Heat Exchangers

Energy Engineering Set

coordinated by
Abdelhanine Benallou

Volume 6

Heat Exchangers

Design and Sizing Algorithms

Abdelhanine Benallou

iSTE

WILEY

First published 2024 in Great Britain and the United States by ISTE Ltd and John Wiley & Sons, Inc.

ISTE Ltd
27-37 St George's Road
London SW19 4EU
UK

www.iste.co.uk

John Wiley & Sons, Inc.
111 River Street
Hoboken, NJ 07030
USA

www.wiley.com

Library of Congress Control Number: 2023952195

British Library Cataloguing-in-Publication Data
A CIP record for this book is available from the British Library
ISBN 978-1-78630-286-1

Contents

Introduction

Recent technological developments relate to different areas of industry: telecommunications, computing, avionics, electric cars, railways, new energies, home automation, etc. These developments have led to enormous progress involving the increasing use of ever-more-powerful embedded electronics and computing systems. With the advent of insulated gate bipolar (IGB) transistor technology, the electric car, or the "more electric" aircraft (Boeing 787 Dreamliner), and more generally next-generation transport equipment, are becoming extremely demanding in terms of energy performance, ergonomics and dimensions: the ultimate performance needs, from now on, to adapt to the smallest of sizes.

However, the operation of power electronics, increasingly used in this type of equipment, releases a great amount of heat. As a matter of fact, the heat density released by the operation of a computer chip can reach 450 to 550 kW/m^2. For comparison, the average power of a semi-trailer truck is 300 kW and the heat density released at the nose of a space shuttle upon entering the atmosphere is in the order of 500 kW/m^2. This means that the heat released by an electronic circuit in operation can be enormous.

However, operation of all electronic components is temperature sensitive. Their performance is defined within very precise temperature ranges. This performance becomes poor outside the prescribed temperature limits. Furthermore, excessively high temperatures can even lead to their deterioration. This constraint means that the cooling efficiency becomes important to maintain the integrity of electronic equipment; importance stressed by an ever-increasing search for small-size designs, compared with the significant power released during operation.

For these particular uses, a specific category of heat exchangers is used to cool electronic boards in computers or power equipment such as those used in new generations of transport vehicles, in mobile telephony or in home automation systems.

Heat exchangers also continue to be present in various industrial processes that require heating or cooling of the products in order to obtain the desired characteristics (chemical reactions, space heating or cooling, etc.). At times, there is even a requirement to vaporize a liquid or condense a vapor (for example, in distillation). All of these types of thermal transformations are carried out in heat exchangers. Indeed, heat exchangers are present almost everywhere; even in the domestic field and long before home automation, where they are the centerpieces of boiler rooms, refrigerators, air conditioners and sanitary water heating systems, whether electric, gas or solar.

However, although several works deal with the conceptual aspects of heat exchangers, sizing and calculation techniques remain the least documented in the literature related to energy engineering. In most cases, specialized books and heat exchanger manufacturers' instruction manuals, which exist in different languages and with different degrees of detail and precision, are either complex, or present different methodologies, linked to a type of manufacture. They are then beyond the reach of users wishing to produce a rapid and independent design.

This book introduces the different technologies available in the industry of heat exchangers in the broadest sense. It presents the most conventional exchangers, but also new-generation exchangers, more specific to the fields of the automotive, avionics or embedded electronics industry.

In addition to illustrating the fundamental concepts governing each of the available technologies, this book specifies the advantages and disadvantages of each type of exchanger, as well as their preferred areas of use. The criteria that need to govern the choice of the exchanger type to be used, according to the task to be carried out and according to the fluids treated, are thus highlighted.

We have made the choice to start with a detailed description of the available technologies. Subsequently, the general design principles that apply to all heat exchanger types are outlined. Nevertheless, technology diversities impose different sizing methods. For this reason, we have reserved Chapter 6 for the presentation of calculation algorithms for each type of exchanger. These algorithms are implemented in the various practical illustrations that accompany Chapter 6.

1

Available Technologies

1.1. Introduction

A heat exchanger is a device designed to transfer energy from one fluid to another, without any mixing between the fluids. When these energy transfers occur without any phase change, these devices are referred to as *heat exchangers*. When, on the other hand, heat transfer is accompanied by a phase change (liquid to vapor or vice versa), they are known as *evaporators, reboilers or condensers*, as appropriate.

Heat exchangers are frequently used in industry to heat or cool gases, liquids and even solids. In the automotive industry, heat exchangers are used for engine cooling; in this case, we refer to radiators. However, in the embedded electronics field, we know them as *heat sinks*, which are essential for cooling power components in circuit boards.

This diversity in designations also covers a diversity of technologies and above all a variety of calculation methods. As a result, several heat-exchanger design and production technologies exist. The purpose of Chapter 1 is to present the different technologies available on the market and to identify the parameters used to define each type of device and the corresponding fields of use.

1.2. The simplest form of heat exchanger: single-tube or coaxial

The simplest device is the *single-tube heat exchanger*, consisting of two coaxial tubes: one of the fluids circulates in the central tube, and the second fluid circulates in the annulus between the two coaxial tubes. Figure 1.1 shows a laboratory embodiment where the outer tube (the sleeve or shell) is made of glass, enabling the inner tube to be viewed. In practice, however, the two concentric tubes are metallic (usually made of steel or copper).

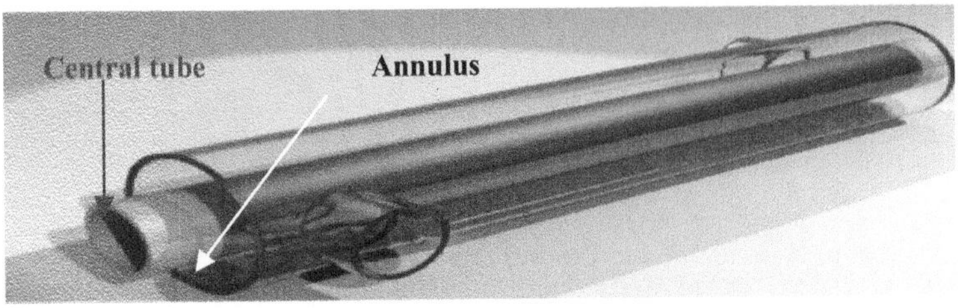

Figure 1.1. *"Single-tube" exchanger*[1]*. For a color version of this figure, see www.iste.co.uk/benallou/energy6.zip*

This exchanger belongs to the category of tubular heat exchangers also known as "shell-tube heat exchangers". By its simplicity, it offers high reliability and relatively low costs.

It is, however, suitable only for limited power applications, and presents risks of hammering for small diameters (<50 mm). It did, however, recently gain new interest for use as a *recuperative heat exchanger in the liquid-hydrogen zero-boil-off storage system.* (Qiu et al. 2023)

1.3. Shell-and-tube heat exchangers

The slightly more elaborate shape of shell-tube exchangers makes it possible to have larger transfer areas in a given volume: the volume of the sleeve. Instead of the heat exchanger consisting of a central tube placed in a sleeve (the *shell)*, a *bundle of parallel tubes* is placed in the shell (see Figure 1.2). One of the fluids (generally, the hot fluid) circulates inside the tubes, while the other flows shell-side. Heat exchange then takes place between the fluid passing through the shell and that circulating in the tubes.

The transfer area depends on the number of tubes integrated into the shell.

This type of heat exchanger is by far the most widely used in the industry, thanks to its relative ease of production and, above all, thanks to its robustness and low cost. Table 1.1 summarizes the advantages and disadvantages of this type of exchanger.

1 Source: https://encryptedtbn0.gstatic.com/images?q=tbn:ANd9GcSugFAO8zsdsBKlTpKM4D4vUyU94O48m75FG)à_4B2qEVtaSv8xVJ.

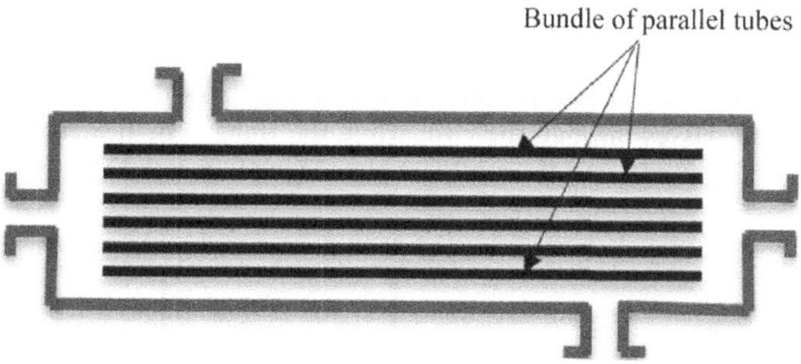

Figure 1.2. *Shell-and-tube heat exchanger. For a color version of this figure, see www.iste.co.uk/benallou/energy6.zip*

Advantages	Disadvantages	Usage
Ease of manufacture Robustness Few breakdowns	Prone to tube fouling	High pressure vapor/water Superheated water/water Thermal fluid/water Fumes/water
Good thermal performance Very good overall heat transfer coefficient	Not sufficiently compact	Liquid/Liquid Liquid/Gas Gas/Gas
Good price/performance ratio Reliable and simple	Difficult to clean	Evaporation
Withstands high pressures Accepts large temperature differences Can be used under boiling conditions	Limited power	Can be used for large flowrates, but must undergo pressure inspection if the volume exceeds 100 liters
Wide range of applications Can be used in partial condensation	Sensitive to vibrations	Oils Liquid/fumes Water/thermal fluid Refrigerants

Table 1.1. *Advantages and disadvantages of shell-and-tube heat exchangers*

1.4. Coil-in-tank heat exchangers

By far the easiest to produce, coil-in-tank heat exchangers are widely used to perform tasks of preheating by heat recovery on available fluids.

They belong to the category of *shell-tube heat exchangers.*

A coil-in-tank heat exchanger consists of a coil immersed in a tank. Usually, the heating medium passes through the coil, while the fluid to be heated is in the tank, which is generally stirred and thermally insulated (see Figure 1.3).

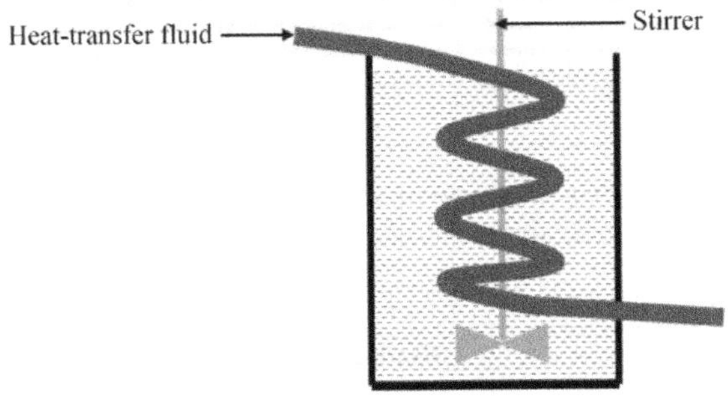

Figure 1.3. *Coil-in-tank heat exchanger. For a color version of this figure, see www.iste.co.uk/benallou/energy6.zip*

Advantages	Disadvantages	Usage
Ease of production Low cost	Low thermal efficiencies Heat loss through the walls of the tank Heat loss by evaporation if the tank is not covered	Good solution to be performed rapidly to meet an immediate need
Can be built with recovery equipment		Thermal tasks to be completed rapidly for process modifications or to conduct tests
Wide range of applications		Liquid/Liquid Water/low-pressure vapor
Ease of tank cleaning	Inside of the coil difficult to clean	Limit circulation in the helical tube to fluids that are clean

Table 1.2. *Advantages and disadvantages of coil-in-tank heat exchangers*

1.5. Compact heat exchangers

Exchangers presenting a large heat transfer surface area per unit volume are generally referred to as compact heat exchangers[2]. The interest placed in this category stems from the fact that, in practice, the aim is always to use devices presenting the most m^2 available for heat exchange per m^3 of device.

This type of heat exchanger includes cross-flow exchangers, finned exchangers used as car radiators or refrigeration evaporators (which generally have surface densities of around 1,000 m^2/m^3), or plate exchangers, plate-fin exchangers, spiral or lamella exchangers and printed-circuit exchangers.

1.5.1. *Finned-tube cross-flow heat exchangers*

These exchangers are generally used to transfer heat between a liquid and a gas. To compensate for the low transfer coefficients on the gas side, the transfer area in contact with the gas flow is increased by attaching fins to the tubes (see Figure 1.4).

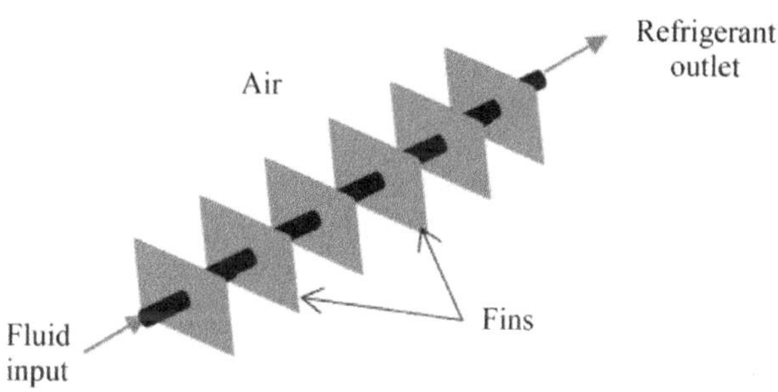

Figure 1.4. *Finned tube. For a color version of this figure, see www.iste.co.uk/benallou/energy6.zip*

Thus, cross-flow exchangers are most often composed of a tube or a set of tubes that pass through metal plates that act as fins, always with a view to increasing the transfer area (see Figure 1.5). The liquid generally circulates inside the tubes and the gas circulates fin-side. The directions of the two fluids are perpendicular, hence the name "cross-flow". Note that the fluids cross each other inside the exchanger, but without mixing.

2 Generally between 700 and several thousand m^2/m^3 (see Chapter 3).

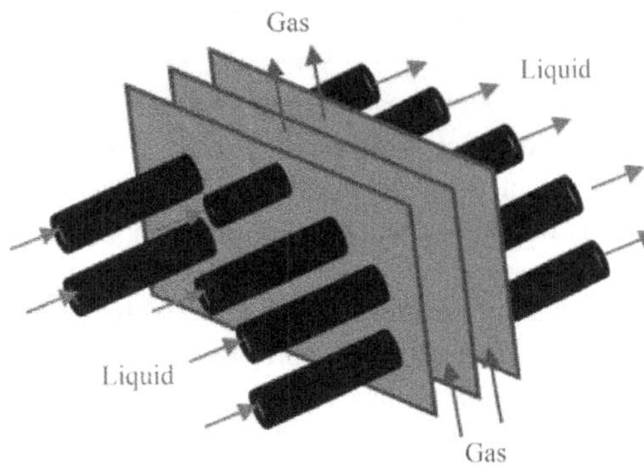

Figure 1.5. *Finned-tube cross-flow heat exchanger. For a color version of this figure, see www.iste.co.uk/benallou/energy6.zip*

1.5.2. *Car radiators*

The car radiator is a common version of the finned-tube cross-flow heat exchanger. It consists of tubes passing through a honeycomb of fins with a density between 450 and 1,100 fins/m, enabling compactnesses of between 950 and 1,700 m^2/m^3.

Car radiators are used to cool the water circulating to remove the heat produced by the engine (see Figure 1.6). The cooling water flows through the tubes and the air passes fin-side.

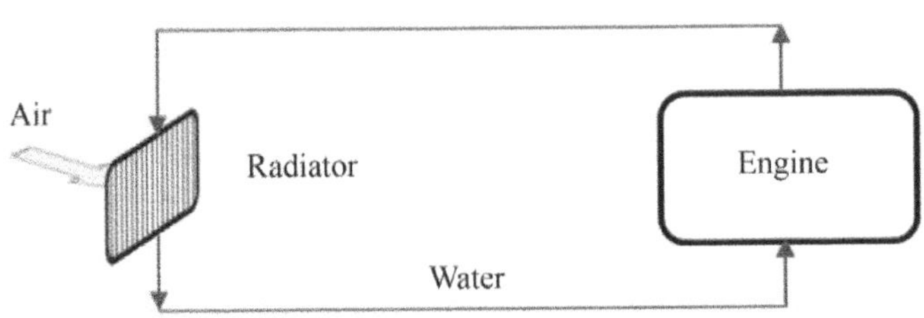

Figure 1.6. *Principle of the car engine cooling system. For a color version of this figure, see www.iste.co.uk/benallou/energy6.zip*

Therefore, the heat is first transferred by convection from the hot fluid circulating in the tubes toward the tube walls. It is then conveyed by conduction toward the metal fins, then by convection between these fins and the air. In this way, heat is extracted from the cooling liquid (which passes through the tubes) to the outside air crossing fin-side (see Figure 1.7).

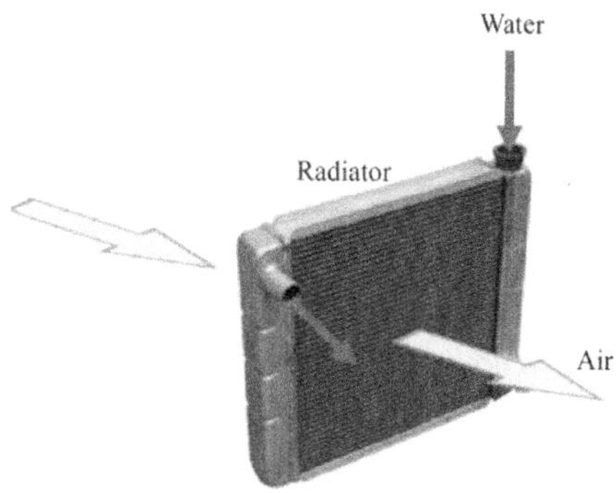

Figure 1.7. *Car radiator. For a color version of this figure, see www.iste.co.uk/benallou/energy6.zip*

Advantages	Disadvantages	Usage
Good performance Can take on precise shapes	Sensitive to shocks	Water/air Oil/air Solid/air

Table 1.3. *Advantages and disadvantages of finned-tube heat exchangers*

1.5.3. *Plate-fin heat exchangers*

Commonly used in gas/liquid applications, this type of exchanger consists of a stack of modules, each comprising of two metal plates between which fins of various shapes have been fixed. Figure 1.8 shows an exploded view of a module, while Figure 1.9 indicates that fluids can circulate counter-flow or cross-flow.

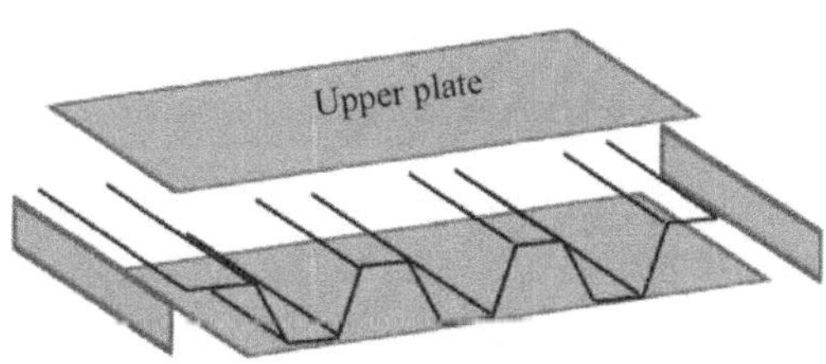

Figure 1.8. *Plate-fin heat exchanger. For a color version of this figure, see www.iste.co.uk/benallou/energy6.zip*

Plate-fin heat exchangers can reach compactnesses beyond 5,500 m^2/m^3.

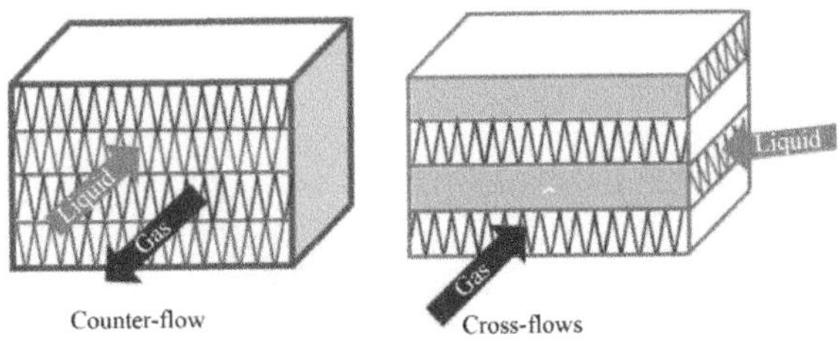

Figure 1.9. *Common flow configurations in a plate-fin heat exchanger. For a color version of this figure, see www.iste.co.uk/benallou/energy6.zip*

Table 1.4 shows the advantages and disadvantages, as well as the common uses, of this type of exchanger.

Advantages	Disadvantages	Usage
Good compactness	Difficult to clean Applications limited to relatively clean fluids	Gas/gas exchanges Gas/liquid exchanges
Low weight Heat transfer area per kg more than 25 times that of shell-and-tube heat exchangers	Rigid structure	Embedded exchangers Commercial aircraft Aerospace Locomotives
Good thermal efficiency	Limited to low pressures $P \leq 1$ kPa	Heating Suitable for gas/gas and gas/liquid energy recovery techniques
Good circuit sealing	Relatively high pressure drops	Heat exchanges where there must be no fluid contamination
Good value for money	Possibility of poor flow distribution Relative fragility	Protected areas

Table 1.4. *Advantages and disadvantages of plate-fin heat exchangers*

1.5.4. *Plate heat exchangers*

Another type of compact heat exchanger has received particular attention in recent decades, as it provides very large transfer areas in very small volumes: the *plate heat exchanger*.

Figure 1.10 shows an exploded diagram of this type of exchanger. It is possible to visualize how the fluids circulate in juxtaposed plates arranged alternately: "hot" plates, shown in red, and "cold" plates, shown in blue. Each "hot" plate is in this way framed by two "cold" surfaces and vice versa.

Distribution of the fluids over plates is ensured by a head plate, in such a way that the respective fluid circuits do not allow mixing.

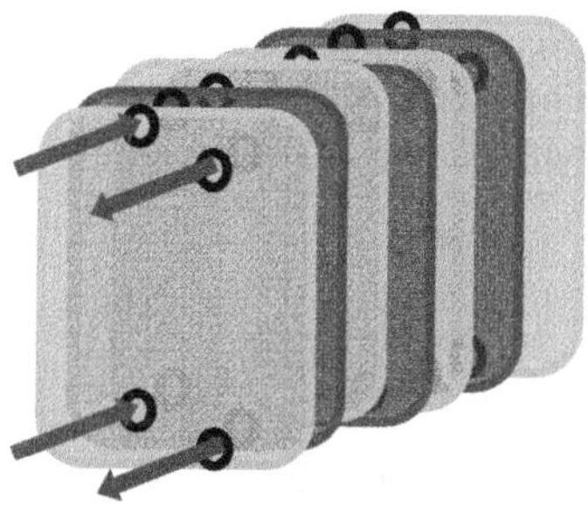

Figure 1.10. *Plate heat exchanger. For a color version of this figure, see www.iste.co.uk/benallou/energy6.zip*

Advantages	Disadvantages	Usage
Good thermal efficiency Very good heat transfer coefficients Low heat loss	Relatively large pressure drops	Water/water exchanges Water/fruit juice exchanges Water/sea water exchanges
Good compactness Fully modular: extensions and reductions possible	Relatively limited working pressures	Low-pressure vapor/oil Oil/water
Suitable for energy recovery techniques	Rather suitable for small temperature differences	Food industry Heating Urban heating Collective networks Solar heating
Good value for money	Poor in gas/gas heat exchanges Poor for high-viscosity fluids	Dairies Marine industry Refrigeration condensation Heat pumps

Table 1.5. *Advantages and disadvantages of plate heat exchangers*

It should be noted that in addition to its high compactness, this type of exchanger offers very good thermal efficiencies and competitive costs.

1.5.5. *Spiral heat exchangers*

Spiral heat exchangers form another category of compact exchangers. They are made from rolling two metal sheets one over the other, using studs to separate the different sheets so as to leave a gap between them. These gaps will constitute channels for the circulation of fluids. The exchanger thus obtained is closed at both ends by covers that ensure both sealing of the device and distribution of the fluids.

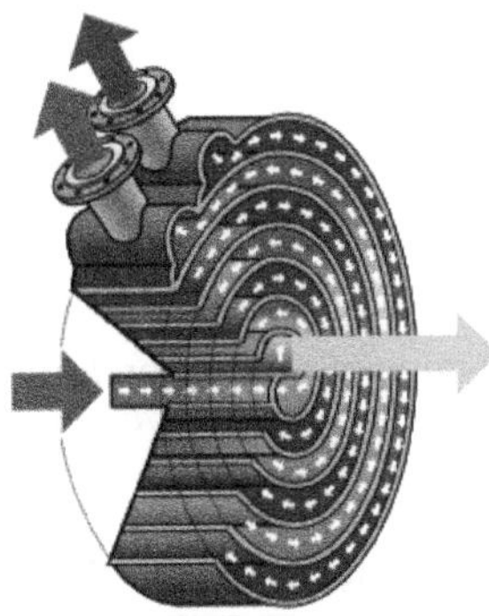

Figure 1.11. *Circulation of fluids inside a spiral[3] heat exchanger. For a color version of this figure, see www.iste.co.uk/benallou/energy6.zip*

Figure 1.11 shows the spiral heat exchanger obtained, in which the "hot" (shown in red) and "cold" (shown in blue) fluids circulate, in counter-flow, in the spiral channels formed by the sheets. This gives an alternation of "hot" and "cold" cylindrical plates inside the exchanger. A spiral heat exchanger could thus be viewed as a specific plate heat exchanger; instead of its plates being flat, they are cylindrical. This type of geometry offers several advantages, in addition to compactness. In particular, spiral heat exchangers can be used for charged fluids, with small temperature differences, and pressures as high as 16 bar.

Moreover, the fact that the channels are accessible (after opening of the covers) offers a definite advantage for easy cleaning of the device. Note that the studs placed between the rolled plates not only make it possible to maintain a gap for fluid passage, but also play an important role in heat transfer efficiency. Indeed, their

3 Source: https://www.alfalaval.com/globalassets/images/products/heat-transfer/plate-heat-exchangers/spiral-heat-exchangers/alshe_flow_hiw_250x338.jpg.

presence in the fluid flow path promotes *turbulent flow* in each channel. They therefore make it possible to obtain larger heat transfer coefficients (see Volume 3).

Note that spiral heat exchangers are very efficient in the case of charged solutions. Indeed, the fact that the fluid flowrate is not shared (as in the case of multitube heat exchangers where the flowrate is divided between the tubes) means that it is forced through the available gap in its entirety. This specific feature imparts a self-cleaning effect on this type of exchanger, since any fouling deposits on the sheets lead to a decrease in the gap available for fluid flow, and therefore an increase in the pressure and flow velocity, which eventually clears these deposits away.

Advantages	Disadvantages	Usage
Good thermal efficiency		Heat recovery Paper industry Mining industry Water treatment Pharmaceutical industry
Good compactness		Aerospace Maritime
Good heat transfer coefficients thanks to the presence of studs, which increase turbulence	Pressure drops can be significant	Steelworks Petrochemical industries Refining
Suitable for energy recovery techniques		All sectors
Suitable for very high-viscosity charged fluids containing solid particles or fibers.		Excellent solution for heat exchange between fouling and dirty fluids Waste heat recovery Suitable for waste-water heat recovery
Good value for money		Low return times for energy-recovery investments
Easy to maintain and clean without the use of special tools		

Table 1.6. *Advantages and disadvantages of spiral heat exchangers*

1.5.6. *Printed-circuit heat exchangers*

Printed-Circuit Heat Exchangers (PCHEs) are manufactured using a process that enables etching (printing) on the inside of solid plates, channels or circuits intended for fluid flow. These channels are created, in a similar way to electronic printed circuits, by chemical etching (Deng et al. 2020). They can also be created by diffusion bonding.

In cross-section, a flow channel generally takes the shape of a semicircle having a diameter of 1 to 2 mm (see Figure 1.12). Other channel shapes (triangles, waves, etc.) can also be encountered, however, depending on the model and the manufacturer; the objective being to obtain the maximum heat transfer area in each channel. The flow configuration may be counter-flow (see Figure 1.12) or cross-flow (see Figure 1.13).

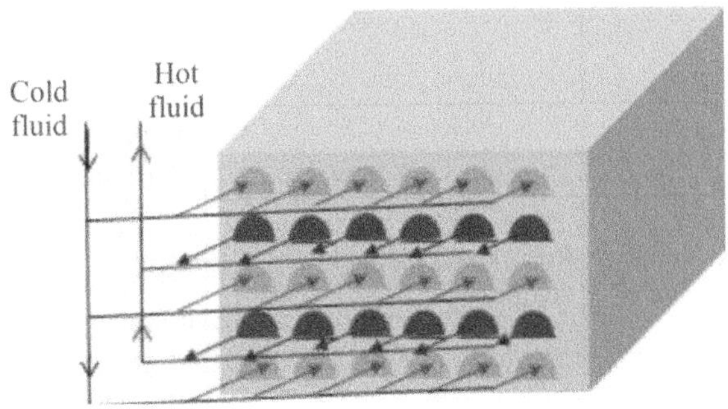

Figure 1.12. *Diagram of a counter-flow PCHE core. For a color version of this figure, see www.iste.co.uk/benallou/energy6.zip*

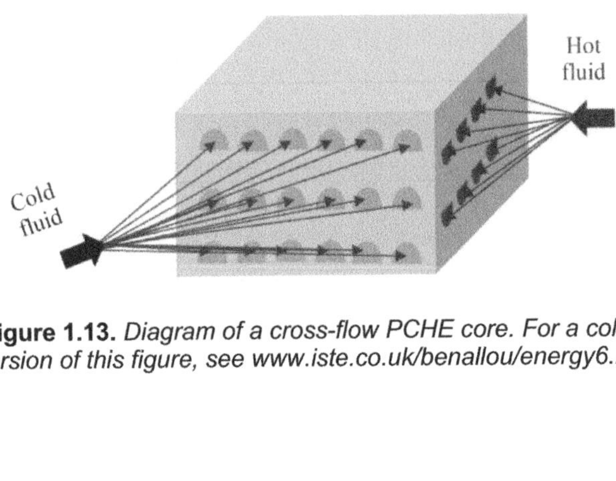

Figure 1.13. *Diagram of a cross-flow PCHE core. For a color version of this figure, see www.iste.co.uk/benallou/energy6.zip*

The fact that the flow channels are etched in the mass enables gasketless and weldless heat exchangers, called "exchanger cores", which have high resistance and integrity.

This technology provides ultracompact heat exchangers that offer very high performance (Oh and Kim 2008): up to 85 % smaller and lighter than traditional heat exchangers (see comparison, Table 1.7). It also enables the development of heat-exchanger cores that are in line with the constraints and detailed specifications of the customer. Figure 1.14 shows an image of a PCHE core custom-made to occupy a specific space.

Type of heat exchanger	Channel diameter (mm)	Compactness (m^2/m^3)
Shell-and-tube	10–50	100
Plates	5	200
Plate-fins	2	1,000
PCHE	1	2,000

Table 1.7. *Characteristics of the PCHE compared to other heat exchangers*

Figure 1.14. *Image of a custom-made[4] PCHE core. For a color version of this figure, see www.iste.co.uk/benallou/energy6.zip*

4 Source: https://www.alfalaval.com/globalassets/images/products/heat-transfer/plate-heat-exchangers/pche/pche_explosion.png.

Unless otherwise specified for embedded exchangers, PCHE cores are usually mounted in bodies made of stainless steel or composite materials (see Figures 1.15 and 1.16).

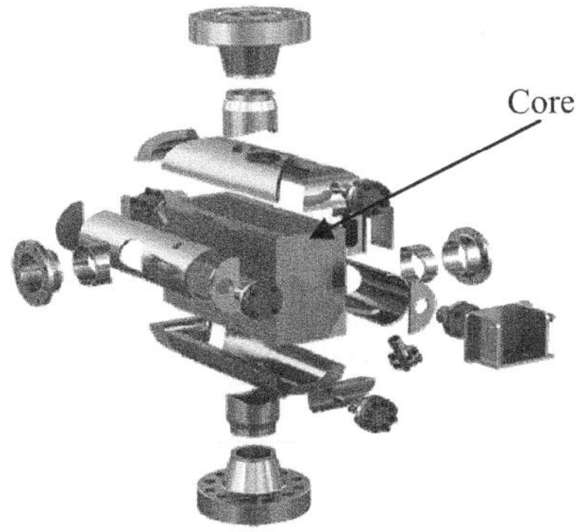

Figure 1.15. *Mounting of a PCHE core*[5]*. For a color version of this figure, see www.iste.co.uk/benallou/energy6.zip*

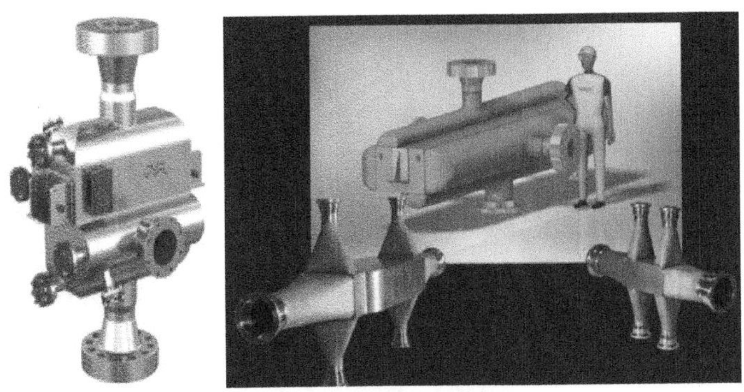

Figure 1.16. *Commercial PCHE*[6] *models. For a color version of this figure, see www.iste.co.uk/benallou/energy6.zip*

5 Source: https://www.alfalaval.com/globalassets/images/products/heat-transfer/plate-heat-exchangers/pche/pche_explosion.png.

6 Sources: https://www.alfalaval.com/globalassets/images/products/heat-transfer/plate-heat-exchangers/pche/pche-640x360-large.jpg; https://images.squarespace-cdn.com/content/v1/557894aae4b002ee535e742a/1434650380584-Q59TYGWFZSZAEBID09EO/ke17ZwdGBToddI8pDm48kFxJkVPJUnwhrx3TZKdnUXV7gQa3H78H3Y0txjaiv_0fDoOvxcdMmMKkDsyUqMSsMWxHk725yiiHCCLfrh8O1z5QPOohDIaIeljMHgDF5CVlOqpeNLcJ80NK65_fV7S1UUTtlt0KybxjZ2Y8OsLd7N6ggbdnJVjo911wfKe77wUrOpYghpI-Ha_TwZsqqmJXng/image-asset.png?format=50).

Developed only recently but presenting great potential, printed-circuit heat exchangers are on course to occupy an important place in the heat-transfer equipment industry. However, the poor availability of thermohydraulic data can make sizing difficult. Recent studies (Ma et al. 2021) make it possible to fill this gap in certain situations.

Advantages	Disadvantages	Usage
Excellent thermal efficiency and great robustness	Fluids must be very clean Difficult to clean Chemical cleaning required	Aerospace Maritime equipment: LNG regasification on ships
Exceptional compactness Up to 650 to 1,300 m^2/m^3	Transients can lead to material fatigue	High-pressure vaporization in fuel gas supply systems
Suitable for very low and very high pressures	Small channel size Pressure drops can be significant for low- and medium-pressure applications	Wide pressure interval: from vacuum to 1,000 bar Petrochemical industries: offshore gas compression systems Pharmaceutical industry
Suitable for very low and very high temperatures	Low availability of thermo-hydraulic data	Wide temperature interval: Up to 1,200°C
Energy recovery		Combustion gas treatment systems Electricity generation: preheating of turbine fuel gas
Various materials, as required	Carbon steels not suitable	Stainless steel, chrome, Cu-Ni-Ti alloys

Table 1.8. *Advantages and disadvantages of printed-circuit heat exchangers*

1.6. Rotary regenerative heat exchangers

Rotary heat exchangers are mainly used in gas/gas applications.

A rotary heat exchanger (see Figure 1.17) consists of a carousel of large metal fins that store heat as the hot gas passes through. The carousel is driven by a rotational motion that allows the fins to be in contact, sometimes with the "hot" gas, and sometimes with the "cold" gas. This makes it possible to transport the heat stored during passage through the hot compartment, to the compartment through which the gas to be heated passes. The rotational velocity of the carousel is usually low (<1 rpm) to allow heat to transfer from the fins to the gases.

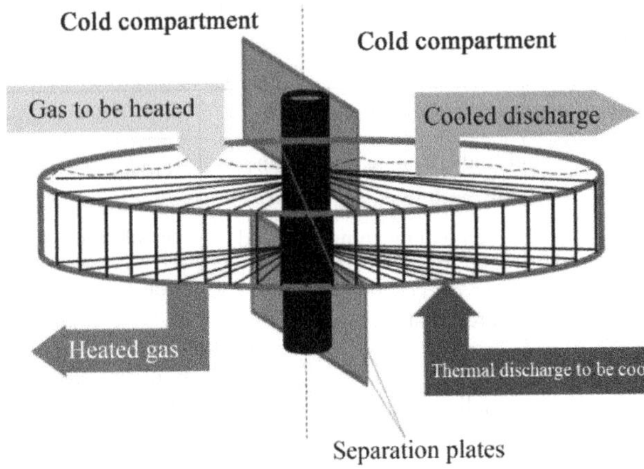

Figure 1.17. *Rotary regenerative heat exchanger. For a color version of this figure, see www.iste.co.uk/benallou/energy6.zip*

Advantages	Disadvantages	Usage
Vast transfer areas	Low heat transfer coefficients	Energy recovery on gas streams
Exploitation of latent heat	Contamination risks between the two flows	Gas/gas exchanges
Ordinary construction materials	Low price/efficiency ratio	Waste heat cooling

Table 1.9. *Advantages and disadvantages of rotary heat exchangers*

1.7. Phase-change heat exchangers

1.7.1. *Condensers*

Condensers are specific heat exchangers where the fluid to be cooled undergoes a *phase change*: it enters the exchanger in the vapor state and exits it in the liquid state.

There are several types of condensers; the most widely used in industry are the shell-and-tube type. Generally, the fluid to be condensed circulates in the shell, while the cooling liquid circulates in the tubes (see Figure 1.18).

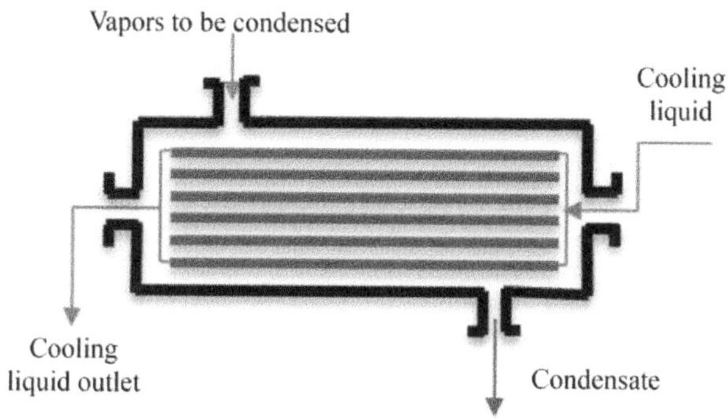

Figure 1.18. *Multitube condenser. For a color version of this figure, see www.iste.co.uk/benallou/energy6.zip*

Other categories of condensers take the form of a coil tube fitted with fins. This type of condenser is mainly used in refrigeration. It is air-cooled (see Figure 1.19), generally by natural convection.

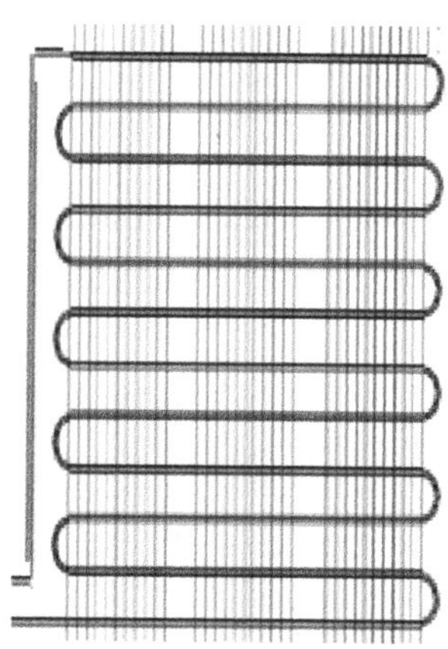

Figure 1.19. *Finned-tube condenser*

In addition, another type of condenser is also often used in refrigeration, simply consisting of a coaxial single-tube exchanger, usually water-cooled (see Figure 1.20). The water generally circulates in the annulus, while the refrigerant circulates in the central tube.

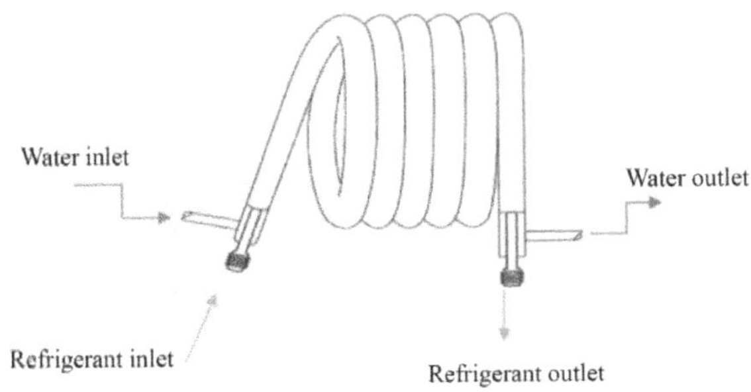

Figure 1.20. *Single-tube condenser. For a color version of this figure, see www.iste.co.uk/benallou/energy6.zip*

1.7.2. *Reboilers and evaporators*

These exchangers make it possible to boil or evaporate a fluid, the objective being to concentrate a solution or produce vapor. Therefore, they are heat exchangers where a phase change occurs. They are widely used in the refining industry (reboilers) and in sugar factories (evaporators). The most widespread technology is that of multitube heat exchangers, but plate heat exchangers are increasingly used.

1.7.2.1. *Reboilers*

Reboilers are generally the heat sources for distillation columns.

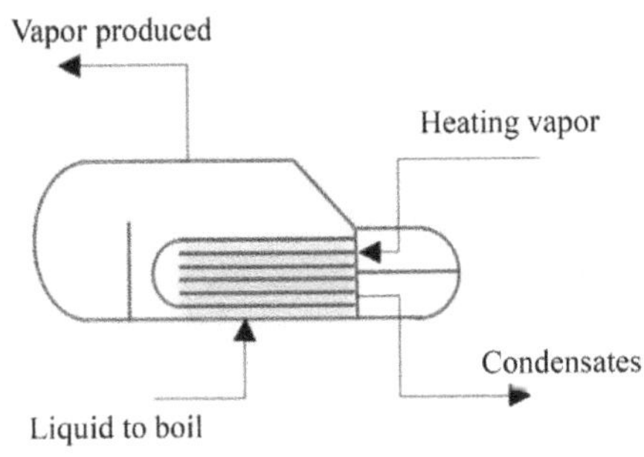

Figure 1.21. *Example of a reboiler. For a color version of this figure, see www.iste.co.uk/benallou/energy6.zip*

These are composed of several tubes placed in a shell with an enlarged part providing space for the vapor produced (see Figure 1.21). The liquid to be boiled is heated by the vapor flowing through the tubes, or (although this is less common) by an electrical resistor dipping inside the reboiler.

By releasing its latent heat, the heating vapor leaves the tubes in the liquid state.

1.7.2.2. *Single-effect evaporators*

A single-effect evaporator is composed of a multitube heat exchanger (see Figure 1.22). This type of exchanger is used for concentrating solutions in different industries: sugar factories, manufacture of concentrated fruit juice, etc.

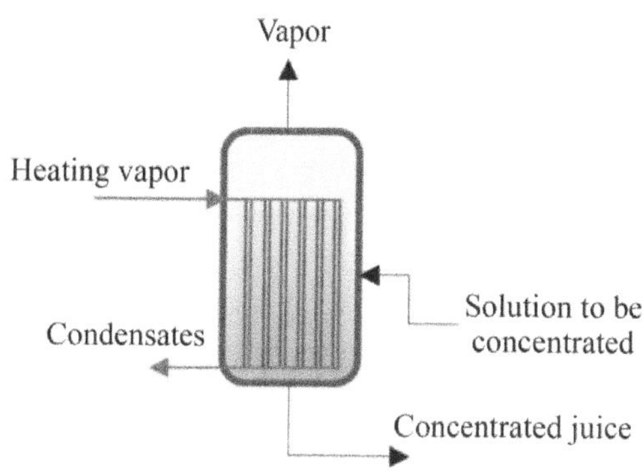

Figure 1.22. *Single-effect evaporator. For a color version of this figure, see www.iste.co.uk/benallou/energy6.zip*

The evaporator is usually heated using a heating vapor that condenses in the tubes, releasing its latent heat. The concentrated juice is recovered at the evaporator outlet.

1.7.2.3. *Multiple-effect evaporators*

Multiple-effect evaporators consist of a set of multitube exchangers connected in series. The concentrated solution leaving the first effect is fed to the second effect to be concentrated further, and so on. Figure 1.23 shows a three-effect evaporator.

The use of multiple effects is generally dictated by energy saving requirements. In fact, the heating vapor, V_0, makes it possible to produce the concentrated juice, J_1, but also produces vapor V_1, which is used in the second effect. Likewise, vapor V_1 produces J_2 and V_2, etc. Therefore, in a battery of *n* multiple-effect evaporators, the consumed vapor, V_0, makes it possible to produce *n* times V_0 and therefore makes it possible for all other effects beyond the first to operate.

Multiple-effect evaporators are widely used in the sugar industry to concentrate sugar juices.

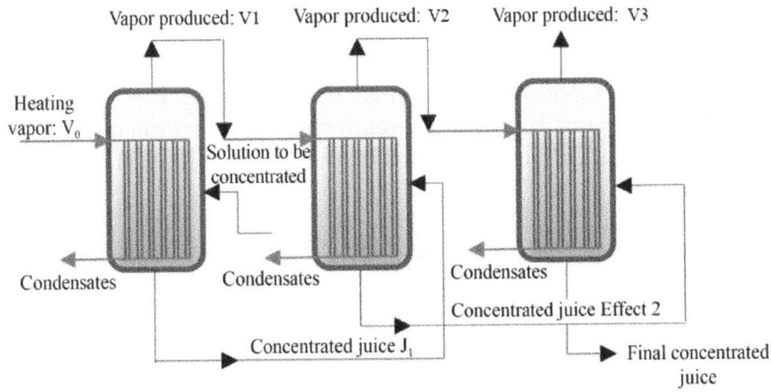

Figure 1.23. *Multiple-effect evaporators. For a color version of this figure, see www.iste.co.uk/benallou/energy6.zip*

1.7.2.4. *Refrigeration evaporators*

Refrigeration evaporators are used in cooling cycles to evaporate a refrigerant. Cooling is generally provided by air in natural convection; hence the name *natural circulation evaporators* (also known as *static evaporators*).

Note that the convective heat transfer coefficients on the gas side are low (see Volume 3), both in natural convection and in forced mode. This weakness is compensated for by an increase in the transfer area that is in contact with the gas flow. This is achieved by attaching fins onto the tubes (see Figures 1.4 and 1.5).

Natural circulation evaporators are used to refrigerate meat or fish display cases, bakery presentation cabinets and cheese-shop display counters. They consist of a multitude of finned tubes, which are mounted on manifolds that ensure distribution of the fluid between the tubes (see Figure 1.24).

For large systems, air circulation can be forced using one or more fans. We then refer to forced-circulation refrigeration evaporators. The latter are usually mounted on walls (see Figure 1.25) or ceilings (see Figure 1.26) of cold rooms.

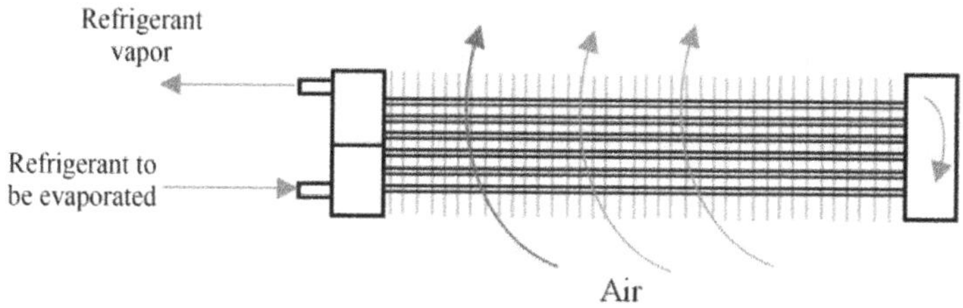

Figure 1.24. *Refrigeration evaporator composed of finned tubes. For a color version of this figure, see www.iste.co.uk/benallou/energy6.zip*

Figure 1.25. *Wall evaporator. For a color version of this figure, see www.iste.co.uk/benallou/energy6.zip*

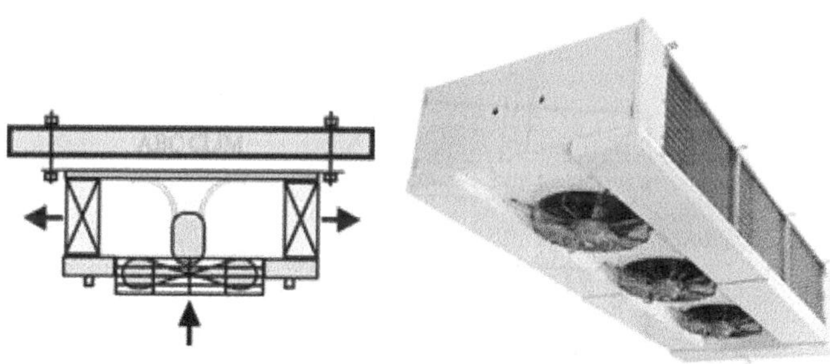

Figure 1.26. *Ceiling evaporator. For a color version of this figure, see www.iste.co.uk/benallou/energy6.zip*

1.7.3. *Heat-pipe heat exchangers*

A heat pipe consists of a sealed tube, containing a heat-transfer fluid, without any other gas (see Figure 1.27). On one of its zones (zone 1), the tube is subjected to a hot fluid whose heat is to be recovered. The temperature of the heat-transfer medium in this zone increases, causing it to evaporate. The heat-transfer fluid vapors then accumulate in zone 2 of the heat pipe, where they condense on the inner walls of the tube, releasing their latent heat to the "cold" fluid (see Figure 1.27). The condensates flow in droplets on the walls of the tube and return to zone 1, where they are again subjected to the heat flux and evaporate.

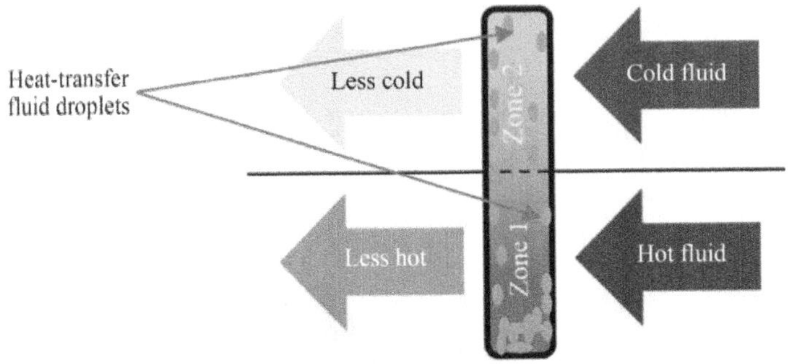

Figure 1.27. *Principle of the heat pipe. For a color version of this figure, see www.iste.co.uk/benallou/energy6.zip*

Zone 1 of the heat pipe is generally known as the "evaporator", while zone 2 is referred to as the "condenser". The use of latent heat instead of sensible heat for energy transfer allows large amounts of heat to be exchanged over reduced surfaces, leading to high compactness.

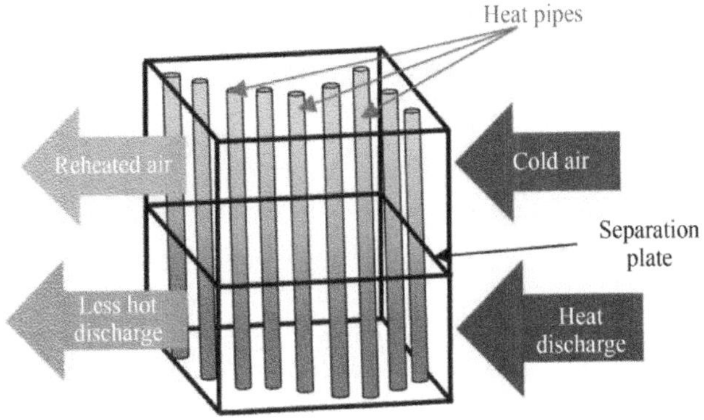

Figure 1.28. *Heat-pipe heat exchanger. For a color version of this figure, see www.iste.co.uk/benallou/energy6.zip*

Figure 1.28 shows a heat exchanger composed of several heat pipes. These are generally formed of finned tubes. This type of exchanger is generally used for *energy recovery from waste heat*, especially for heating premises. It is also used to cool electronic power systems (see section 1.8.7).

Advantages	Disadvantages	Usage
Good compactness	Production complexity	Energy recovery Gas/gas exchanges
Exploitation of latent heat	Relatively high costs	Waste heat cooling Heating of premises

Table 1.10. *Advantages and disadvantages of heat-pipe heat exchangers*

1.8. Heat exchangers for power electronics

Operation of electronic components (transistors, integrated circuits, microprocessors, etc.) generates significant amounts of heat. If this heat is not extracted, it leads to an increase in the temperature of these components, which can alter their properties and operation. This alteration can induce degradation of the bonding wires, delamination of solders and the appearance of leakage currents, etc. Heat removal is therefore a major challenge for the reliability of electronic systems and components.

Several types of heat exchangers are used to remove this heat; the simplest are electronic radiators, also called heat sinks.

1.8.1. *Electronic radiators and heat sinks*

Electronic heat sinks are a specific type of heat exchanger; they ensure the evacuation into the environment of the heat generated by electronic component operation. They are mainly of the compact type (finned surfaces, printed-circuit heat exchangers, etc.).

Figure 1.29 shows examples of finned exchangers used as heat sinks on electronic components.

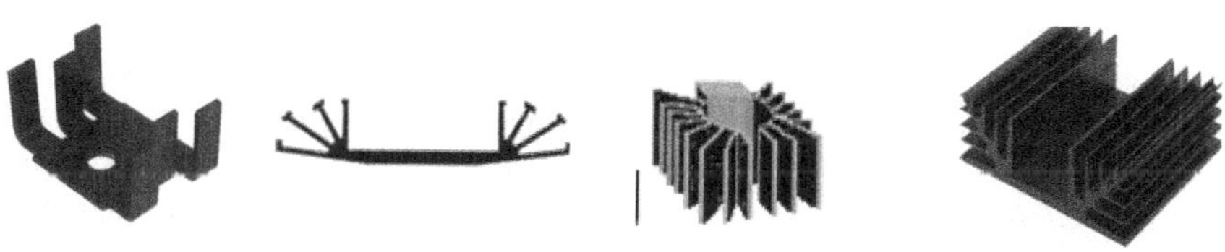

Figure 1.29. *Different models of heat sinks composed of fins*

The sink is usually mounted on the electronic part component that generates heat-typically transistors, integrated circuits and central processing units (CPUs) – thus increasing the heat transfer area by convection with ambient air (see Figure 1.30).

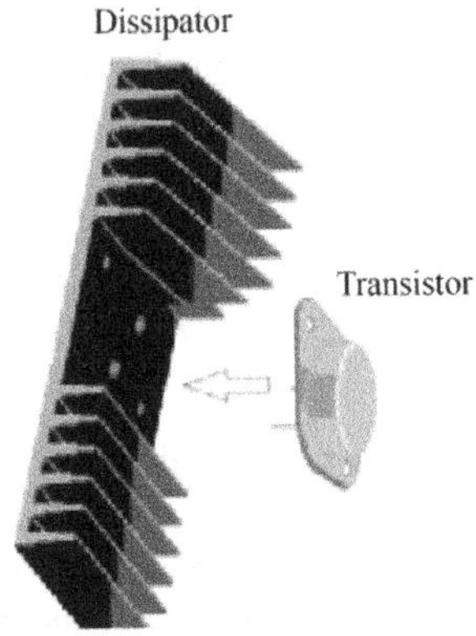

Figure 1.30. *Mounting of a transistor on a heat sink. For a color version of this figure, see www.iste.co.uk/benallou/energy6.zip*

In the majority of simple electronic circuits, heat is evacuated by *natural convection*. However, for power electronics circuits, whose operation generates large amounts of energy, it is necessary to use *forced-convection* cooling systems, with or without refrigeration (Zhang et al. 2019a). Electronic radiators operating by natural convection have been extensively covered in Volume 3 of this series (sections 5.17 to 5.19). Forced-convection systems are presented in section 1.8.2.

1.8.2. *Forced-convection heat sinks*

The technological developments recorded over the last decade have led to unprecedented information fluxes (data or multimedia). This has generated important data transfers (Bluetooth and WiFi) and the need for ever-increasing computing capabilities (Benallou, 2021).

Furthermore, the advent of the *Internet of Things* means that the amount of connected devices is constantly increasing. Meanwhile, *embedded electronics* and *mobile telephony* have imposed ergonomics and space limits (volume) that are often too rigid. In systems engineering, this implies that *ever larger power circuits need to be integrated into ever smaller volumes*. Under such conditions, the amount of energy generated by electronic circuits is so large that natural convection is no longer sufficient to extract them. As soon as the flux generated by electronic equipment operation exceeds 70 W/cm² (Priatherm 2018), a *forced convection* exchange must be used: fans are employed to force air through the fins of the electronic radiators. This type of assembly can be miniaturized in cooling modules.

Using the fan to force air through the heat sink allows *significant increases in convective heat transfer coefficients*: up to 3,000 $W.m^{-2}°C^{-1}$ (Phelan et al. 2002). By way of comparison, this coefficient is between 25 and 500 $Wm^{-2}°C^{-1}$ in the case of natural convection.

1.8.3. *Cold-plate heat exchangers*

The introduction of forced convection has certainly helped improve thermal management of electronic systems. However, as technology becomes more and more demanding, the thermal power released (mobile telephones, electric vehicles, embedded avionics systems, high-speed trains, etc.) is becoming so great that systems to evacuate heat by air in forced convection are no longer sufficient (Milnes 2011).

Cooling systems using so-called cold plates are then used (Baviere 2005). These plates are fitted with pipes (see Figure 1.31) or with microchannels through which a cooling fluid flows (see Figure 1.32). These are attached to the walls to be cooled. The cooling fluid is conveyed using a pump and is cooled in turn by means of a compact heat exchanger (Meysenc 1998).

1.8.4. *Microchannel coolers*

Microchannel coolers can be considered as a specific subcategory of printed-circuit heat exchangers. These are, in fact, exchangers that are very small in size (a few millimeters), allowing thermal fluxes of around 800 W/cm² to be absorbed (Tuckerman and Pease 1981).

In these *micro-heat-exchangers*, the cooling fluid (generally air) circulates in *microchannels*, of microscopic diameters (around 10 to 60 μ), formed by etching on plates made of metal or composite materials (Ohadi et al. 2013). These microchannels are around 0.5 mm in height. The modules thus formed are placed beneath the electronic power components, such as microprocessors (Colgan et al. 2007). Figure 1.32 illustrates such an assembly, in which the electronic circuit to be cooled is shown in semi-transparency, above the microchannels

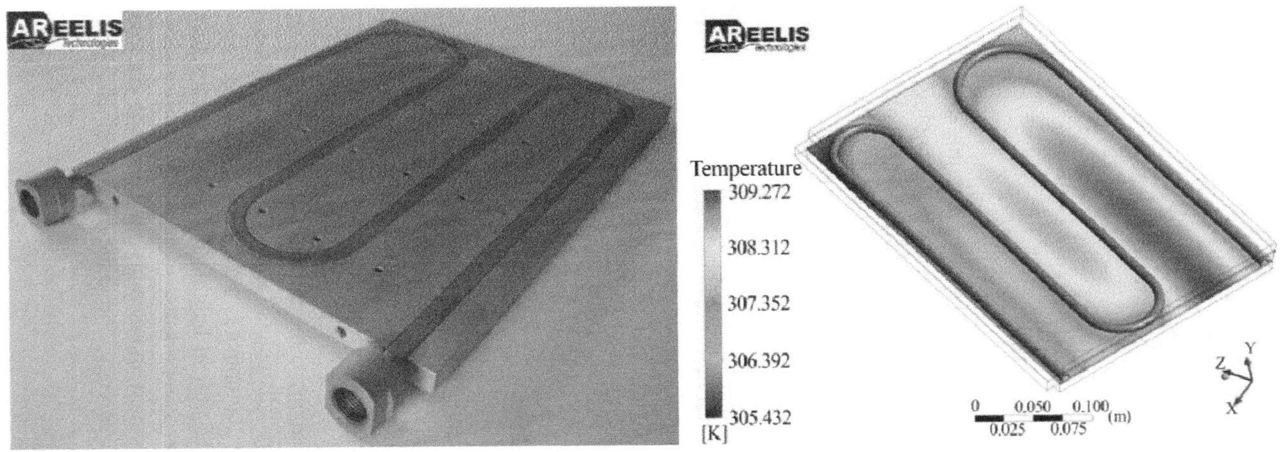

Figure 1.31. *Cold-plate liquid cooling (sources: https://www.areelis.fr/wp-content/uploads/2019/05/plaques_froides-495x400.jpg; https://www.areelis.fr/wp-content/uploads/2019/05/FRE-0814-01-2-495x400.png) (AREELIS Technologies). For a color version of this figure, see www.iste.co.uk/benallou/energy6.zip*

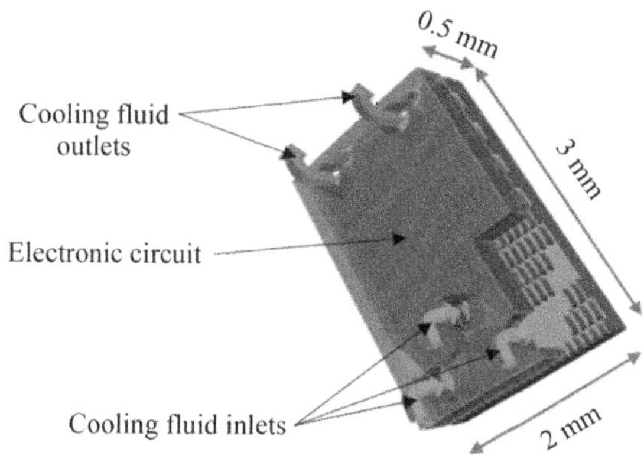

Figure 1.32. *Microchannel heat exchanger. For a color version of this figure, see www.iste.co.uk/benallou/energy6.zip*

1.8.5. *Refrigerated heat exchangers*

Although it improves performance compared to air-cooled systems, *cold-plate* cooling, with pipes or microchannels, presents several drawbacks.

Indeed, in addition to its implementation complexity and the significant pressure drops associated with circulation in the microchannels, it does not fully meet the requirements of thermal management in electronic power systems (International Technology Roadmap for Semiconductors 2007).

For this reason, significant efforts have been recently devoted to the development of plates associated with refrigeration systems, using a refrigerant with a cold production cycle (Webb 2005; Marcinichen et al. 2016). These cooling systems are used in particular to cool power supplies of lasers and their optical systems. They are capable of extracting heat flux densities in excess of 1,000 W/cm^2, while maintaining the electronic chips at temperatures of below 65°C (Shannon et al. 1999; Shannon 2002; Phelan et al. 2010; Barbosa et al. 2012; Burnett 2014).

However, this type of cooling remains difficult to achieve. In fact, all of the components of a *refrigeration cycle* (compressor, expansion valve, evaporator and condenser) need to be assembled in rather minute spaces (see Figure 1.33).

These systems therefore need to meet a major challenge: the miniaturization required by designs favoring small sizes (Shannon 2002; Mancin et al. 2013).

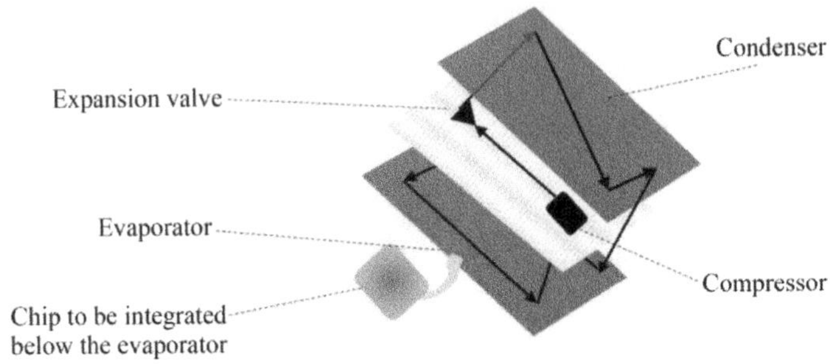

Figure 1.33. *Exploded view of a chip mounted under a refrigerated heat exchanger. For a color version of this figure, see www.iste.co.uk/benallou/energy6.zip*

1.8.6. *Spray heat exchangers: spray coolers*

These are phase-change micro-exchangers. The electronic circuit is cooled by a jet of fluid that partially evaporates as it absorbs heat (see Figure 1.34). The heat-transfer fluid is directly projected onto the surface of the power component to be cooled (Kercher et al. 2006; Thibault 2009; Shicheng et al. 2014). Spraying makes it possible to transfer large heat fluxes, thus facilitating fluid vaporization, even at relatively low temperatures (60 to 75°C, under 1 Atm). It can be carried out using a single injector (Horacek et al. 2005) or multiple injectors (Horacek et al. 2004).

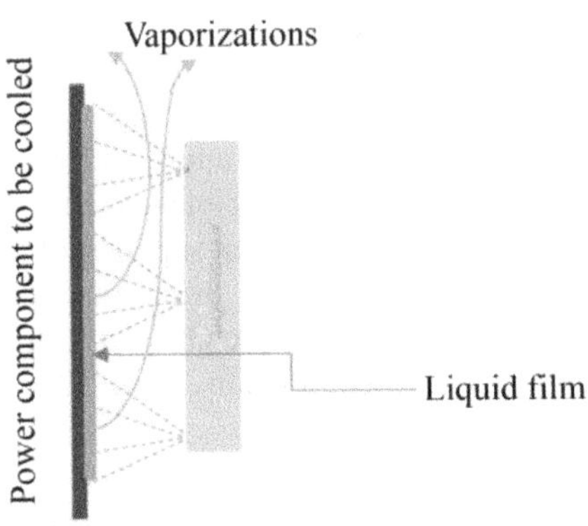

Figure 1.34. *Principle of spray cooling. For a color version of this figure, see www.iste.co.uk/benallou/energy6.zip*

However, as in the case of refrigerated heat exchangers, spray cooling is relatively complex to put in place. In addition to the sprayers, it requires installation of a condenser to collect the generated vapors, and a compressor to supply the sprayer injectors (see Figure 1.35).

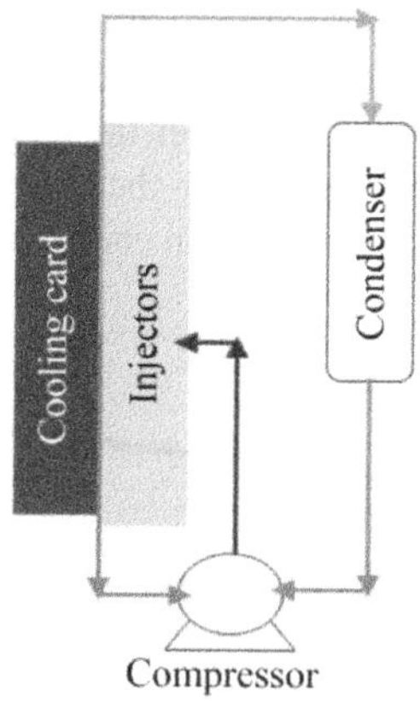

Figure 1.35. *Implementation of spray cooling. For a color version of this figure, see www.iste.co.uk/benallou/energy6.zip*

1.8.7. *Micro heat pipes for electronic systems*

Figure 1.36 shows one way to use a micro heat pipe for cooling electronic systems: six copper heat pipes are mounted on a central processing unit (CPU) to convey heat to a heat sink consisting of fins.

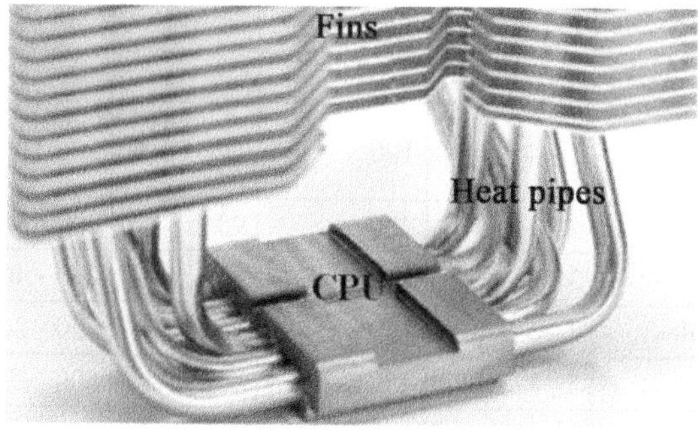

Figure 1.36. *Heat pipes used to cool a central unit. For a color version of this figure, see www.iste.co.uk/benallou/energy6.zip*

1.8.8. *Evaporation-chamber heat exchangers*

Evaporation-chamber heat exchangers are, in reality, evaporation-condensation devices that operate on roughly the same principle as heat pipes: a significant latent-heat fluid is placed in a sealed sleeve, *the chamber*. The latter is assembled on the electronic component to be cooled. Under the action of the heat released by this component, the fluid evaporates. The vapor then condenses on the cold surface of the chamber, which releases the condensation energy to the outside.

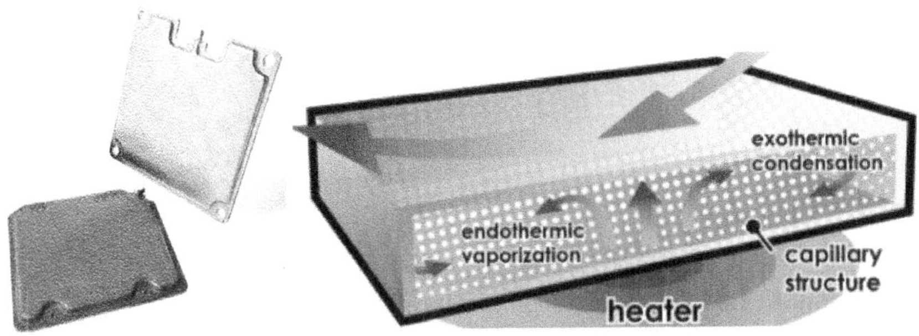

Figure 1.37. *Evaporation-chamber heat exchanger*[7]. *For a color version of this figure, see www.iste.co.uk/benallou/energy6.zip*

Usage	Thickness (mm)	Length (mm)	Width (mm)	Power (W)	Materials
Mobile telephones	0.3 - 0.4	80	50	5	Cu alloy or stainless steel
Laptop computers	0.5	220	60	15 - 25	Cu alloy
Telecom base stations	2 - 4	100	100	50 - 100	Cu or Cu alloy
Graphics cards	2 - 4	200	90	200 - 350	Cu or Ti
Servers	3 - 4	150	80	200 - 400	Cu
Games console	3 - 5	150	150	150 - 200	Cu
Inverters	3 - 8	450	450	500 - 2,000	Cu

Table 1.11. *Current uses and construction materials of evaporation chambers*

7 Source: http://tglobalcorp.com/storage/upload/product/image/2019-02-11/2PWZVcZfzLV2iTLzDkwC4tNZO1AaceUiSYD9Kgjb.jpe.

Figure 1.37 shows this type of heat exchanger. It should be noted that the dimensions are extremely small to meet the miniaturization requirements of embedded electronics and mobile telephony. The thickness of such an exchanger is from 2 to 8 mm. Current developments are focused on the ultra-miniaturization of these exchangers by introducing ultra-thin evaporation chambers: thicknesses from 0.3 mm to 2 mm using walls made of titanium, stainless steel or copper alloys. Table 1.11 shows the current uses of these types of exchangers, as well as their typical dimensions and the thermal power conveyed.

2

Tubular Heat Exchanger Design

2.1. Introduction

Thanks to the simplicity of their construction, tubular heat exchangers are the most commonly encountered in industrial applications. They cover a wide range of uses and come in different forms. As we saw in Chapter 1, their simplest form is *the coaxial tubular heat exchanger, also referred to as a single-tube exchanger*; it consists of two coaxial tubes. *The coaxial tubular heat exchanger* is characterized by the dimensions of the pipes that constitute it (see section 2.2). These determine the values of the heat transfer coefficients, and, consequently, the transfer area.

Another heat exchanger that is very simple and commonly encountered in industry, especially for liquid cooling, is *the coil-in-tank heat exchanger*. This is quite simply composed of a coil that is immersed in a tank through which a cooling (or heating) fluid passes. Section 2.3 is dedicated to presenting the parameters that characterize this type of exchanger.

But by far the most common form of tubular exchangers is *the multitube heat exchanger*. Frequently used in different industries, the multitube heat exchanger takes the form of several tubes placed in a sleeve (shell). The heating or cooling fluid passes through the shell, and the second fluid circulates in the tubes. The data defining the construction technologies for this category of heat exchangers are presented in section 2.4.

Lastly, it should be noted that the parameters and nomenclatures presented in this chapter are important for *determining the exchange area* and for the *general sizing procedure*, which are presented, respectively, in Chapters 3 and 4 of this book.

2.2. Coaxial tubular heat exchangers

A coaxial heat exchanger consists of two concentric tubes (see Figure 2.1). This type of exchanger is characterized by the following parameters:

– the outer diameter of the inner tube: D_1;

– the inner diameter of the outer tube: D_2;

– the hydraulic diameter for calculations relating to flow in the annulus: $D_H = D_1 - D_2$;

– the Reynolds number for the annulus: $\mathrm{Re_H} = \frac{D_H v \rho}{\mu}$;

– D_{H1}, the thermal hydraulic diameter for the inner tube: $D_{H_1} = \frac{D_2^2 - D_1^2}{D_1}$;

– D_{H2}, the thermal hydraulic diameter for the outer tube: $D_{H_2} = \frac{D_2^2 - D_1^2}{D_2}$.

These different parameters determine the convective heat transfer conditions between the inner tube and the fluid circulating in the annulus (see Appendix 2). They thus define the corresponding heat transfer area (see Chapter 3).

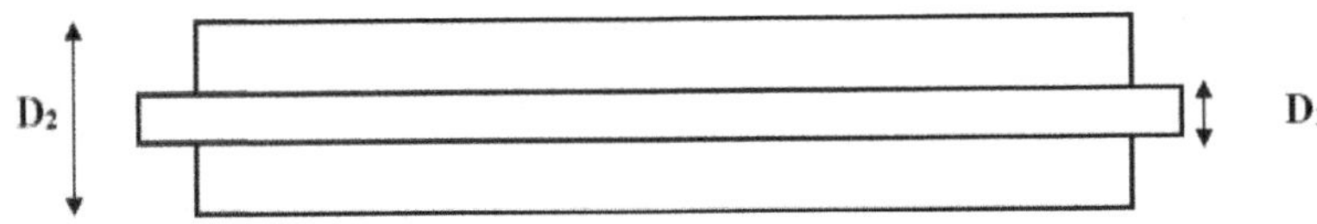

Figure 2.1. *Coaxial heat exchanger*

2.3. Tubular coil-in-tank heat exchangers

Coil-in-tank heat exchangers are commonly used to cool liquids as they leave their production unit. They are also used for heating a liquid load contained in a tank. The liquid load is generally stirred to increase heat transfer. The heat-transfer fluid circulates in the coil, which may be immersed inside the tank (see Figure 2.2a) or envelope it (see Figure 2.2b).

The coil generally consists of a tube, often made of copper, that is wound into a helical form. Using coils makes it possible to obtain a large transfer area between the coil and the liquid located in the tank.

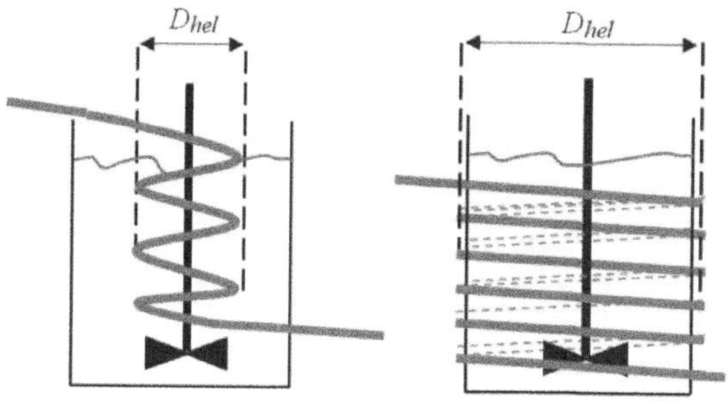

a) b)

Figure 2.2. *a) Immersed coil and b) outer coil. For a color version of this figure, see www.iste.co.uk/benallou/energy6.zip*

Different exchangers of this type are distinguishable by the turbines used to ensure the mixing of the fluid in the tank. Indeed, the quality of this mixing determines, to a large extent, the efficiency of the helix-tank heat exchange (see Appendix 2). As a matter of fact, it determines the convective transfer coefficient in this medium. Thus, coil-in-tank heat exchangers can be distinguished based on the type of turbine used. Figure 2.3 shows the two main types.

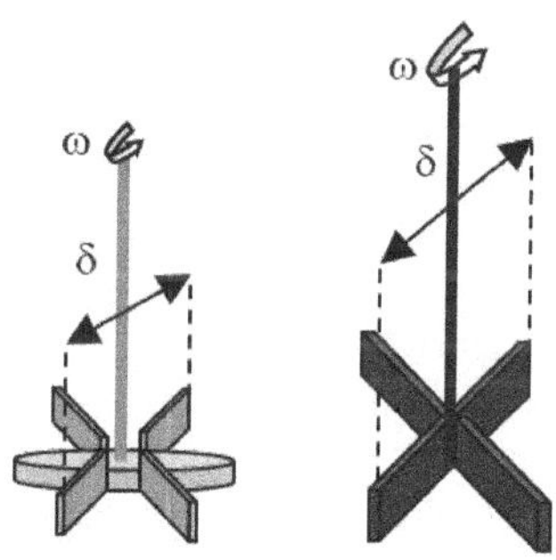

a) b)

Figure 2.3. *a) Blade-disc turbine and b) straight-blade turbine. For a color version of this figure, see www.iste.co.uk/benallou/energy6.zip*

Coil-in-tank heat exchangers are characterized by the following parameters:

d_i: inner diameter of the tube and d_o: outer diameter of the tube;

– D_{tank}: tank diameter;

– D_{hel}: helix diameter;

– δ: turbine diameter;

– ω: turbine rotational velocity;

– Re_m: mixer Reynolds number, defined by: $Re_m = \frac{\delta^2 \omega \rho}{\mu}$.

2.4. Multitube heat exchanger technology

Multitube heat exchangers are used both for heat transfers between two fluids not subjected to a phase change, and for condensing a vapor or evaporating a fluid.

A multitube heat exchanger of course contains several tubes. Figure 2.4 shows this type of exchanger. It particularly shows the tube bundle, the fluid inlets and outlets (in the tubes and in the shell), and a set of elements defined in sections 2.4.1 to 2.4.6.

2.4.1. *Tube plates*

Each end of the tube is attached to a disc, called a tube plate, which is quite simply a plate onto which the tubes are generally either welded or stamped.

Tube plates, together with the sleeves that protect them, are often called *shell heads* or *exchanger heads*. These heads serve both to support the tubes via the tube plates, and to *ensure the orientation of the fluid* toward the inlet or outlet of the tubes.

2.4.2. *Fixed-head heat exchangers*

The head of a multitube heat exchanger can be welded onto the shell or bolted to it after inserting an expansion joint between the two. When the heads are welded, the resulting heat exchangers are said to be *fixed-head*. In these exchangers, the tube bundle cannot be removed, and *no relative motion between the shell and the tube bundle* is permitted. It is the simplest and least-expensive multitube heat exchanger, but it is also an exchanger that *does not allow expansions which may occur due to large temperature differences*. It also does not allow the tubes to be disassembled for cleaning. It is therefore *to be used in cases where there are no large temperature differences* and where the fluids are clean, that is to say, their circulation does not yield deposits (scale or suspended matter) on the tubes. Indeed, these deposits can significantly reduce heat exchange performance (see Appendix 4).

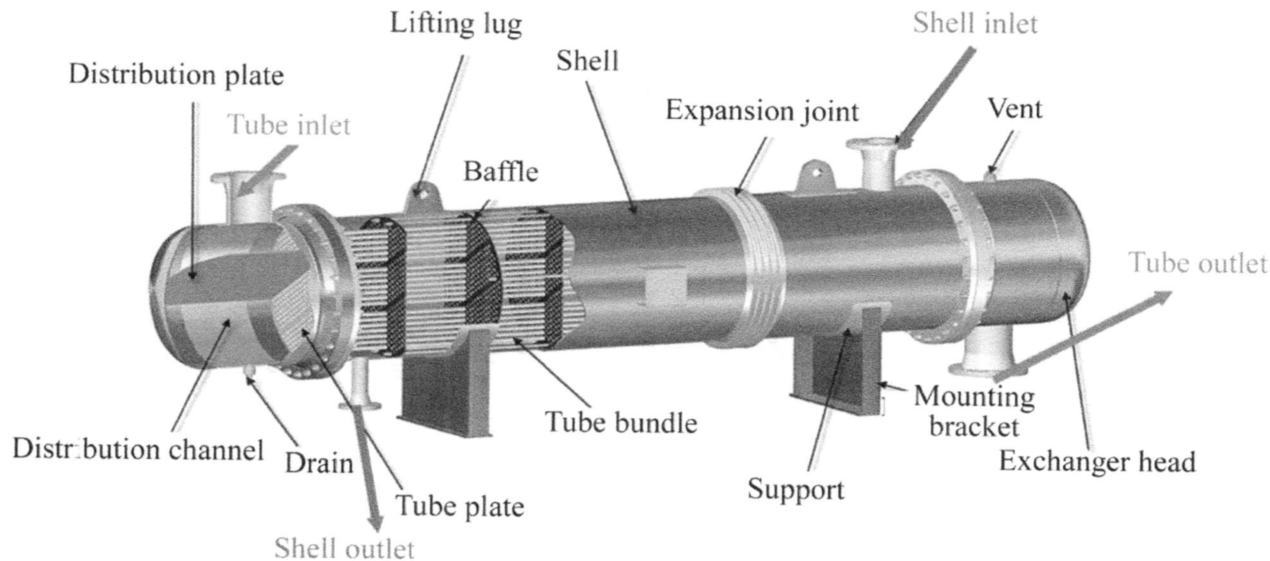

Figure 2.4. *Tubular heat exchangers (source: https://thermofin.net/wp-content/uploads/2016/10/caleos-echangeur-a-faisceau-tubulaire-detail. Courtesy of Thermofin, www.thermofin.net). For a color version of this figure, see www.iste.co.uk/benallou/energy6.zip*

2.4.3. *Floating-head heat exchangers*

By contrast, the heads of heat exchangers can sometimes be required to have a certain flexibility in order to *be able to absorb expansions induced by large temperature variations*. To meet this requirement, multitube heat exchangers need to have *flexible or even removable heads*. This can be realized thanks to the use of joints that withstand thermal expansions, without exceeding the mechanical stresses admissible by the materials constituting the exchanger. They are generally referred to as *floating-head heat exchangers*. *U-tube heat exchangers* also belong to this category. Floating-head heat exchangers are thus *to be used in cases where there are large temperature differences between the hot and cold fluids*. Another advantage of this type of exchanger is that the tubes can be removed, *facilitating their cleaning or replacement*.

2.4.4. *Multipass heat exchangers*

When the two fluids pass through the heat exchanger only once, the exchanger is said to be *one-pass, single-pass or mono-pass*.

In order to increase the effective transfer area per unit length of the device, it is sometimes envisaged for the fluid circulating in the tubes, and possibly that flowing in the shell, to be passed through the heat exchanger several times. This type of device is called a *multipass heat exchanger*.

In the case where several passes are considered on the shell side, the flow of the fluid is delimited by longitudinal baffles. Figure 2.5 shows a shell where the exchanger tubes are not depicted in order to clearly illustrate fluid circulation in such a shell. This figure shows that in the case of a two-pass shell, a longitudinal baffle separates the available volume into two parallel spaces, allowing the fluid to make a round trip within the device before exiting.

Given that the tubes can also make several passes within the vessel, multipass heat exchangers are characterized by two figures. The first indicates the number of shell-side passes. The second corresponds to the number of tube-side passes. We will therefore refer to *m-n heat exchangers*, where:

– *m* is the number of shell-side passes;

– *n* is the number of tube-side passes.

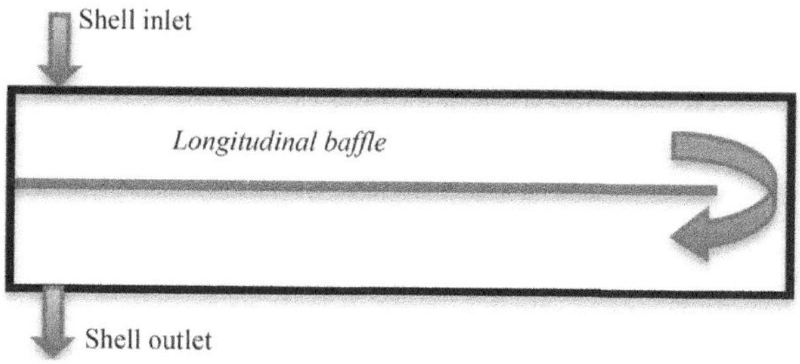

Figure 2.5. *Separation of the shell volume by a longitudinal baffle. For a color version of this figure, see www.iste.co.uk/benallou/energy6.zip*

Thus, a 2-4 heat exchanger allows the fluids to make two shell passes and four tube passes.

The geometric configurations and the technical implementation schematics become complex as the number of passages multiplies. However, the gains in compactness and efficiency justify these technical complications.

Figure 2.6 shows, for illustrative purposes, a 1-2 heat exchanger (one shell-side pass and two tube-side passes) and a 2-2 heat exchanger with two shell-side passes and two tube-side passes.

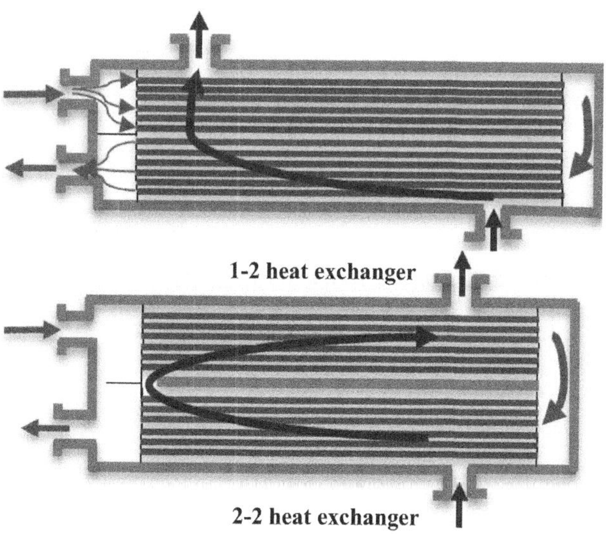

Figure 2.6. *Multipass heat exchangers. For a color version of this figure, see www.iste.co.uk/benallou/energy6.zip*

2.4.5. *Transverse baffles*

It should be noted that, inside the shell, the tubes generally pass through *vertical plates* called *transverse baffles*, which make it possible to support the tubes inside the shell, on the one hand, and to improve the shell-side flow characteristics, on the other hand.

The transverse baffles are attached to the inside of the shell, perpendicular to the direction of flow (see Figure 2.4). The tubes cross the baffles through holes that leave a passage for the fluid circulating in the shell.

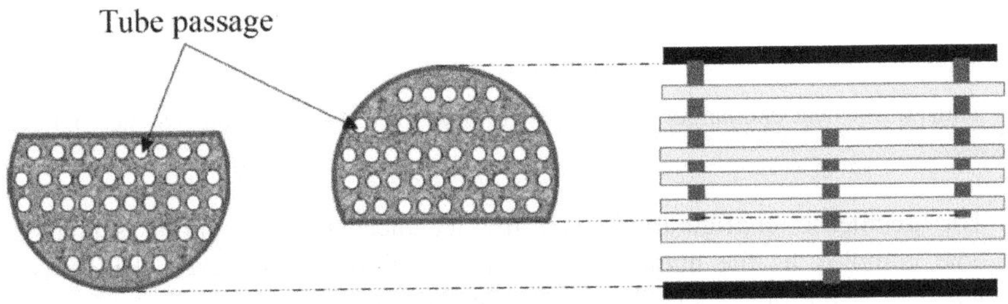

Figure 2.7. *Transverse baffles in the form of single segments. For a color version of this figure, see www.iste.co.uk/benallou/energy6.zip*

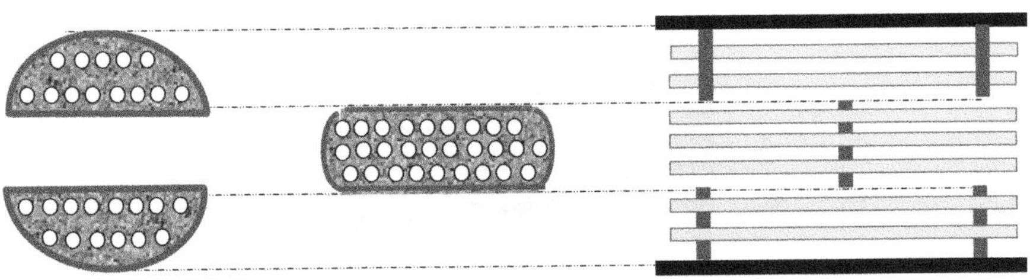

Figure 2.8. *Transverse baffles in the form of discs and segments. For a color version of this figure, see www.iste.co.uk/benallou/energy6.zip*

The baffles can be of various shapes. The most common are those *in the form of discs*, on which segments have been cut out. They may be *single-segment* (see Figure 2.7), *segment and disc* (see Figure 2.8), or *disc and crown* (see Figure 2.9).

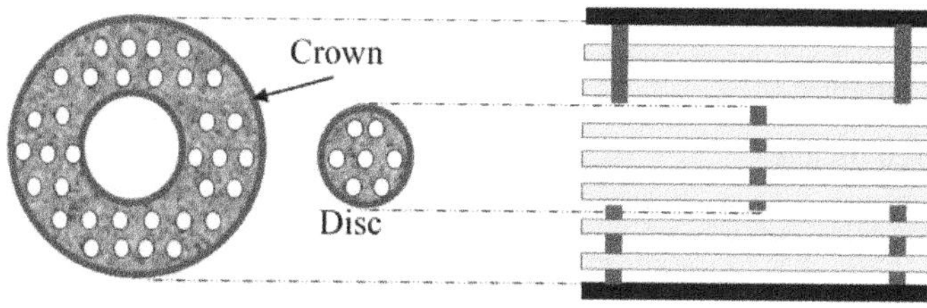

Figure 2.9. *Baffles in the form of discs and crowns. For a color version of this figure, see www.iste.co.uk/benallou/energy6.zip*

2.4.6. *Arrangement of tubes inside tubular heat exchangers*

2.4.6.1. *Tube layers*

In industrial heat exchangers, the position of the tubes is precisely defined. In fact, the tubes are arranged equidistantly from one another to form parallel *layers* (see Figure 2.10). Each layer consists of a given number of tubes. The distance between the layers is itself well defined. The set of layers forms a tube bundle.

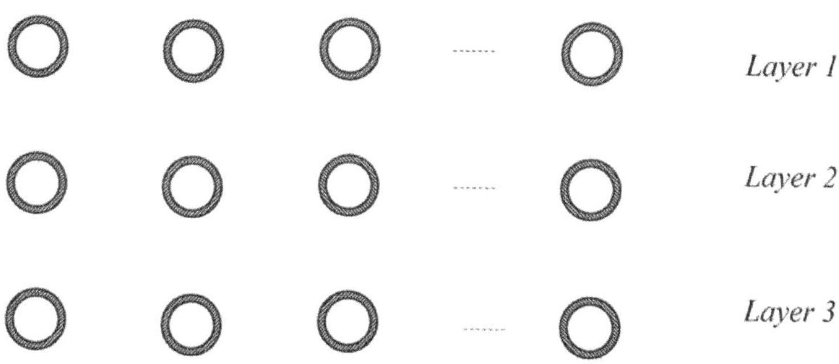

Figure 2.10. *Bundle of three layers of tubes*

2.4.6.2. *Tube bundles*

A tube bundle generally consists of several layers. It is characterized by:

– the number of layers;

– the bundle *array*. This can be square or triangular. The array is the *elementary geometric configuration of the tubes*, based on which the bundle is formed. It is a kind of period that allows the bundle to be constructed repetitively.

There are two types of bundles (see Figures 2.11 and 2.12):

– in-line bundles;

– staggered bundles.

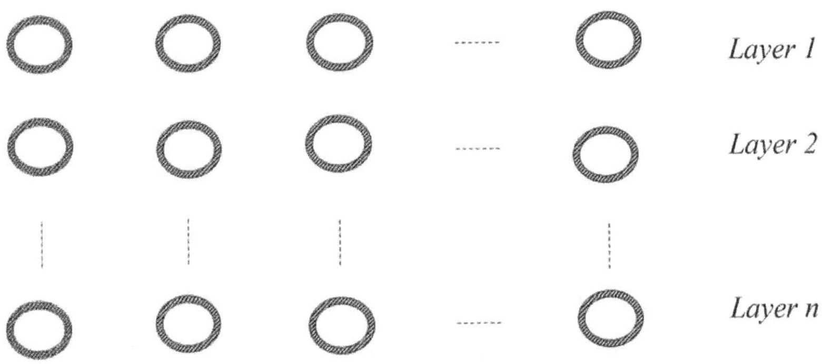

Figure 2.11. *Bundle of n in-line layers*

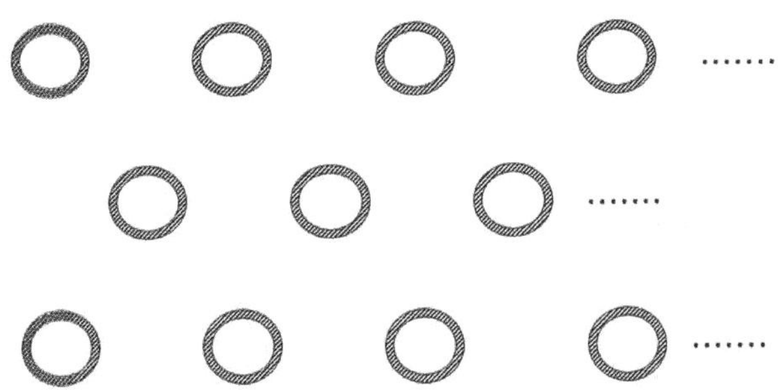

Figure 2.12. *Bundle of three staggered layers*

2.4.6.3. *Square and triangular arrays related to flow patterns*

Tube bundles are defined by a *square* or *triangular* array. Two types of square arrays are encountered (see Figure 2.13, the *in-line square array* and the *staggered square array*). The staggered square array is obtained by rotating the in-line square array 45°.

The in-line square array is often referred to as a *90° array*. In this array, the flow forms a 90° angle with one side of the array and is parallel to the other side. In the staggered square array (*45° array*), meanwhile, the flow forms a 45° angle with both sides of the array.

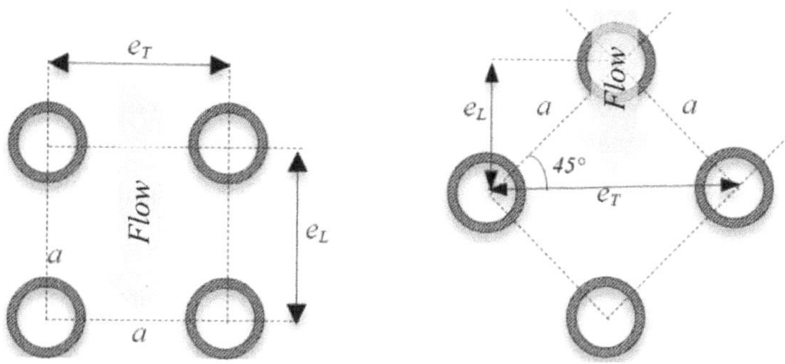

a) In-line, 90° flow (b) Staggered, 45° flow

Figure 2.13. *Square array. For a color version of this figure, see www.iste.co.uk/benallou/energy6.zip*

Likewise, there are two different types of triangular array; the in-line triangular array (also called *30° array*) and the staggered triangular array (*60° array*).

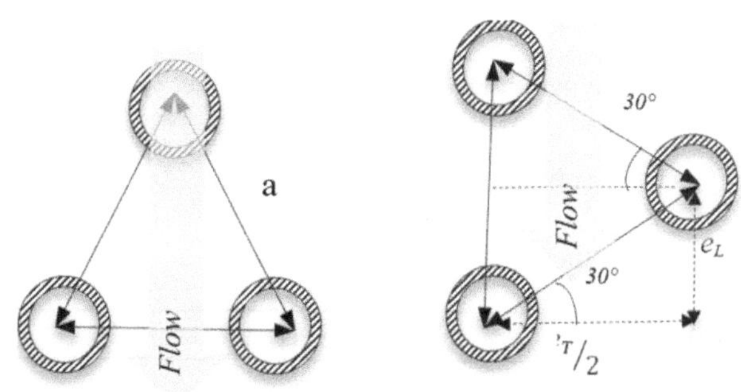

a) In-line, 30° flow b) Staggered, 60° flow

Figure 2.14. *Triangular array. For a color version of this figure, see www.iste.co.uk/benallou/energy6.zip*

The in-line triangular array is constructed from an equilateral triangle of side *a*. The staggered triangular array is obtained by rotating the in-line triangular array 30°.

2.4.6.4. *Array pitch and spacings*

The different arrays (square and rectangular) are characterized by the following three parameters:

– the pitch of the array: this is the side, a, of the square or the equilateral triangle. It determines the distance between the centers of two adjacent tubes inside the square or the basic triangle;

– the transverse spacing, e_T, which defines the distance between the centers of two adjacent tubes of the same layer;

– the longitudinal spacing, e_L, which defines the distance between two consecutive layers.

Thus, for an in-line square array of side a, we have (see Figure 2.13):

$$e_T = e_L = a$$

But, for a staggered square array of side a, we have:

$$e_T = 2\,a\cos\left(\frac{\pi}{4}\right) \quad \text{and} \quad e_L = a\sin\left(\frac{\pi}{4}\right)$$

Likewise, for an in-line triangular array of side a, we have (see Figure 2.14):

$$e_T = a \quad \text{and} \quad e_L = a\sin\left(\frac{\pi}{3}\right)$$

But, for a staggered triangular array of side a, we have:

$$e_T = 2\,a\cos\left(\frac{\pi}{6}\right) \quad \text{and} \quad e_L = a\sin\left(\frac{\pi}{6}\right)$$

Table 2.1 summarizes these different cases.

Array type	e_L	e_T
In-line triangular (30°)	$e_L = a\,\frac{\sqrt{3}}{2}$	$e_T = a$
Staggered square (45°)	$e_L = a\,\frac{\sqrt{2}}{2}$	$e_T = a\sqrt{2}$
Staggered triangular (60°)	$e_L = \frac{a}{2}$	$e_T = a\sqrt{3}$
In-line square (90°)	$e_L = a$	$e_T = a$

Table 2.1. *Expressions of tube spacings according to array type*

2.4.6.5. *Outer limit of the tubes in the shell*

When placing the tubes in the shell, we will take care to leave a distance, Δ, between the inner wall of the shell and the circumference of diameter D_{lim}, defined by the outer limit of the last tubes in the bundle (see Figure 2.15).

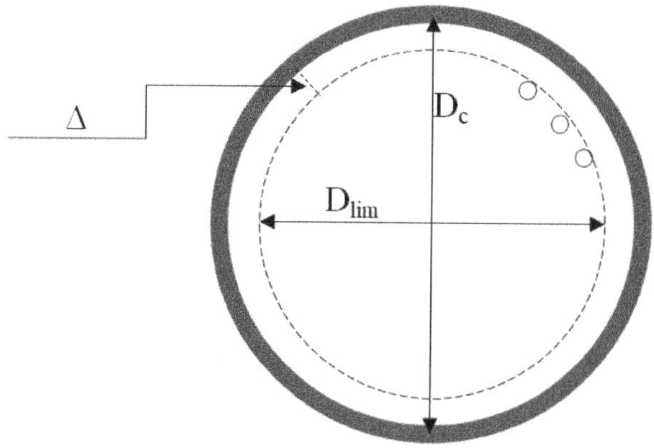

Figure 2.15. *Outer limit of tubes in the shell. For a color version of this figure, see www.iste.co.uk/benallou/energy6.zip*

$$\Delta \geq D_{sh} - D_{lim}$$

where D_{sh} is the shell inner diameter and D_{lim} is the limit diameter defined by TEMA standards, as follows:

$$D_{lim} = \sqrt{\frac{C_1 N_t a^2}{0.78}} + d_o$$

where:

– $C_1 = 0.86$ for 30° tube arrangements, = 1 for 45° or 90° arrangements (see Figures 2.13 and 2.14);

– N_t is the total number of tubes in the shell;

– a is the pitch;

– d_o is the outer diameter of the tubes.

2.5. Flow in a tubular heat exchanger

When specifying a heat exchanger, the direction of flow of the fluids must be given, as these can either flow in the same direction, or in opposite directions. If the two fluids flow in the same direction, the heat exchanger is said to be *parallel flow* or *co-flow*. If, on the other hand, the two fluids move in opposite directions, the exchanger is said to be *counter-flow*. Figure 2.16 offers a schematic view of the flow circuits in a counter-flow multitube heat exchanger.

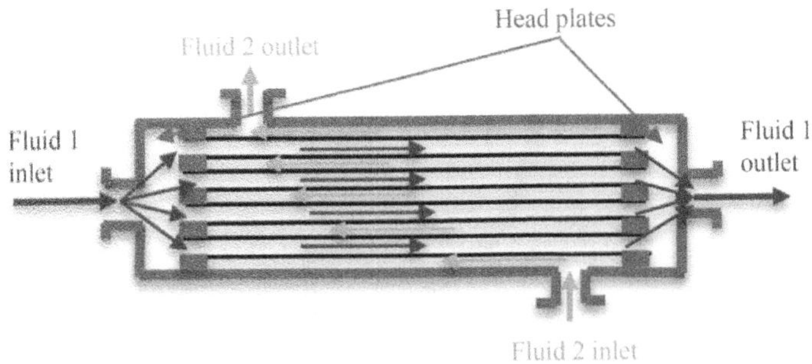

Figure 2.16. *Fluid circuit in the shell (Fluid 2) and in the tubes (Fluid 1). For a color version of this figure, see www.iste.co.uk/benallou/energy6.zip*

This diagram shows that fluid 1 circulates in the tubes and fluid 2 circulates in the shell, of course with no physical contact between the two fluids (no mixing). It also shows, in a simplified manner, how the head plates ensure the distribution of fluid 1 on the tubes and its collection at the other end of the heat exchanger.

Figure 2.17 illustrates how the two types of heat exchanger are schematically represented in relation to the flow.

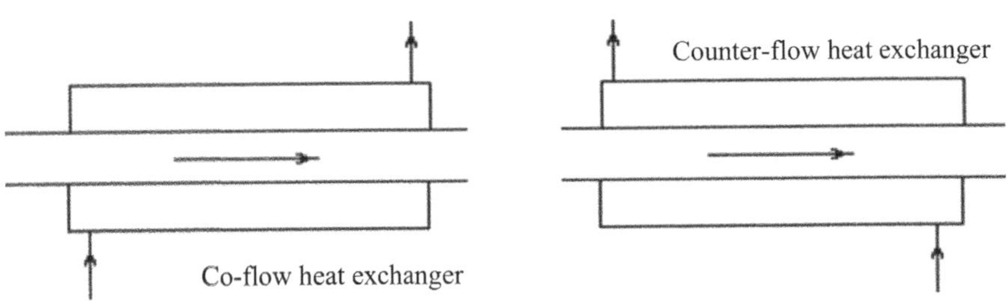

Figure 2.17. *Co-flow and counter-flow*

In reality, the flow inside the heat exchanger is a little more complex than the schematics of Figure 2.17. Indeed, while the flow in the tubes is generally straight and could correspond to the schematic, that in the shell is generally far from being straight. The reason is that the transverse baffles change the directions of fluid flow; the fluid is thus returned lengthwise and widthwise by means of these baffles, which block part of the passage. Figure 2.18 shows a shell fitted with transverse baffles where the tubes are not represented to allow a clear representation.

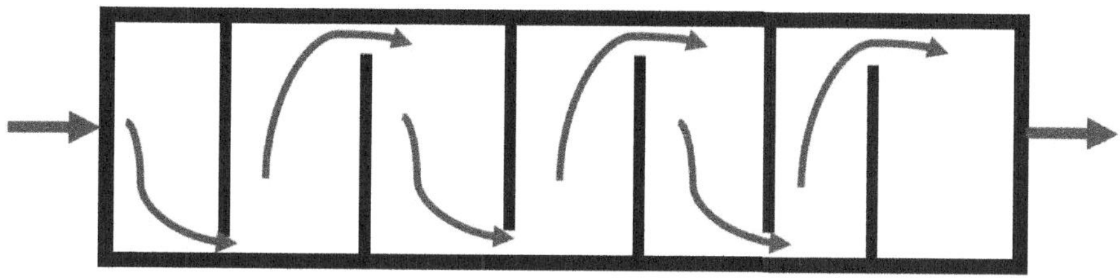

Figure 2.18. *Fluid flow around the baffles in the shell. For a color version of this figure, see www.iste.co.uk/benallou/energy6.zip*

Thus, in the shell, circulation therefore sometimes occurs parallel to the tubes, and sometimes perpendicularly. It should be noted, however, that the cross-section area available for the flow is different, depending on the case. It is considerably reduced when passing (below or above) the baffles. This results in increases in flow velocity at these passages, inducing an increase of turbulence.

Thus, by improving turbulence, transverse baffles should lead to an improvement of heat exchange between the tubes and the shell fluid. Figure 2.19 shows a shell fitted with transverse baffles in which the tubes are represented. It should be recalled that while allowing an increase in turbulence, transverse baffles also serve as mechanical supports for the tubes, allowing them to not bend inside the shell.

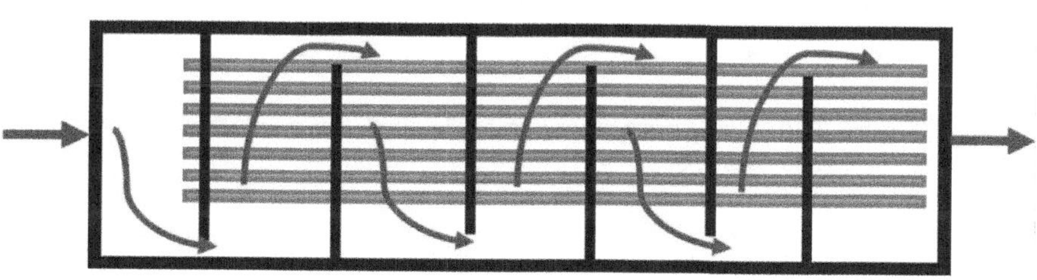

Figure 2.19. *Fluid flow around transverse baffles in the shell. For a color version of this figure, see www.iste.co.uk/benallou/energy6.zip*

2.6. Position of transverse baffles and flow parameters

While transverse baffles play a dual role in the shell (mechanical support of the tubes, and improvement of heat transfer), they render the shell-side flow complex. This flow is actually difficult to describe; it is sometimes transverse (perpendicular to the tubes) when passing between two baffles, and sometimes longitudinal (parallel to the tubes), when passing (above or below) a baffle. This complex flow configuration has a significant impact on the definition of the flowrate to be used in hydraulic and thermal calculations of tubular heat exchangers, and in particular for determining the Reynolds number.

Thus, in order to correctly describe the flow inside the shell, it is necessary to use a number of parameters, which will have a key role in calculating Reynold's numbers inside the heat exchanger shell. These parameters are defined in sections 2.6.1 to 2.6.6.

2.6.1. *Position of baffles in the shell*

Consider a heat-exchanger shell of inner diameter D_{sh}, comprising N_t tubes of outer diameter d_o, placed inside a circumference of diameter D_{lim} (see Figure 2.20). To support the tubes and improve turbulence, transverse baffles are arranged in the shell with a spacing, e_c. This spacing is generally not set arbitrarily.

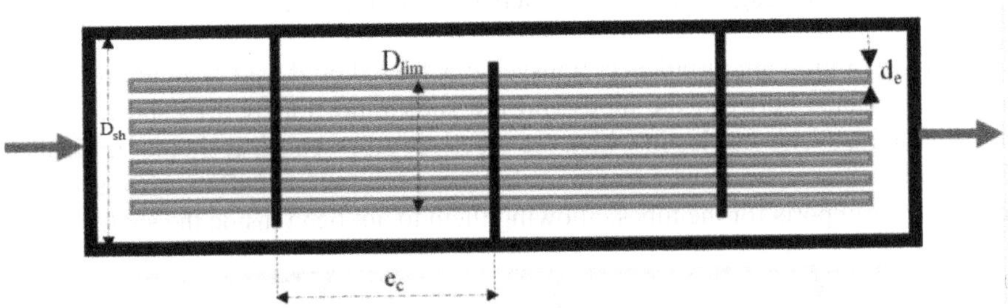

Figure 2.20. *Flow parameters in a baffle shell. For a color version of this figure, see www.iste.co.uk/benallou/energy6.zip*

The Tubular Exchanger Manufacturers Association (TEMA) recommends that e_c be greater than 1/5th of the shell diameter and never less than 2 inches.

In practice, we will use the following constraint to fix e_c:

$$\frac{D_{sh}}{5} \leq e_c \leq D_{sh}$$

2.6.2. *Baffle geometry*

Figure 2.21 shows a cross-section of a shell where the tubes are not represented for simplicity of representation, but where we show a segmental baffle. Note that the baffle leaves above it a free area, A_F, and a free height, δ_F. A_F is often referred to as the *baffle window*.

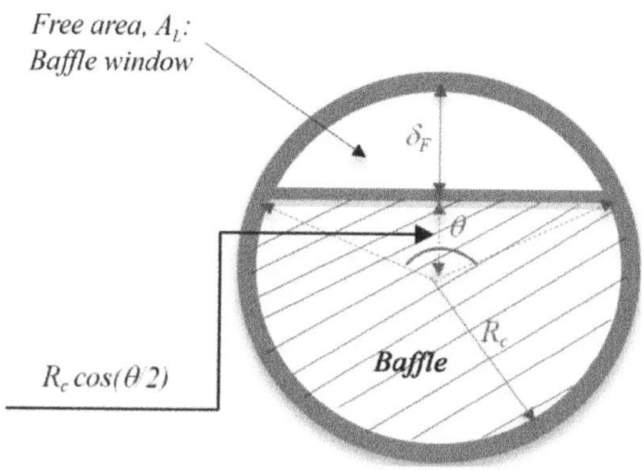

Figure 2.21. *Free area, A_F, and height, δ_F, above the baffle. For a color version of this figure, see www.iste.co.uk/benallou/energy6.zip*

In practice, the baffle is characterized by one of the following geometric parameters:

– the free height, δ_F, above the baffle;

– the central angle, θ, which intercepts the arc of circle defined by the baffle segment;

– the cutting ratio, ε, which is the ratio between the free area, A_F, and the transverse cross-section area of the shell, A_T: $\varepsilon = \frac{A_F}{A_T} = \frac{4A_F}{\pi D_{sh}^2}$; D_{sh} being the shell diameter (radius R_{sh}).

These different parameters are linked by the following relations:

$$\delta_F = R_{sh}\left[1 - \cos\left(\frac{\theta}{2}\right)\right] \text{ or } \theta = 2\cos^{-1}\left(1 - \frac{\delta_F}{R_{sh}}\right) \text{ and } A_F = \frac{R_{sh}^2}{2}(\theta - \sin\theta)$$

Thus, knowledge of one of these parameters makes it possible to determine the others, but it should be noted that knowing ε does not make it possible to directly determine δ_F. This first requires determination of the central angle, θ.

2.6.3. *Illustration: A_F, δ_F and θ from ε*

Baffles are generally produced by carrying out cuts on discs of diameter D_{sh}. These cuts are generally characterized by the ratio (ε), which represents the cut surface area with respect to the total surface area of the disc.

The cutting ratios are between 15 and 45 % in practice. Cuts of 20 to 25% generally allow flows in the shell that ensure good energy transfer, while maintaining pressure drops within acceptable limits.

1) Calculate δ_F and θ corresponding to ε ratios of 15%, 20%, 25%, 30%, 40%, 45% and 50%.

2) Show that for cutting ratios greater than or equal to 25 %, the free height above the baffle, δ_F, can be approximately related to ε by: $\delta_F \cong \varepsilon D_{sh}$ with a relative error below 15%.

Given datum: D_{sh} = 520 mm.

Solution

1) Calculation of δ_F and θ for different ratios

We have: $\varepsilon = \frac{4A_F}{\pi D_{sh}^2}$.

Hence: $A_F = \frac{\varepsilon\pi D_{sh}^2}{4}$.

Moreover: $A_F = \frac{R_{sh}^2}{2}(\theta - \sin\theta)$.

θ is thus the first solution to the equation: $\theta - \sin\theta = \frac{8A_F}{D_{sh}^2}$.

Once the value of θ is known, δ_F is calculated from: $\delta_F = R_{sh}\left[1 - \cos\left(\frac{\theta}{2}\right)\right]$.

Table 2.2 presents the results obtained for the different values of ε.

ε	A_F (m²)	$\frac{8A_{LF}}{D_{sh}^2}$	θ	δ_F (in m)
15%	0.032	0.94	1.89	0.11
20%	0.042	1.26	2.11	0.13
25%	0.053	1.57	2.31	0.15

30%	0.064	1.88	2.49	0.18
35%	0.074	2.20	2.66	0.20
40%	0.085	2.51	2.83	0.22
45%	0.096	2.83	2.98	0.24
50%	0.106	3.14	3.14	0.26

Table 2.2. *Heights above the baffle for different values of ε*

2) Approximation of δ_F

The approximation $\delta_F \cong \varepsilon D_{sh}$ leads us to commit a relative error: $\frac{\varepsilon D_{sh}-\delta_F}{\delta_F}$.

Table 2.3 shows the values of this relative error.

ε	15%	20%	25%	30%	35%	40%	45%	50%
δ_F	0.11	0.13	0.15	0.18	0.2	0.22	0.24	0.26
εD_{sh}	0.08	0.10	0.13	0.16	0.18	0.21	0.23	0.26
$\frac{\varepsilon D_{sh} - \delta_F}{\delta_F}$	-29%	-20%	-13%	-13%	-9%	-5%	-2%	0%

Table 2.3. *Comparison of δ_F and εD_{sh}*

We can see that for $\varepsilon \geq 25\%$, we can approximate δ_F by: εD_{sh} with a relative error less than or equal to 13%.

2.6.4. *Shell-baffle and baffle-tube clearances*

Installation of transverse baffles in the shell should observe the following clearances (spaces):

– a clearance between each baffle and the inner wall of the shell;

– a clearance between the tubes and their passages through the baffle (see Figure 2.22).

These clearances are set by TEMA standards as follows:

– The shell-baffle clearance, D_{SB}, is given by: $D_{SB} = 3.1 + 0.004D_{sh}$.

– The tube-baffle clearance, L_{TB}, should be set to: $L_{TB} = 0.0008$ m.

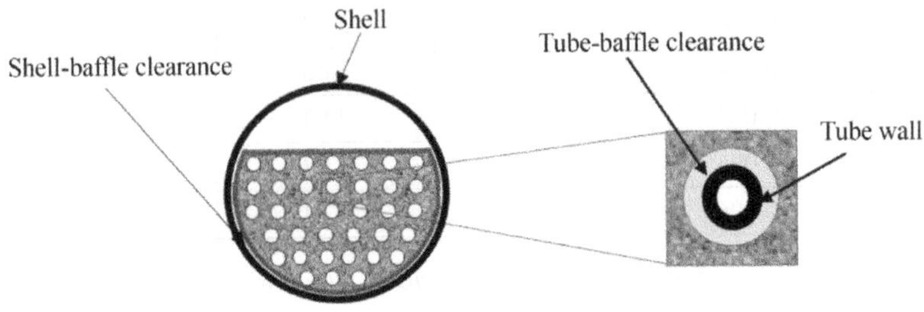

Figure 2.22. *Tube-baffle clearance. For a color version of this figure, see www.iste.co.uk/benallou/energy6.zip*

It should be noted that these clearances induce flow leakages that are characterized by the passage areas at these clearances.

2.6.5. *Longitudinal free area*

Assuming that the transversal cross-section area, A_T, is passed through by N_t tubes of outer diameter d_o, the area that would be available for longitudinal flow, in the absence of a baffle would then be:

$$A_{T_{\text{Without baffles}}} = \frac{\pi}{4}(D_{sh}^2 - N_t d_o^2)$$

The longitudinal free area is defined as the transverse cross-section surface area of the shell that is available for longitudinal flow. It is given by:

$$a_L = \varepsilon A_{T_{\text{Without baffles}}} = \frac{\varepsilon\pi}{4}(D_{sh}^2 - N_t d_o^2)$$

2.6.6. *Transverse free area*

As with longitudinal flow, the transverse free area is defined by:

$$a_T = \frac{e_c D_{sh}(\sqrt{e_T e_L} - d_o)}{\sqrt{e_T e_L}}$$

where:

– e_c is the spacing between baffles;

– D_{sh} is the shell diameter;

– d_o is the outer diameter of the tubes;

– e_T is the transverse spacing of the tubes;

– e_L is the longitudinal spacing of the tubes.

2.6.7. *Flowrates*

The flow in the shell is thus represented by a longitudinal flow, through the longitudinal free area, a_L, and a transverse flow through the transverse free area, a_T. As a result, two flowrates in the shell are defined: a longitudinal flowrate and a transverse flowrate.

– Longitudinal mass flowrate is defined by: $G_L = \frac{W}{a_L}$, where W is the mass flowrate on the shell side;

– The transverse mass flowrate is then defined by: $G_T = \frac{W}{a_T}$.

The average mass flowrate is thus defined as the geometric average of the mass, longitudinal and transverse flowrates: $G_M = \sqrt{G_L G_T}$.

The maximum mass flowrate is defined as follows for each tube bundle.

For an in-line bundle, G_{Max} is defined by:

$$G_{Max} = \frac{v_{sh}\rho_{sh}e_T}{e_T - d_o}$$

where:

– v_{sh} is the velocity of the fluid in the shell, away from the baffle; $v_{sh}\rho_{sh} = \frac{4W}{\pi D_{sh}^2}$;

– ρ_{sh} is the density of the fluid circulating in the shell;

– d_o is the outer diameter of the tubes.

For a staggered bundle, G_{Max} can be defined either with respect to the minimum area between tubes of the same layer, or with respect to the minimum area between tubes of the same diagonal. Taking the minimum area between the tubes of the same layer, we will have: $G_{Max} = \frac{v_\infty \rho_\infty e_T}{e_T - d_o}$.

The maximum transverse mass flowrate is defined in relation to the minimum clearance area, a_{Min}:

$$G_{TM} = \frac{W}{a_{Min}}$$

W being the mass flowrate on the shell side.

2.7. Calculation of the Reynolds number for flow in the shell

The presence of baffles and tubes inside the shell leads to complex flow patterns. The case of shells without baffles is rather simple. It is treated using a hydraulic diameter, in a way similar to that used for flows in non-cylindrical pipes. For shells with baffles, in contrast, the Bell-Delaware method is used.

2.7.1. *Use of hydraulic diameter for shells without baffles*

For shells without baffles, the Reynolds number is computed using the hydraulic diameter defined in the same way as for non-cylindrical pipes (see Volume 3, section 3.2):

$$D_H = \frac{4S}{P}$$

where *S* is the free cross-section area perpendicular to the flow and *P* is the wetted perimeter through which the heat exchange takes place.

For each basic array (see Figure 2.23, for example), the free cross-section area is the surface, $e_T\, e_L$, minus the cross-section of a tube. The wetted perimeter is that of a tube.

That is: $S = e_T e_L - \pi \frac{d_o^2}{4}$ and $P = \pi d_o$.

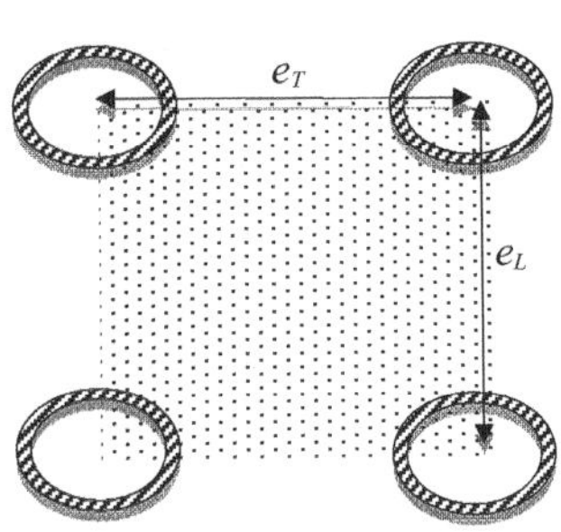

Figure 2.23. *Free cross-section area and wetted perimeter*

Hence: $D_H = 4\,\dfrac{e_T e_L - \pi\frac{d_o^2}{4}}{\pi d_o}$.

This diameter is sometimes called (in some works) the *equivalent diameter*.

Depending on the retained flowrate, two Reynolds numbers are usually used in the correlations recommended for computation of the shell-side heat transfer coefficients: the average Reynolds number, Re_{avg}, and the maximum Reynolds number, Re_{max}.

The average Reynolds number is defined with respect to the average mass flowrate, G_{avg}, as follows:

$$Re_{avg} = \frac{G_{avg} D_H}{\mu}$$

The maximum Reynolds number is defined relative to the maximum flowrate, G_{Max}, as follows:

$$Re_{max} = \frac{G_{max} D_H}{\mu}$$

In the Bell-Delaware method, the shell-side Reynolds number is calculated with respect to the maximum transverse flowrate, G_{TM}:

$$Re = \frac{d_o G_{TM}}{\mu}$$

μ being the shell-side viscosity.

2.7.2. *The Bell-Delaware method for shells with baffles*

Developed for segmental-baffle shells, the Bell-Delaware method considers a flow description that allows deviations with respect to a perfect parallel-flow to be taken into consideration.

In particular, this method takes into account the leakages through the spaces (clearances) between the tubes and the orifices of the baffles on the one hand, and between the baffles and the shell, on the other hand.

These leakages are characterized by the shell-baffle leakage area, a_{sb}, and the tube-baffle leakage area, a_{tb}, defined in sections 2.7.2.2 and 2.7.2.3. It also takes into account the flow's bypass around the space between the tube bundle and the shell.

This bypass is quantified by calculating the minimum clearance area at the center of the tube bundle.

2.7.2.1. *Calculation of the minimum free area at the center of the bundle*

This area is defined by:

$$a_{Min} = e_c\left[D_{sh} - D_{lim} + \frac{D_{lim}-d_o}{e_E}(a - d_o)\right]$$

where:

– e_c is the baffle spacing;

– D_{sh} is the shell diameter;

– D_{lim} is the limit diameter for placement of the tubes in the shell, as defined in section 2.4.6.5;

– a is the pitch and e_E is the effective spacing, given in Table 2.4, as a function of the array arrangement, e_T being the transverse spacing given by Table 2.1 as a function of the array arrangement.

Arrangement	Expression of e_E
30° array	$e_E = e_T$
45° array	$e_E = \frac{e_T}{2}$
90° array	$e_E = e_T$

Table 2.4. *Effective spacing according to the array arrangement*

2.7.2.2. *Leakage free area between shell and baffle*

It is noted a_{sb}. It is defined by: $a_{sb} = D_{sh}D_{SB}\left(\pi - \frac{\theta}{2}\right)$

where:

– θ is the central angle given by (see section 2.6.2): $\theta = 2\cos^{-1}(1 - 2\varepsilon)$;

– ε is the cutting ratio of the baffle; 30 % cut ⇨ $\varepsilon = 0.3$;

– D_{SB} is the shell-baffle clearance (see section 2.6.4): $D_{SB} = \frac{3.1+0.004D_{sh}}{1000}$.

2.7.2.3. *Leakage free area between tubes and baffle*

It is noted a_{tb}.

It is defined by: $a_{tb} = \frac{\pi}{4}\left[(d_o + L_{TB})^2 - d_o^2\right]N_t(1 - F_w)$

where:

– d_o is the outer diameter of the tubes;

– L_{TB} is the tube-baffle clearance (see section 2.6.4): $L_{TB} = 0.0008$ m;

– N_t is the total number of tubes in the shell;

– $F_W = \frac{\theta_{lim} - \sin(\theta_{lim})}{2\pi}$;

– $\theta_{lim} = 2\cos^{-1}\left[\frac{D_{sh}}{D_{lim} - d_o}(1 - 2\epsilon)\right]$; ε being the cutting ratio;

– ε is the cutting ratio (30 % cut ⇨ ε = 0.3);

– $D_{lim} = \sqrt{\frac{C_1 N_t p^2}{0.78}} + d_o$ (see section 2.4.6.5);

– $C_1 = 0.86$ for 30° arrangements and $C_1 = 1$ for 45° or 90° arrangements.

2.7.2.4. *Bypass free area*

This is the cross-section area through which leakage flows occur by bypass. It is given by: $a_b = e_c\left(D_{sh} - D_{lim} - \frac{d_o}{2}\right)$

where e_c is the spacing of the baffles.

2.7.2.5. *Effective number of cross-flow layers between baffles*

It is given by: $N_{tcc} = \frac{D_{sh}}{e_L}(1 - 2\varepsilon)$

where:

– D_{sh} is the shell diameter;

– e_L is the longitudinal spacing given by Table 2.1, according to the layer arrangement.

3

Compact Heat Exchanger Design

3.1. Introduction

As established in Chapter 1, compact heat exchangers designate a category of devices with a compactness of over 700 m^2/m^3.

This category of exchangers is typically used for liquid-gas heat transfer. It has undergone several developments in recent decades, with applications such as embedded equipment on aircraft, locomotives and space shuttles, where space available for equipment installation is one of the main design parameters.

More recently, new uses of compact heat exchangers have emerged; notably the evacuation of heat produced by electronic power equipment operation, by concentrated solar power plants or by fourth-generation nuclear reactors. In all of these new applications, high thermal flux densities are conveyed through minimal volumes and at very high temperatures, requiring the development of compact heat exchangers that utilize new materials.

Chapter 3 is dedicated to presenting technologies and design parameters of this category of exchangers, including:

– cross-flow heat exchangers;

– plate-fin heat exchangers;

– plate heat exchangers;

– spiral heat exchangers;

– printed-circuit heat exchangers.

3.2. General parameters of compact heat exchangers

For the description and calculations of compact exchangers, the commonly-used parameters (e.g. hydraulic diameter, free flow area, etc.) are broken down according to the specific geometry of this category of vessels. Moreover, parameters specific to compact heat exchanger design are also defined in the following sub-sections (Bannon et al. 1965; Kuppan 2000).

3.2.1. *Compactness*

The compactness of any heat exchanger is generally evaluated using a ratio β, defined by the heat transfer area, *A*, divided by the volume of the device. This parameter can be defined with respect to the total volume of the vessel, *V*, or with respect to the volume of each side or flow path of the heat exchanger, V_i.

Therefore, for a compact heat exchanger, we can define three parameters relative to compactness:

– the total compactness defined by: $\bar{\beta} = \frac{A}{V}$; *A and V* being, respectively, the overall transfer area and volume of the vessel;

– the specific compactness of side *i*, defined by: $\beta_i = \frac{A_i}{V_i}$; A_i *and* V_i being, respectively, the transfer area and volume relative to side i; $i = 1; 2$;

– the compactness of side *i* with respect to the total volume, *V*, defined by: $\bar{\beta}_i = \frac{A_i}{V}$; *V* and A_i being, respectively, the total volume and the transfer area relative to side i.

Sometimes referred to as *surface density*, compactness is an important parameter in the specification of compact heat exchangers. Compact heat exchangers have compactnesses exceeding *700* m^2/m^3 and up to *several thousand* m^2/m^3.

3.2.2. *Hydraulic diameter*

The concept of hydraulic diameter extends to compact heat exchangers:

$$D_H = \frac{4\,A_0\,L}{A}$$

where:

– A_0 is the free flow area;

– A is the transfer area;

– L is the length of the flow.

In the same way as for compactness, the hydraulic diameter can be defined for each side or flow path of the heat exchanger. Thus, the hydraulic diameter on side i is defined by:

$$D_{H_i} = \frac{4\, A_{0_i}\, L_i}{A_i};\quad i = 1;\, 2$$

where:

– A_{0i} is the free flow area on side i;

– A_i is the transfer area on side i;

– L_i is the length of the flow on side i.

Compact heat exchangers are characterized by hydraulic diameters of under 6 mm in gas circuits.

3.2.3. *Porosity*

To characterize the extent of the free flow area with respect to the front surface of the exchanger, we use the porosity σ, defined for each side of the heat exchange by:

$$\sigma_i = \frac{A_{0_i}}{A_{f_i}};\ i = 1;\, 2$$

where A_{0_i} is the free flow area of side i, and A_{f_i} is the frontal area of side i.

3.2.4. *Relationship between compactness and hydraulic diameter*

Using the expression of A_{0i} as a function of the hydraulic diameter, we have, in the most general case:

$$A_{0i} = \frac{D_{Hi} A_i}{4\, L_i}$$

Hence: $\sigma_i = \frac{D_{Hi}}{4\, L_i} \frac{A_i}{A_{fi}}$.

Yet: $\frac{D_{Hi}}{4L_i} = \frac{A_{0i}}{A_i} = \frac{\sigma_i A_{fi}}{\beta_i V_i} = \frac{\sigma_i A_{fi}}{\beta_i A_{fi} L_i} = \frac{\sigma_i}{\beta_i L_i}$.

Hence: $D_{Hi} = 4\frac{\sigma_i}{\beta_i}$.

3.3. Finned-tube compact heat exchanger technology and design

Finned-tube heat exchangers constitute an important category of compact heat exchangers. They are very commonly used in industry, especially for liquid/gas exchanges. Their great compactness allows them to also be used as embedded heat exchangers in vehicles (radiators), locomotives (space heating) and aircraft (equipment cooling). They generally consist of individually -or collectively- finned tubes.

3.3.1. *Individually-finned tube heat exchangers*

The primary objective of the fins is the *increase in transfer area* by geometric extensions attached to the tubes. Figure 3.1 shows the most commonly-used fins.

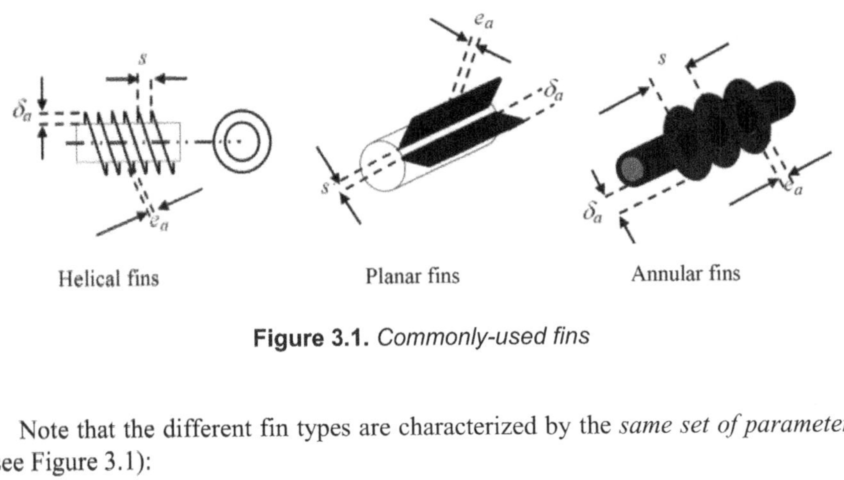

Figure 3.1. *Commonly-used fins*

Note that the different fin types are characterized by the *same set of parameters* (see Figure 3.1):

– the distance, s, at the base between two fins; $s = a_a - e_a$;

– the pitch, a_a, of the fins;

– the height of the fin da;

– the thickness, e_a, of a fin.

In fact, a similar set of parameters is used to characterize the fins of collectively-finned tube heat exchangers, covered in section 3.3.2.

3.3.2. *Collectively-finned tube heat exchangers*

Figure 3.2 shows that this type of device consists of a series of tubes that collectively pass through sheets of fins.

The liquid usually circulates in the tubes and the gas circulates on the outside.

3.3.2.1. *Flow configuration*

As with shell-and-tube heat exchangers, collectively-finned tube heat exchangers can be used in parallel flow (*co-flow*), *counter-flow* or *cross-flow* configurations. However, it is the *cross-flow* scheme that is most common in industry.

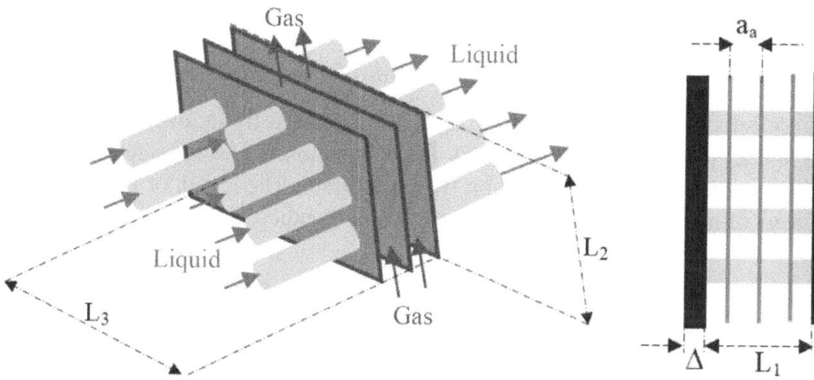

Figure 3.2. *Collectively-finned tube heat exchanger. For a color version of this figure, see www.iste.co.uk/benallou/energy6.zip*

3.3.2.2. *Flux splitting*

DEFINITION.– A flux is said to be split if its fluid flows through individual pipes or ducts without being able to cross in the transverse direction.

Typically, when the hot (or cold) fluid flows through tubes, its flux will be said to be split: this is the case with the liquid in the exchangers shown in Figures 3.2 and 3.3. Note that in Figure 3.2, the gas flux is separated as well, since it flows through individual ducts without crossing from one duct to another. By contrast, in the exchanger shown in Figure 3.3, whereas the liquid flux is split, the gas flux is able to circulate in any direction before leaving the exchanger, and *it is therefore not split*. This device is therefore *a split/mixed-flux cross-flow heat exchanger*. In comparison, the heat exchanger in Figure 3.2 is *a split/split-flux cross-flow* exchanger.

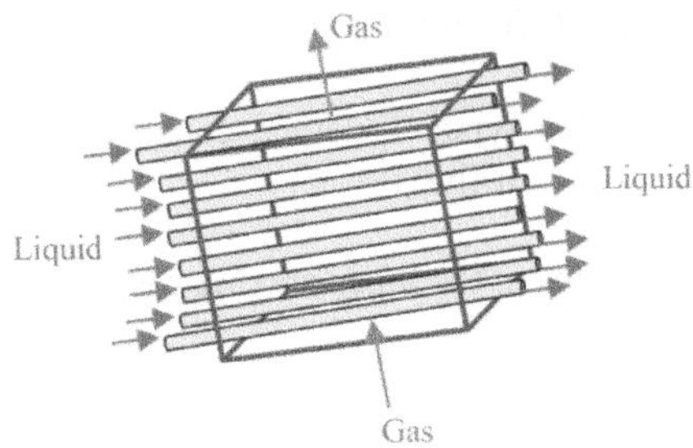

Figure 3.3. *Finned-tube heat exchanger with gas flux not split (fins not shown). For a color version of this figure, see www.iste.co.uk/benallou/energy6.zip*

3.3.2.3. *Overall characteristics of collectively-finned tube heat exchangers*

A collectively-finned tube heat exchanger is defined by its overall dimensions, L_1, L_2 and L_3 (see Figures 3.2 and 3.4). By convention, L_2 *is taken in the gas flow direction* and L_1 in the liquid direction (Kuppan 2000). The length of the tubes is L_1. The layers of tubes are counted in the direction perpendicular to flow.

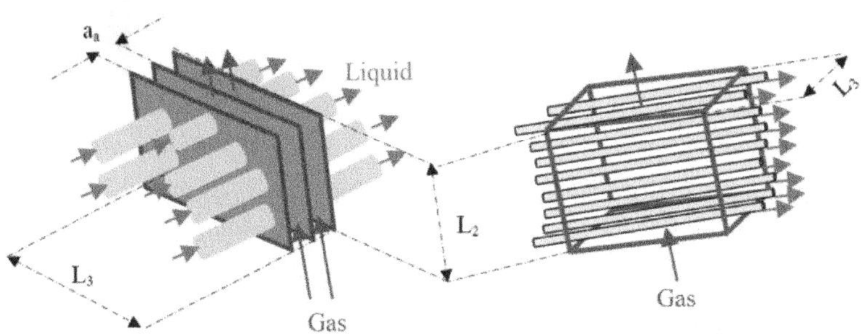

Figure 3.4. *Dimensions of cross-flow heat exchangers. For a color version of this figure, see www.iste.co.uk/benallou/energy6.zip*

3.3.2.4. *Collectively-finned tube heat exchanger specifications*

Fin-and-tube heat exchanger specifications shall therefore specify the following parameters:

– the overall dimensions of the device: L_1 (liquid flow direction), L_2 (gas direction), L_3;

– the flow configuration (parallel flow, cross flow or counter flow);

– the flow mode (split or mixed flows);

– the tube arrangement: array type (square or rectangular, in-line or staggered), array pitch, longitudinal and transverse spacings;

– the fin characteristics (continuous sheet passed through by the tubes or individual fins on each tube: height, pitch, a_a, thickness, e_a).

The parameters describing the tube arrangement in the shell space (cylindrical or parallelepipedic) are the same as those defined for shell-and-tube heat exchangers (see section 2.3.6). The longitudinal (e_L) and transverse (e_T) spacings can therefore be calculated from Table 2.1. Moreover, Figure 3.5 shows the various parameters relating to the tube arrangement, for in-line and staggered bundles of cross-flow finned-tube heat exchangers.

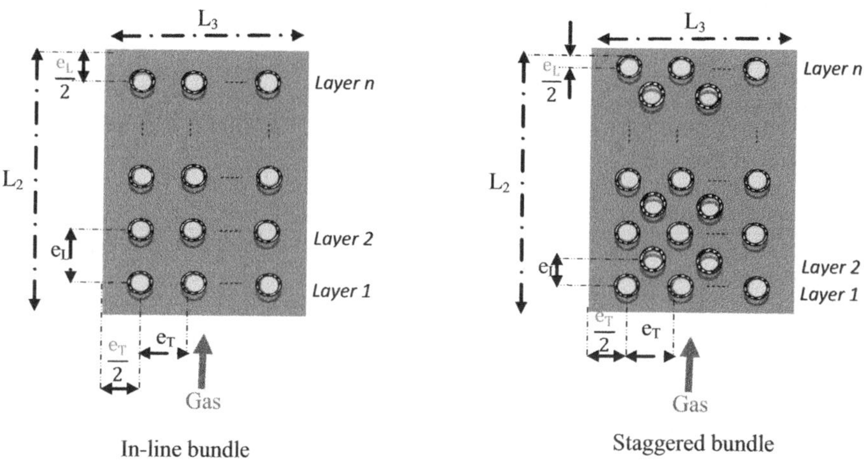

Figure 3.5. *Dimensions of cross-flow heat exchangers. For a color version of this figure, see www.iste.co.uk/benallou/energy6.zip*

The tube-side calculations are the same as those given for shell-and-tube heat exchangers. Conversely, the shell-side sizing shall specify, in particular:

– the number of fins and their geometric and thermal characteristics;

– the number of layers perpendicular to the direction of flow;

– the number of tubes per layer;

– the free flow area;

– the hydraulic diameter;

– the efficiency of the fins and the effective heat transfer area.

These parameters shall be calculated using the relationships presented in Table 3.1.

Parameters	In-line bundles	Staggered bundles
Number of layers[1]	$N_n = \frac{L_2 - e_L}{e_L} + 1 = \frac{L_2}{e_L}$	$N_n = \frac{L_2}{e_L}$
Number of tubes per layer	$n_T = \frac{L_3 - e_T}{e_T} + 1 = \frac{L_3}{e_T}$	For layers 1, 3…; $n_{T_i} = \frac{L_3}{e_T}$ For layers 2, 4…: $n_{T_p} = \frac{L_3}{e_T} - 1$
Total number of tubes	$N_T = N_n n_T$	$N_T = N_i n_{T_i} + N_p n_{T_p}$
Number of fin-sheets	$N_a = \text{Ent}\left(\frac{L_1}{a_a + e_a} - 1\right)$	
Free flow area	$A_0 = (e_T - d_o)L_1 N_T - N_a e_a L_3$	
Heat transfer area	$A = N_T \pi d_o (L_1 - N_a e_a) + 2N_a\left(L_2 L_3 - N_T \frac{\pi d_o^2}{4}\right)$	
Effective transfer area	$A_e = N_T \pi d_o (L_1 - N_a e_a) + 2N_a\left(L_2 L_3 - N_T \frac{\pi d_o^2}{4}\right)\eta_e$	
Plate-fin efficiency	$\eta_e = \frac{\tanh(m_e L_3)}{m_e L_3}$, with: $m_e = \sqrt{\frac{2h_e}{e_a \lambda_a}}$	
Hydraulic diameter	$D_H = \frac{4\,A_0 L_2}{A}$	

Table 3.1. *Calculation of the characteristics of collectively-finned tube heat exchangers*

It should be noted that the *effective heat transfer area* in fact corresponds to the total external exchange area, including the area of the fins, corrected to take account of the efficiency, η_e, of the latter.

1 e_L and e_T are the longitudinal and transverse spacings, respectively. They are calculated from Table 2.1. N_I: number of odd-numbered layers; N_p: number of even-numbered layers. a_a is the pitch of the fin-sheets; e_a is the thickness of one fin-sheet. λ_a is the heat conductivity of the fin-sheet material; h_e is the fin-side convection heat transfer coefficient, given in section A2.5.2.3.

Note that for embedded exchangers (car and locomotive radiators, aircraft heat exchangers, etc.), the overall dimensions, L_1, L_2 and L_3, are generally determined by spatial constraints; the volume available for the heat exchanger installation being generally limited. These space constraints need therefore to be taken into account in designing the exchanger.

3.4. Plate-fin heat exchanger technology and design

Plate-fin heat exchangers are mainly used for gas/gas energy transfers, with compactnesses of around 6,000 m^2/m^3.

3.4.1. *Flow configuration*

Figure 3.6 shows a plate-fin heat exchanger with a cross-flow configuration. Plate-fin heat exchangers can be used in parallel flow (*co-flow), counter-flow* or *cross-flow*, but the cross-flow and counter-flow configurations are most commonly encountered in industry.

3.4.2. *Fin type and arrangement*

Note that Figure 3.6 shows an exchanger with triangular fins. Of course, different technologies use different fin types. These may be corrugated or simply have a square, rectangular or triangular cross-section area or, in principle, be of any shape likely to increase the transfer area of the base plate.

Some technologies offer perforated fins to increase turbulence in gas flows.

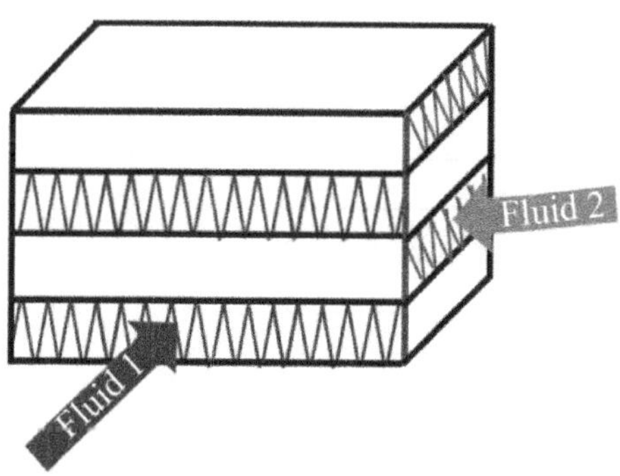

Figure 3.6. *Plate-fin heat exchanger. For a color version of this figure, see www.iste.co.uk/benallou/energy6.zip*

Figure 3.7 shows the various types of fins used in plate-fin heat exchangers.

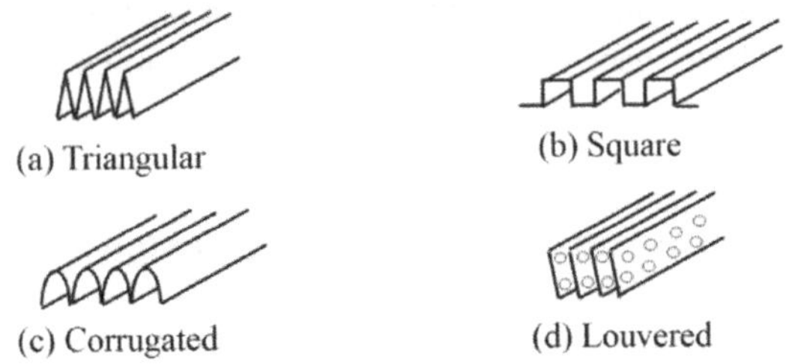

Figure 3.7. *Fins used in plate-fin heat exchangers*

3.4.3. *Fin-plate attachments and exchange modules*

The fins are attached to the plates by welding, brazing or cold pressing. To a large extent, operating temperature ranges are determined by the type of fin-plate bond and the materials used to manufacture the plate-fin structure. However, depending on these materials, it is possible to have cryogenic plate-fin heat exchangers operating at negative temperatures (all-aluminum manufacture), or exchangers made of composite materials that can process fluids at very high temperatures (up to 800°C).

A layer of fins attached between two sheets constitutes an *exchange module.* A plate-fin heat exchanger consists of several modules superposed to form a matrix. The exchanger matrix shown in Figure 3.6 consists of four stacked modules: two for fluid 1 and two for fluid 2. However, as Figure 3.8 shows, an additional module could have been stacked on top for fluid 1, giving three passages for fluid 1 and two for fluid 2.

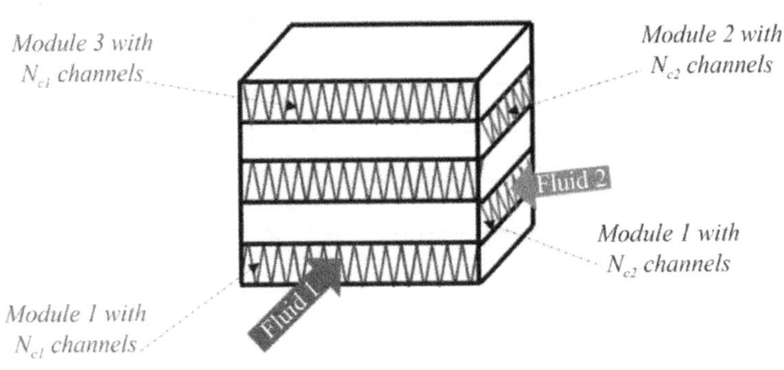

Figure 3.8. *Plate-fin heat exchanger with unequal numbers of modules. For a color version of this figure, see www.iste.co.uk/benallou/energy6.zip*

Each module makes it possible to have a clearance zone for fluid flow. This is why the number of modules on each side of the heat exchanger is often referred to as the number of passes. Moreover, each module comprises a number of channels, N_c. The number of channels in each side is the same: N_{c1} for all the passes on side 1 and N_{c2} for those on side 2.

Therefore, similarly to shell-and-tube heat exchangers, a plate-fin heat exchanger is designated by two pairs of digits. We will then refer to an exchanger (N_1; N_{c1})-(N_2; N_{c2}). The first pair defines the number of passes, N_1, and the number of channels, N_{c1}, on the hot side (fluid 1). The second does the same for the cold side (fluid 2).

3.4.4. *Geometric parameters*

The geometric parameters usually used for the design and description of plate-fin heat exchangers are defined in sections 3.4.4.1 to 3.4.4.9. These parameters include (see Figure 3.9):

– the number of passages (modules) on each side: N_i: $i = 1; 2$;

– the number of channels per module: N_{ci}: $i = 1; 2$;

– the height of a module; this is the clearance height between plates: δ_i : $i = 1; 2$;

– the flow lengths, L_1 and L_2, and the dead height, L_3;

– the available volume on each side of the exchanger: V_i;

– the thickness of each plate: e;

– the compactness of the circuit: β_i;

– the exchange area: A;

– the frontal area: A_f;

– the free area: A_o;

– the porosity: σ;

– the thickness, e_a, of the fin[2];

– the width, s, at the base of the fin;

– the length, l_a, of the fin; only in the case where the latter does not continuously cover the entire flow length. In the case of Figure 3.9, $l_{a1} = L_1$ and $l_{a2} = L_2$.

2 Not to be confused with the thickness, e, of the plates of the matrix.

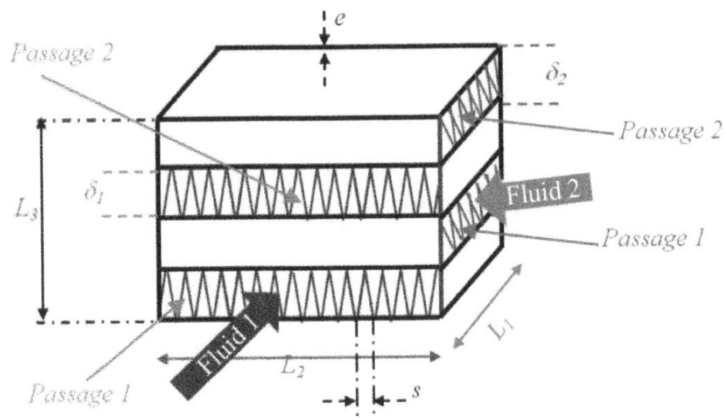

Figure 3.9. *Geometric parameters of plate-fin heat exchangers. For a color version of this figure, see www.iste.co.uk/benallou/energy6.zip*

3.4.4.1. *Flow lengths*

The flow length, L_i, on side i is the length (distance) traveled by the fluid, i, in the exchanger (see Figure 3.9): L_1 and L_2 are the flow lengths on sides 1 and 2, respectively. L_3 is referred to as the *dead height*.

3.4.4.2. *Number of modules*

This is the number of passages available for flow on each side of the exchanger. The numbers of modules, N_1 and N_2, are design degrees of freedom, but with the constraint that $N_2 = N_1$ or $N_2 = N_1 + 1$.

3.4.4.3. *Number of channels*

The numbers of channels per pass, N_{c1} and N_{c2}, are also design degrees of freedom. Note that they are related to the flow lengths. This relationship depends on the shape of the fin used. For triangular fins such as those shown in Figures 3.7a, 3.8 and 3.9, Nc_1 and Nc_2 are related to L_1 and L_2 by:

$$N_{c_i} = \frac{L_i}{s_i + 2e_i}$$

3.4.4.4. *Height of a module*

This is the height between two plates. Most often, this height is the same on either side of the exchanger. However, generally, and particularly if one of the fluids is charged with impurities, we can choose to have a different height on one side of the exchanger. We will then have a height for the hot side (fluid 1), δ_1, and another for the cold side (fluid 2): δ_2.

3.4.4.5. *Available volume*

Since the modules can have a different height on either side of the exchanger, it follows that the volumes between plates are not necessarily equal for each of the circuits. Moreover, they are not equal to the total volume, V, of the exchanger.

Indeed, the volume of a module on side i is: $v_i = L_1 L_2 \delta_i$; i = 1; 2.

With side i of the exchanger having N_i passes, the *available volume on side i* is given by: $V_i = N_i v_i = N_i\, L_1 L_2 \delta_i$: i = 1; 2.

Thus, the available volumes, for each side of the exchanger, will be calculated based on the following relationships: $V_1 = N_1\, L_1 L_2 \delta_1$ and $V_2 = N_2\, L_1 L_2 \delta_2$.

3.4.4.6. *The compactness on side i*

The compactness, β_i, of circuit i is the ratio of the exchange area of circuit i, divided by the volume available in the same circuit: $\beta_i = \frac{A_i}{V_i}$.

3.4.4.7. *The exchange area*

The exchange area of each circuit is calculated based on the defining equation for compactness: $A_i = \beta_i V_i$; (i = 1; 2); where β_i is the compactness of circuit i.

3.4.4.8. *The frontal area*

This is the cross-section area perpendicular to flow. It is defined by: $A_{f_i} = \frac{V_i}{L_i}$.

It follows that: $A_{f_1} = N_1\, L_2 \delta_1$ and $A_{f_2} = N_2\, L_1 \delta_2$.

3.4.4.9. *Porosity*

This is the ratio, σ_i, defined by: $\sigma_1 = \frac{A_{0_1}}{A_{f_1}}$

where A_{0_i} and A_{f_i} are, respectively, the free flow area and the frontal area on side i.

$A_{0_i} = N_i\left(\mathrm{L_d \delta_i} - 2\mathrm{N_{c_i} s_i e_{c_i}}\right)$ and $A_{f_i} = N_i\, L_j \delta_i$; j = 2 if i = 1 and j = 1 if i = 2.

Hence: $\sigma_i = 1 - 2\frac{N_{c_i} e_{c_i} s_i}{L_j \delta_i}$ with j = 2, if i = 1 and j = 1, if i = 2.

Therefore, a plate-fin heat exchanger is described by a series of parameters for fluid 1 and another for fluid 2, in a manner reminiscent of shell-and-tube heat exchangers, where a series of parameters describes the tube-side arrangements and another relates to the shell side. These parameters are considered degrees of freedom in the design.

3.4.5. *Calculation of the hydraulic diameter for plate-fin heat exchangers*

The hydraulic diameter of each circuit can be inferred directly from the compactnesses and porosities of the latter, where these are available: $D_{Hi} = 4\frac{\sigma_i}{\beta_i}$

In general, σ_i and β_i are supplier data; they are available for each model. However, in the case where only the compactnesses, β_i, are known, the hydraulic diameter can be calculated directly from the equation of its definition and the geometric parameters.

Indeed, by definition, we have:

$$D_{H_i} = \frac{4\,A_{0_i}\,L_i}{A_i}$$

$$A_{0_1} = A_{f_1} - N_1 N_{c_1}(2 s_i e_{c_1}) = N_1\left(L_2\delta_1 - 2N_{c_1} s_i e_{c_1}\right) \text{ and } A_1 = \beta_1 V_1$$

$$D_{H_1} = \frac{4N_1L_1}{\beta_1 V_1}\left(L_2\delta_1 - 2N_{c_1} s_i e_{c_1}\right)$$

And, likewise:

$$D_{H_2} = \frac{4N_2L_2}{\beta_2 V_2}\left(L_1\delta_2 - 2N_{c_2} s_i e_{c_2}\right)$$

with: $V_1 = N_1\, L_1 L_2 \delta_1$ and $V_2 = N_2\, L_1 L_2 \delta_2$.

3.4.6. *Calculation of the Reynolds number*

For plate-fin heat exchangers, the Reynolds number is calculated using the hydraulic diameter: $Re_{E_i} = \frac{D_{H_i} v_i \rho_i}{\mu_i}$; i = 1; 2 with: $D_{H_i} = \frac{4\sigma_i}{\beta_i}$; i = 1; 2.

By substituting the expression of the hydraulic diameter, we obtain:

$$Re_{E_i} = \left(\frac{v_i \rho_i}{\mu_i}\right)\left(\frac{4\sigma_i}{\beta_i}\right).$$

For each side, i, the velocity, v_i, can be expressed as a function of the mass flow rate, W_i, namely:

$$v_i = \frac{W_i}{\rho_i A_{0_i}}; \text{ with: } A_{0_i} = \sigma_i A_{f_i}; \ A_{f_i} = N_i \, L_j \delta_i; \ j = 2 \text{ if } i = 1 \text{ and } j = 1 \text{ if } i = 2$$

Hence: $v_i = \frac{W_i}{\rho_i \sigma_i N_i \, L_j \delta_i}$; $j = 2$, if $i = 1$ and $j = 1$, if $i = 2$.

It follows: $\frac{v_i \rho_i}{\mu_i} = \frac{W_i}{\sigma_i N_i \, L_j \delta_i \mu_i}$.

Thus: $Re_{E_i} = \frac{4W_i}{N_i \, \delta_i \beta_i \mu_i L_j}$; j = 2, if i = 1 and j = 1, if i = 2.

3.5. Plate heat exchanger technology and design

In this type of exchanger, the fluids circulate inside the plates. A plate consists of two metal sheets, generally crimped on all four sides, leaving a free distance between the two sheets for fluid circulation (see Figure 3.10). There are circular openings in the corners of the plates, commonly referred to as corner holes.

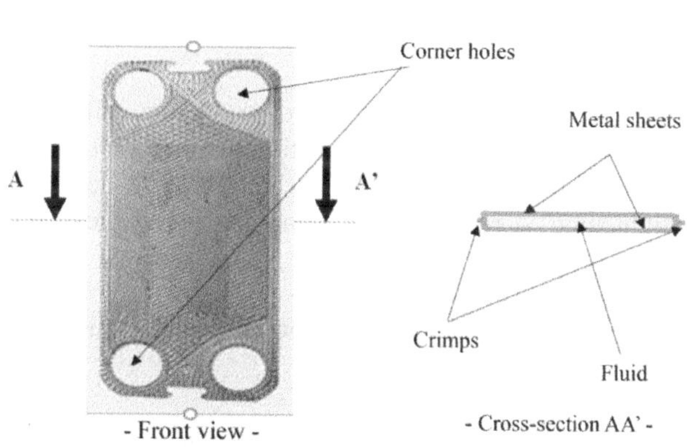

Figure 3.10. *Exchanger plate. For a color version of this figure, see www.iste.co.uk/benallou/energy6.zip*

The materials generally used to manufacture the plates are stainless steel, titanium and titanium-palladium. The plates are juxtaposed to form the exchanger (see Figure 3.11).

Figure 3.11. *Plates juxtaposed to form the exchanger*[3]. *For a color version of this figure, see www.iste.co.uk/benallou/energy6.zip*

The plates are held clamped together by two head plates called cover plates (one fixed and the other removable), by means of tie rods (see Figure 3.12). Two guide bars, one at the top and the other located at the bottom of the device, keep the plates aligned during tightening. A rear foot or column constitutes the attachment element of the movable cover plate.

Once the sheets are assembled, the corner holes, which are superposed, constitute channels for fluid flow.

The plate surfaces are often corrugated (chevron-shaped). This makes it possible, firstly, to promote turbulence during the passage of the fluid achieving high thermal performance, and, secondly, to avoid clogging of the plates. In addition, by multiplying the points of contact, corrugated plates offer better resistance to pressure.

3 Source: https://upload.wikimedia.org/wikipedia/commons/thumb/e/e6/Plate_heat_exchanger_-_dismantled_pic02.jpg/220px-Plate_heat_exchanger_-_dismantled_pic02.jpg.

Two plate assembly technologies exist, defining two distinct categories of exchangers: gasketed-plate heat exchangers and brazed-plate heat exchangers.

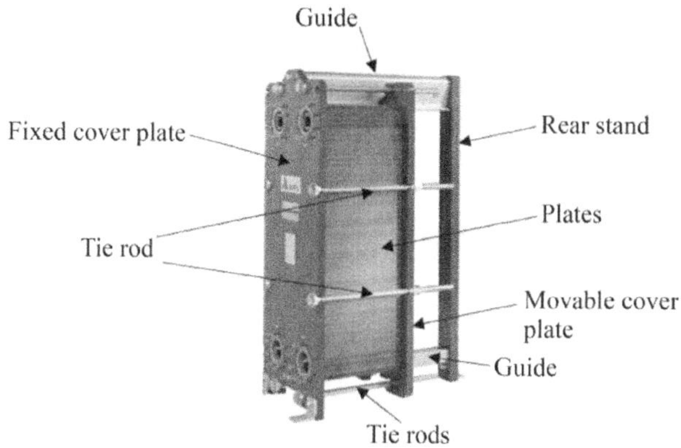

Figure 3.12. *Plate heat exchanger plates, guide and tie rods*[4]*. For a color version of this figure, see www.iste.co.uk/benallou/energy6.zip*

3.5.1. *Gasketed-plate heat exchangers*

In gasketed-plate heat exchangers, each plate is fitted with a peripheral seal that both guarantees tightness towards the outside and between the fluids, and ensures fluid distribution within the exchanger. This type of assembly presents two essential advantages:

– the flexibility of the transfer area, since plates can be added to or removed from the exchanger;

– the ease of cleaning, since the exchanger can be disassembled.

However, seals are among the exchanger components with the shortest service life. Replacing them can prove to be a relatively recurrent task. Recent developments in this field have nevertheless led to the development of non-bonded seals (snap-on and clip-on processes), facilitating their changing by means of a system of interlocking plates, therefore avoiding having to bond the seals and saving a lot of time when disassembling and reassembling the plates.

4 Source: https://encryptedtbn0.gstatic.com/images?q=tbn:ANd9GcSCndSttLhAi6Gx8gZ8MyUEu5u94pIDXUt00ZxgrQ1n7EWec9HaFw.

3.5.2. *Brazed-plate heat exchangers*

Due to their construction, brazed-plate heat exchangers have a rigid structure and a fixed transfer area. Indeed, the plates are stacked and then stamped and brazed together to the end plates to form a compact assembly (see Figure 3.13). The end plates, exchange plates and connections are vacuum brazed at a high temperature, and the exchanger is permanently assembled.

Figure 3.13. *Brazed[5]-plate heat exchangers. For a color version of this figure, see www.iste.co.uk/benallou/energy6.zip*

Therefore, while gasketed-plate heat exchangers present the drawback of being difficult to assemble and disassemble, brazed-plate heat exchangers simply cannot be disassembled at all. They do not offer the flexibility, through the addition or removal of one or more plates on their transfer area.

Moreover, since they consist of a welded, rigid structure, brazed-plate heat exchangers also present poor resistance to large temperature differences, since welded contacts do not offer the possibility of large expansions, and are unable to absorb large temperature variations. Furthermore, they are difficult to clean. The exchanger can nevertheless be cleaned if necessary, with the use of specific products and techniques.

Therefore, brazed-plate heat exchangers may be used for clean liquids' heating and cooling. They are particularly suitable for use as condensers and evaporators in a refrigeration setting.

5 Source: https://www.barriquand.com/dam/hd/I_1444033872432.JPG.

3.5.3. *Sizing parameters*

3.5.3.1. *Number of passes*

The number of passes is one of the degrees of freedom that are available to the designer. Figure 3.14 shows a single-pass, counter-flow plate heat exchanger, whereas Figures 3.15 and 3.16 show 2-2 and 3-1 heat exchangers, respectively.

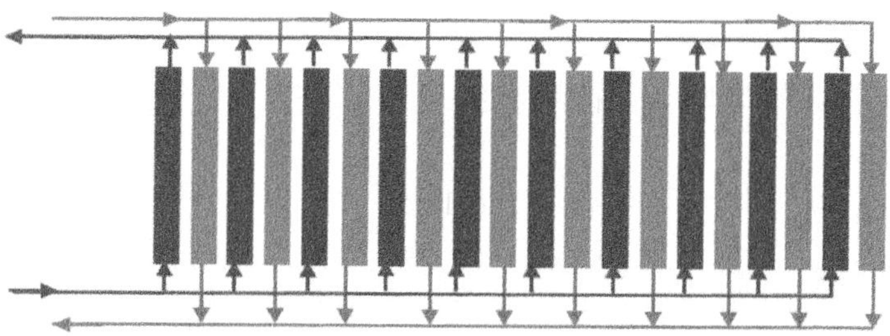

Figure 3.14. *Counter-flow 1-1 plate heat exchanger. For a color version of this figure, see www.iste.co.uk/benallou/energy6.zip*

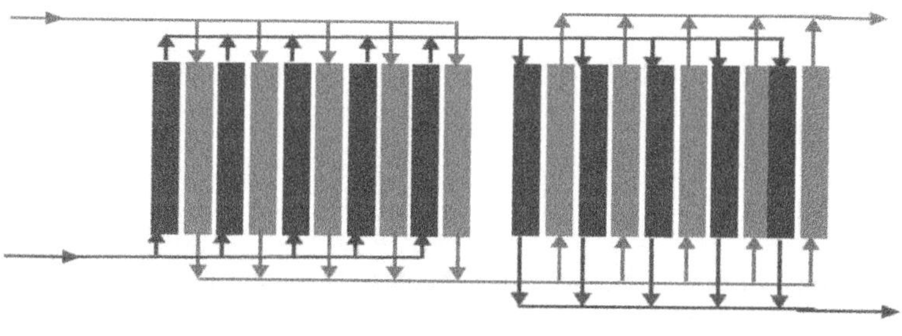

Figure 3.15. *Counter-flow 2-2 plate heat exchanger. For a color version of this figure, see www.iste.co.uk/benallou/energy6.zip*

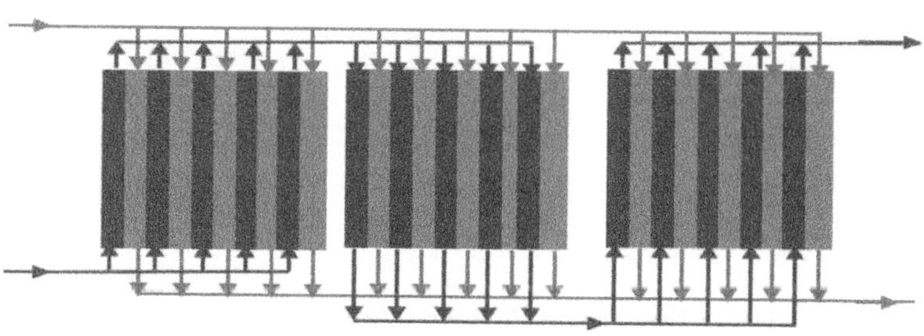

Figure 3.16. *3-1 plate heat exchanger. For a color version of this figure, see www.iste.co.uk/benallou/energy6.zip*

The number of passes shall not be chosen arbitrarily, since it is linked to the number of transfer units, which in turn determines the transfer area. Moreover, the number of passes also influences pressure drop.

In general, one-pass exchangers are preferred, but in order to get a sense of the number of passes needed for a given case, we will use the transfer unit (NTU) approach. Note that for each side, hot and cold, the number of transfer units is given by:

$$NTU_h = \frac{(\theta_{in} - \theta_{out})_h}{\Delta\theta_{ML}}; \text{ for the hot side}$$

$$\text{and } NTU_c = \frac{(\theta_{out} - \theta_{in})_c}{\Delta\theta_{ML}}; \text{ for the cold side}$$

with $\Delta\theta_{mL} = \frac{(\Delta\theta_1 - \Delta\theta_2)}{Ln\frac{\Delta\theta_1}{\Delta\theta_2}}$; $\Delta\theta_1 = \theta_{h_1} - \theta_{c_1}$ and $\Delta\theta_2 = \theta_{h_2} - \theta_{c_2}$; the index '1' being assigned to the end of the exchanger corresponding to the hot fluid inlet.

The number of passes, N_p, depends on the average NTU, defined by: $NTU_m = \frac{NTU_h + NTU_c}{2}$.

N_p is given by: $N_p = \frac{NTU_i}{NTU_{Max}}$; where NTU_{Max} is given in Tables A1.24 to A1.27, for each type of plate.

RULE OF THUMB.–

For every choice of plate type, will take the nearest integer given by Tables A1.24 to A1.27 and will attempt, as far as possible, to keep to a single-pass exchanger.

3.5.3.2. *Intraplate flow rate*

For each side, this is the flow rate of fluid that passes through each plate.

This is given by:

$$F_i = \frac{W_i}{N_i}$$

where:

– W_i is the total mass flow rate of the fluid on side i;

– N_i is the number of plates per pass on side i.

3.5.3.3. *Plate parameters*

Figure 3.17 shows the characteristic dimensions of an exchanger plate, notably:

– the height, a, and width, b;

– the diameter of the connections, δ;

– the longitudinal (e_L) and transverse (e_T) gaps;

– the spacing, e, for fluid flow[6];

– the wall thickness of the plates Δx;

– the length of the flow, L, as defined in Figure 3.17.

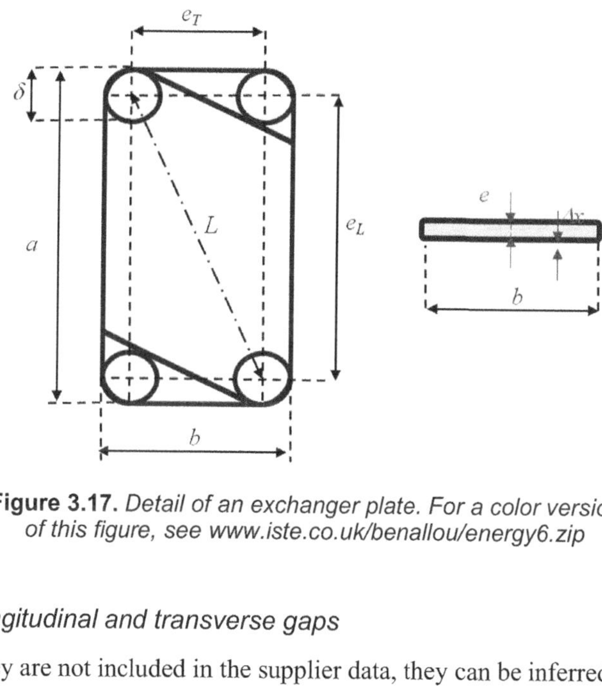

Figure 3.17. *Detail of an exchanger plate. For a color version of this figure, see www.iste.co.uk/benallou/energy6.zip*

3.5.3.4. *Longitudinal and transverse gaps*

When they are not included in the supplier data, they can be inferred as follows:

$$e_L = a - \delta \text{ and } e_T = b - \delta$$

3.5.3.5. *Calculation of the flow length in the plate*

The flow length (see Figure 3.17) is defined by: $L = \sqrt{e_L^2 + e_T^2}$

6 Manufacturer datum, in general, e =1.5 to 5 mm (Kuppan 2000).

3.5.3.6. *Calculation of the hydraulic diameter*

By definition (see section 3.2.2): $D_H = \frac{4\,A_0\,L}{A}$.

If *e* designates the spacing for fluid circulation, the free flow cross-section area is given by (see Figure 3.16): $A_0 = b\,e$.

The transfer area is: $A = 2\,(a\text{-}\delta)b$; (two faces).

Hence:

$$D_H = \frac{4\,b\,e\,L}{2\,(a-\delta)b}$$

$$L = \sqrt{e_L^2 + e_T^2}$$

Thus, the hydraulic diameter is given by: $D_H = \frac{4\,b\,e\,\sqrt{e_L^2+e_T^2}}{2\,(a-\delta)\,b}$.

Thus: $D_H = \frac{2\,e\,\sqrt{e_L^2+e_T^2}}{a-\delta}$.

Often, we equate *a* to *a*-δ and the length of the flow, *L*, to the height, *a*, of the plate.

We then have: $D_H \cong \frac{4\,bea}{2ab}$.

Thus: $D_H \cong 2e$.

3.5.3.7. *Calculation of the Reynolds number*

For plate heat exchangers, the Reynolds number is calculated for each side. It is related to the hydraulic diameter:

$$Re_{E_i} = \frac{D_{H_i} v_i \rho_i}{\mu_i};\ \mathrm{i} = 1; 2$$

$$D_{H_i} = \frac{2\,e_i\,\sqrt{e_{L_i}^2+e_{T_i}^2}}{a_i-\delta_i};\ \mathrm{i} = 1; 2$$

$$v_i = \frac{F_i}{\rho_i A_{0_i}}$$

where:

– $F_i = \frac{W_i}{N_i}$;

– W_i is the total mass flow rate of the fluid on side i;

– N_i is the number of plates per pass on side i.

Thus: $v_i = \frac{W_i}{N_i \rho_i b_i e_i}$.

Hence: $Re_{E_i} = \frac{2}{\mu_i} \frac{W_i}{N_i b_i} \frac{\sqrt{e_{L_i}^2 + e_{T_i}^2}}{a_i - \delta_i}$; i = 1; 2.

3.6. Spiral heat exchanger technology

Spiral heat exchangers are mainly used for heat transfer between two liquids. They are used secondarily in condensation and for gas cooling.

Figure 3.18 illustrates a spiral heat exchanger where the hot (red) and cold (blue) fluids circulate counter-flow.

Figure 3.18. *Spiral heat exchanger[7]. For a color version of this figure, see www.iste.co.uk/benallou/energy6.zip*

Figure 3.19 shows that a spiral heat exchanger is characterized by:

– the large and small diameters of the spiral, D_s and d_s;

– the width of the spiral, b;

– the number of revolutions, n;

7 Source: https://www.alfalaval.com/globalassets/images/products/heat-transfer/plate-heat-exchangers/spiral-heat-exchangers/spiral_heat_exchanger_flow.jpg.

– the gap, e, of the channels;

– the wall thickness, Δx.

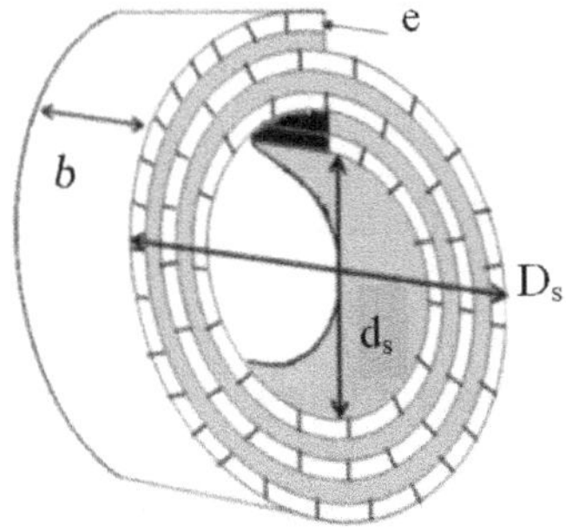

Figure 3.19. *Geometric parameters of a spiral heat exchanger*

3.6.1. *Gap*

For each circuit (hot and cold), the *gap between the channels, e,* is an important design parameter. It shall be set with respect to the allowable pressure drops and flow rates.

RULE OF THUMB.–

In the case of charged fluids, the gap shall be larger than the size of the particles suspended in the fluid.

In general, the gap between the channels is between 5 and 25 mm.

3.6.2. *Hydraulic diameter*

By definition (see section 3.2.2):

$$D_H = \frac{4\,A_0\,L}{A}$$

where:

– A_0 is the free flow cross-section area;

– A is the exchange area;

– L is the length of the flow.

For a spiral heat exchanger:

– The free flow area on side i, A_{0i}, is: $A_{0i} = e_i b$.

– The transfer area is: $A = 2\, b\, L$; L being the length of the flow.

Hence: $D_{H_i} = \frac{4\, e_i b\, L}{2bL} = 2\, e_i$.

3.6.3. *Reynolds number*

By definition, the Reynolds number is given by: $Re_i = \frac{D_{H_i} v_i \rho_i}{\mu_i}$.

Yet: $v_i\, \rho_i = \frac{W_i}{A_{0_i}}$.

Hence: $Re_i = \frac{D_{H_i} W_i}{\mu_i A_{0_i}}$.

The standard dimensions of commercial spiral heat exchangers are given in Table A1.28 (see Appendix 1).

3.6.4. *Relationship between number of revolutions and length*

An n-revolution spiral heat exchanger can be considered as a succession of semi-cylindrical metal sheets, of different radii (see Figure 3.19).

Table 3.2 gives the radii of the semicircles, each constituting revolutions made by the metal sheet.

Revolution number	**Radius of the first semicircle**	**Radius of the second semicircle**
1st revolution	$r_s+\Delta x+(e+\Delta x)$	$r_s+\Delta x+2(e+\Delta x)$
2nd revolution	$r_s+\Delta x+2(e+\Delta x)$	$r_s+\Delta x+3(e+\Delta x)$
⋮	⋮	⋮
***n*th revolution**	$r_s+\Delta x+n(e+\Delta x)$	$r_s+\Delta x+(n+1)(e+\Delta x)$

Table 3.2. *Radii of the semicircles constituting each revolution of the spiral*

The length, L, is the sum of the perimeters of the different semicircles.

Revolution number	Perimeters of the first semicircles	Perimeters of the second semicircles
1st revolution	$\pi[r_s+\Delta x+(e+\Delta x)]$	$\pi[r_s+\Delta x+2(e+\Delta x)]$
2nd revolution	$\pi[r_s+\Delta x+2(e+\Delta x)]$	$\pi[r_s+\Delta x+3(e+\Delta x)]$
⋮	⋮	⋮
***n*th revolution**	$\pi[r_s+\Delta x+n(e+\Delta x)]$	$\pi[r_s+\Delta x+(n+1)(e+\Delta x)]$
Sum	$\pi\left[n(r_s+\Delta x)+\frac{n(n+1)}{2}(e+\Delta x)\right]$	$\pi\left[n(r_s+\Delta x)+\left(\frac{(n+1)(n+2)}{2}-1\right)(e+\Delta x)\right]$

Table 3.3. *Sum of the semicircle perimeters*

Hence: $L=\pi\left\{2n(r_s+\Delta x)+\left[\frac{n(n+1)+(n+1)(n+2)-2}{2}\right](e+\Delta x)\right\}$.

Thus: $L=\pi[(e+\Delta x)n^2+2(r_s+e+2\Delta x)n]$.

Thus, L and n shall satisfy the following second-degree equation:

$$(e+\Delta x)n^2+2(r_s+e+2\Delta x)n-\frac{L}{\pi}=0$$

This equation can be used to calculate the number of revolutions, n_0, for a given transfer area, A_0.

Indeed, the length needed for A_0 is: $L_0=\frac{A_0}{b}$.

The discriminant of the equation is: $\delta'=[r_s+e+2\Delta x]^2+(e+\Delta x)\frac{L_0}{\pi}$.

n_0 is then given by: $n_0=\frac{\sqrt{(r_s+e+2\Delta x)^2+(e+\Delta x)\frac{A_0}{\pi b}}-(r_s+e+2\Delta x)}{e+\Delta x}$.

3.7. Printed-circuit heat exchanger technology and design

As a result of their high compactness, high pressure resistance and high thermal efficiency, printed-circuit heat exchangers have generated great interest in recent years in the field of nuclear, solar and power electronics cooling.

As already observed (see Chapter 1), a printed-circuit heat exchanger (PCHE) consists of a superposition of modules (layers) to form a core. Each layer comprises a set of channels "printed" in the material, for the circulation of fluids. In the simplest configurations, the superposed modules form a parallelepiped, which can be used counter-flow or cross-flow (see Figure 3.20). However, current technological developments (Wang et al. 2019; Saeed et al. 2020; Tang et al. 2020) converge toward configurations where the geometry of the fluid circulation channels is much closer to that of electronic printed circuits (see Figure 3.22).

Carbon steels are generally not used, as the small diameters required make the exchanger channels vulnerable to corrosion. The materials used to manufacture PCHEs are often alloys with high heat conductivity. New materials are currently being developed to have high mechanical strength at the very high temperatures encountered in applications using liquid metals or molten salts as heat-transfer fluids (see Table 3.4).

Material	Essential composition	ρ (kg/m^3)	λ (W/m°K)	C_p (J/kg°K)
Alloy 617 (Kim 2008)	52Ni-22Cr-13Co-9Mo-1.2Al	8,360	23.9	586
Alloy 800H (Kim 2012)	33Ni-42Fe-21Cr	7,940	22.8	460
Hastelloy N (Sabharwall et al. 2011)	72Ni-7Cr-16Mo-5Fe	8,860	23.6	523

Table 3.4. *Properties of new materials used in PCHEs*

Printed-circuit heat exchangers are defined by a multitude of geometric parameters usually used for design.

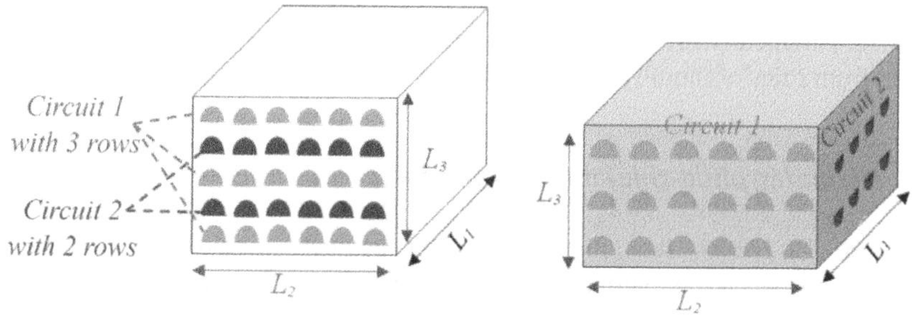

Figure 3.20. *Counter-flow and cross-flow PCHE. For a color version of this figure, see www.iste.co.uk/benallou/energy6.zip*

These parameters relate to the dimensions and arrangement of the different flow channels. They include, for a given circuit (hot or cold, see Figure 3.20):

– the thickness of each module, t_p;

– the diameter of the channels in each circuit, d_i; $i = 1\text{-}2$;

– the pitch of circuit i, a_i; $i = 1\text{-}2$;

– the gap between the channels of a module, t_f, which is the thickness of the wall separating the fluid circuits;

– the number of channels in a layer of circuit i: n_i; $i = 1\text{-}2$;

– the number of layers in each circuit: N_i; $i = 1\text{-}2$;

– the overall dimensions of the device: L_1, L_2 and L_3.

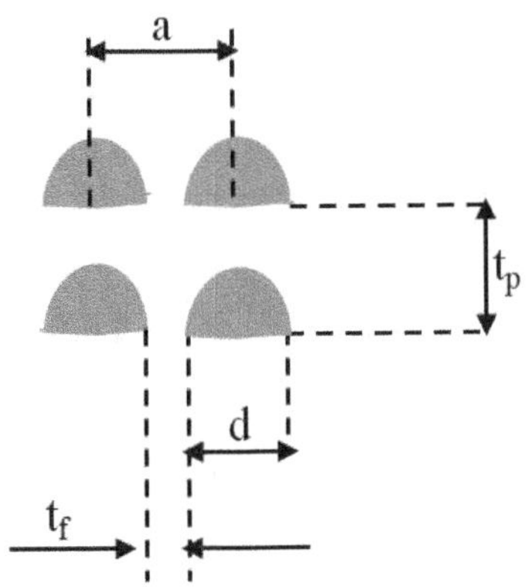

Figure 3.21. *Parameters of the channels a printed-circuit heat exchanger*

These parameters are not independent; they are constrained by relationships derived from rules of thumb or imposed by the device's geometry.

3.7.1. *Diameter-pitch-thickness relationship*

Figure 3.19 shows that the construction geometry imposes the following relationship between the diameter of the channels, the pitch and the thickness, t_f, of the walls separating the channels:

$$t_f = a - d$$

3.7.2. *Constraints on the number of channels per layer*

The number of channels in the layers of circuit 1, n_1, and that of circuit 2, n_2, shall satisfy the following relationships with L_1 and L_2:

– $n_1d + (n_1\text{-}1)t_f = L_2$;

– $n_2d + (n_2\text{-}1)t_f = L_1$ for a cross-flow;

– $n_2d + (n_2\text{-}1)t_f = L_2$ for a counter-flow.

For a counter-flow, these first two relationships impose $n_1 = n_2$; unless we opt for different channel diameters. Conversely, for a cross-flow, these relationships impose the following constraints on n_1 and n_2 as a function of L_1 and L_2:

$$n_1 = \frac{L_2+t_f}{d+t_f} \text{ and } n_2 = \frac{L_1+t_f}{d+t_f}.$$

Thus, for a given diameter d and wall thickness t_f, the choice of L_1 and L_2 determines n_1, n_2 and, conversely, setting n_1 and n_2 determines L_1 and L_2. It should be noted that L_1 and L_2 are often set by space limitations.

3.7.3. *Constraints on numbers of layers*

Furthermore, the numbers of layers, N_1 for circuit 1 and N_2 for circuit 2, shall satisfy the following relationship with L_3: $L_3 = (N_1 + N_2)t_p$.

The total transfer area is then given by:

– for a counter-flow: $A = N_1 n_1 \left(\frac{1}{2}\pi d + d\right) L_1 = N_2 n_2 \left(\frac{1}{2}\pi d + d\right) L_1$

hence $N_1 = N_2$, since $n_1 = n_2$;

– for a cross-flow: $A = N_1 n_1 \left(\frac{1}{2}\pi d + d\right) L_1 = N_2 n_2 \left(\frac{1}{2}\pi d + d\right) L_2$

hence $N_2 = N_1 \left(\frac{L_2+t_f}{L_1+t_f}\right)\left(\frac{L_1}{L_2}\right)$.

By combining with the relationship giving L_3, we obtain:

– for a counter-flow: $L_3 = 2\,N_1 t_p$;

– for a cross-flow: $L_3 = N_1 \left[1 + \left(\frac{L_2+t_f}{L_1+t_f}\right)\left(\frac{L_1}{L_2}\right)\right] t_p$.

Thus, for given values of L_1, L_2, t_f and t_p, the choice of N_1 determines N_2 and L_3.

3.7.4. *Hydraulic diameter*

For each of the two semicircular circuits of the PCHE, the hydraulic diameter is given by (see Volume 3, section 3.2):

$$D_{H_i} = \frac{4A_i}{P_i}; \; i = 1.2$$

In this case, the cross-section area perpendicular to flow is given by:

$$A_i = \frac{\pi d_i^2}{8} \text{ and } P_i = \frac{\pi}{2} d_i$$

The hydraulic diameter of channel i is then: $D_{H_i} = \frac{4A_i}{P_i} = 4\frac{\pi d_i^2}{8}\frac{2}{\pi d_i} = d_i$.

3.7.5. *The Reynolds number*

The Reynolds number for each channel is given by:

$$Re_i = \frac{D_{H_i} v_i \rho_i}{\mu_i} = \frac{d_i v_i \rho_i}{\mu_i}; \; i = 1.2$$

3.7.6. *Mechanical strength*

The channels of a printed-circuit heat exchanger are usually subjected to great forces, due to the pressures of the fluids on the walls. The mechanical strength of a PCHE requires compliance with the rules of material resistance, which impose conditions both on the thickness of the modules, t_p, and on the gap between the channels, t_f, of a module (Oh and Kim 2008):

$$t_f \geq \frac{a}{\left(\frac{\sigma_{max}}{\Delta P}+1\right)} \text{ and } t_p \geq \begin{cases} \frac{d}{2}\sqrt{\frac{\sigma_{max}+P_p}{\sigma_{max}+2P_S-P_p}} & when\ P_p \geq P_s \\ \frac{d}{2}\sqrt{\frac{\sigma_{max}-P_p}{\sigma_{max}-2P_S-P_p}} & when\ P_p < P_s \end{cases}$$

where:

– a is the pitch, in meters;

– ΔP is the pressure difference between the two fluids (Pa);

– σ_{max} is the maximum allowable shear stress (Pa);

– d is the diameter of the channels, in meters;

– P_p is the primary-circuit pressure (Pa);

– P_s is the secondary circuit pressure (Pa).

3.8. New developments in compact heat exchangers

The latest developments recorded in concentrated solar power plants technologies require ever more materials operating at very high temperatures (above 750°C and possibly exceeding 1,400°C). This has enabled the development of new compact heat exchangers that satisfy both the robustness required for such installations and the cost-effectiveness that enables acceptable costs per kWh generated.

The innovation lies in the use of metal ceramics (cermets: a mixture of zirconium carbide and tungsten), leading to higher yields in electricity generation from the sun (Caccia et al. 2018; Wang et al. 2019). In this way, it is possible to achieve improvements of more than 20% in terms of heat-electricity conversion efficiency; this offers hope for a significant reduction in the cost of electricity generated by thermal conversion of concentrated solar radiation.

Compact metal-ceramic heat exchangers can be considered as a subcategory of printed-circuit heat exchangers.

4

Heat Exchanger Selection, and Sizing Parameters

4.1. Introduction

As explored in the previous chapters, several heat-exchanger construction technologies exist: single-pass or multipass tubular exchangers, plate exchangers and cross-flow exchangers, etc. The choice of the type of exchanger to be used in a given situation depends on several parameters, including the heat transfer area needed in order to carry out the desired thermal schedule.

In addition, the surfaces that make up the tubes or plates of the different exchangers are often subjected to fouling, which can be of greater or lesser extent depending on the cleanliness of the fluids involved in the heat transfer. As we show in Appendix 4, this fouling effect introduces an additional thermal resistance, called fouling resistance, R_e, which depends on the type of fluid, the fluid velocity, the surface type and the service life of the heat exchanger. The degree of cleanliness of the treated fluids therefore needs to be taken into account when determining the exchanger type and the necessary transfer area.

Moreover, as we saw previously, in cases where one of the fluids is a gas, fins are often added to surfaces exposed to the gas in the perspective of increasing the transfer area. The calculation of the transfer area will therefore take into account the possible use of fins.

The fluid flow configuration (co-flow, counter-flow or cross-flow) is also part of the parameters to be considered when sizing an exchanger. This means that the parameters to be defined in the design are multiple.

This multiplicity generates complexity in the analysis methods and sizing techniques. This chapter details the parameters to be defined when designing an exchanger. The different sizing methods are subsequently presented in Chapter 5.

4.2. Choosing the type of heat exchanger to be used

The design of the exchangers begins with the specification of the type of exchanger to be used. This choice is guided by the nature of the fluids involved (gases or liquids, clean or charged fluids, corrosive fluids, etc.), the temperature ranges concerned, the operating pressure considered, the heat exchanger's thermal schedule, the space available for its installation, the fouling tendency (see Appendix 4), the ease of inspection, cleaning and repair, the costs, etc. In reality, the choice is made based on a multi-criteria analysis that takes all of these parameters into consideration. But it should be emphasized that, in the majority of practical cases, the choice is driven by the overall cost.

4.2.1. *Nature of fluids and temperature ranges*

This is the starting point of a design problem. As shown in Table 4.1, the maximum allowable temperature depends on the material used to build the heat exchanger. The definition of the thermal schedule determines operational temperature ranges. In turn, these ranges will determine a set of compatible materials which can be used to build the heat exchanger. The final choice of the material will depend on the nature of the fluids concerned. Indeed, construction materials will be compatible not only with the fluids, but also with the temperature ranges of the thermal schedule. For example, stainless steel will be used for exchangers that need to accommodate food fluids.

Construction materials	Allowable fluid temperature (°C)
Carbon steel	540
C-Mo steel	590
Cr-Mo steel	650
Low-alloy steel (< 6% Cr)	590
Alloy steel (< 17% Cr)	590
Austenitic Cr-Ni steel	650
Cast iron	200
Brass	200

Table 4.1. *Construction materials*

As such, the definition of the nature of the fluids and temperature ranges will make it possible to define the construction materials that may be used. However, the wall thicknesses are related instead to the definition of the operating pressure.

4.2.2. *Design pressure*

The operating pressure is important since it is the parameter used to determine the minimum thickness of the exchanger walls (tube and shell in the case of a tubular exchanger, plates for compact exchangers).

4.2.3. *Thermal schedule of a heat exchanger*

Whatever its type (tubular, plate, cross-flow, etc.), a heat exchanger admits two inlets (D_1 at T_1 and D_2 at t_1) and two outlets (D_1 at T_2 and D_2 at t_2), as shown in Figure 4.1. It can thus be represented by a quadrupole connecting the inlet and outlet flow rates of the two fluids.

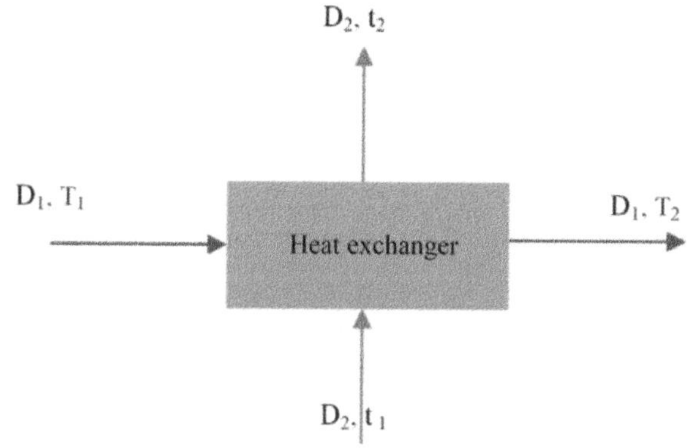

Figure 4.1. *Representation of a heat exchanger by a quadrupole. For a color version of this figure, see www.iste.co.uk/benallou/energy6.zip*

DEFINITION.– A heat exchanger thermal schedule is defined by the following seven parameters:

– mass flow rates D_1 and D_2;

– temperatures T_1, T_2, t_1 and t_2;

– the amount of heat exchanged, Q.

These parameters are related to each other by the following relationships reflecting the heat balances on the two flow rates (C_{p1} and C_{p2} being the specific heats of the hot and cold fluids, respectively):

$$Q = D_1\, C_{p1}\, (T_1 - T_2) \text{ and } Q = D_2\, C_{p2}\, (t_2 - t_1)$$

Thus, the thermal schedule can be defined by only five of the seven parameters mentioned above; the other two can be calculated by the balance equations.

Table 4.2 shows the parameters defined and those calculated by heat balances in several situations typically encountered.

Data description	Typical situation	Calculated parameters
D_2, t_1, t_2, D_1, T_1	We wish to heat a mass flowrate, D_2, from t_1 to t_2. We have a mass flowrate, D_1, at T_1. What will the outlet temperature of D_1 be?	Q and T_2
D_2, t_1, t_2, T_1, T_2	We wish to heat D_2 from t_1 to t_2 using a fluid available at T_1. What is the necessary flowrate, D_1?	Q and D_1
D_1, T_1, T_2, t_1, t_2	What flowrate will be able to be heated by recovering heat from D_1 between T_1 and T_2?	Q and D_2

Table 4.2. *Definition of the thermal schedule*

4.2.4. *The selection itself*

The choice of the type of exchanger to be used to carry out a given thermal schedule therefore poses an optimization problem with several criteria.

Figure 4.2 illustrates that the choice procedure can be organized according to the following four aspects:

– reliability at high temperatures;

– reliability at high pressures;

– the thermal schedule, i.e. the amount of heat to be exchanged;

– the ease of production expressed by the cost.

Figure 4.2 shows that tubular heat exchangers cover wide temperature and pressure ranges for moderate thermal schedules, with relatively low prices.

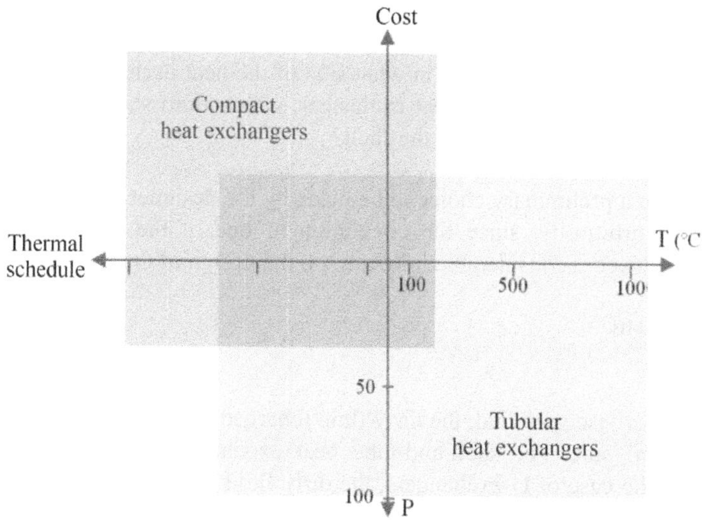

Figure 4.2. *Choosing a heat exchanger. For a color version of this figure, see www.iste.co.uk/benallou/energy6.zip*

However, compact heat exchangers, especially plate heat exchangers, cover a wide range of thermal schedules thanks to their large transfer areas. These therefore appear to be the exchangers to choose for various thermal schedules. Their high prices, due to their production complexity, will nevertheless temper this choice. They are mainly preferred in applications where compactness is sought and where heat treatment is associated with high added value, such as in the food and pharmaceutical industries.

Thus, engineers in charge of choosing an exchanger from manufacturers' catalogs are often faced with an overwhelming choice of different types of vessels off the shelf. But, no matter what choice they make, they will be asked to ensure that the exchanger(s) selected meet the requirements of the predefined thermal schedule.

It should also be emphasized that, often, none of the exchangers in the catalogs perfectly meets all the specifications of a given project. In this case, the design engineers are required, themselves, to make the various choices that lead to the sizing of the appropriate exchanger.

4.3. Selection of fluid circuits: which fluid for what side?

Which fluid needs to circulate in what side of the heat exchanger? For example, in the case of a shell and tube heat exchanger, which fluid should flow inside the tubes, and which fluid will flow in the shell?

This is also a preliminary choice to be made by the designer. But this choice will not be made arbitrarily, since the circulation of one of the fluids, shell-side or tube-side, can exert a considerable influence on the design of the exchanger.

RULES OF THUMB.–

Rule 1

Unless otherwise specified, the dirty fluid (charged with particles) will be on the 'easy to clean' side. For shell-and-tube heat exchangers, this is the tube side. However, in the case of U-exchangers, the dirty fluid should be set to flow outside the tubes.

Rule 2

The most viscous fluid is best suited for the 'easy to flow' side. For shell-and-tube heat exchangers, this is the shell side, since the free flow spaces are larger and the wall thickness is fairly large. This will minimize shell-side pressure drops and any vibrations during operation.

Rule 3

The fluid with the highest pressure is reserved for the tube side.

Table 4.3 gives some general indications regarding the positioning of the fluids. It should be understood that these guidelines are not absolute rules and that the optimal placement of the fluid depends on many service-specific factors.

Tube side	Shell side
– "Hot" fluids – Corrosive liquids – Cooling water – Lower-viscosity fluids – Charged (dirty) fluids – High-pressure flow rates	– "Cold" fluids – Condensation vapors (unless corrosive) – Fluids with a large temperature difference ($\Delta\theta > 50°C$) – Higher-viscosity fluids – Uncharged (clean) fluids – Moderate pressures

Table 4.3. *Choosing the fluid circuits*

4.4. Choosing the flow configuration

The choice between fluid circulation configurations, co-flow or counter-flow, depends on the problem considered.

When there is no constraint on the maximum temperature that one of the fluids could reach, the counter-flow will be preferred since it leads to greater efficiency.

Conversely, when the fluid to be heated runs the risk of degrading at a certain temperature, co-flow will be preferred. In this case we will try to ensure that the fluid concerned enters on the same side as the hot fluid inlet. Under these conditions, the temperature of the fluid to be heated will never exceed the inlet temperature of the hot fluid. If the latter temperature is chosen below the degradation temperature of the fluid to be heated, this fluid will not be degraded. This type of situation is frequently encountered in food industries.

4.5. Illustration: choosing a heat exchanger and determining its thermal schedule

The heat exchanger represented in the form of a quadrupole in Figure 4.3 is intended to recover the energy transported by a flow rate, D_1, of ethanol, which leaves a condenser of a distillation column at temperature T_1. The energy thus recovered will serve to heat a flow rate, D_2, of water for factory workers' showers. This water is available at temperature t_1. It needs to be brought to temperature t_2 before storage.

Figure 4.3. *Energy recovery from ethanol condensates. For a color version of this figure, see www.iste.co.uk/benallou/energy6.zip*

1) What type of exchanger would you recommend?

2) Define the circuits of the two fluids.

3) Calculate the heat flux that needs to be exchanged.

4) Calculate the ethanol outlet temperature.

Given data:

$D_1 = 150$ kg/hr $\quad C_{p1} = 0.75$ kcal/kg°C $\quad T_1 = 78°C$
$D_2 = 100$ kg/hr $\quad C_{p1} = 1$ kcal/kg°C $\quad t_1 = 18°C \quad t_2 = 65°C$

Solution

1) Type of exchanger

The temperatures are in ordinary zones and no constraint for the space reserved for installation has been expressed. A *tubular heat exchanger* will therefore be used.

2) Definition of the circuits of the two fluids

Hot fluid in the tubes and cold fluid in the shell.

3) Calculating the flux

$$Q = D_2\, C_{p2}\, (t_2 - t_1)$$

NUMERICAL APPLICATION

$D_2 = 100$ kg/hr $\quad C_{p1} = 0.75$ kcal/kg°C $\quad t_1 = 18°C \quad t_2 = 65°C$

$$Q = 100 \times 0.75 \times (65 - 18)$$

$$Q = 4{,}700 \text{ kcal/hr}$$

4) Ethanol outlet temperature

We have: $Q = D_1\, C_{p1}\, (T_1 - T_2)$.

Hence: $T_2 = T_1 - \frac{Q}{D_1 C_{p1}}$.

NUMERICAL APPLICATION

$D_1 = 150$ kg/hr $\quad C_{p1} = 0.75$ kcal/kg°C $\quad T_1 = 78°C \quad Q = 28{,}200$ kcal/hr

$$T_2 = 78 - \frac{28{,}200}{150 \times 0.75}$$

$$T_2 = 36°C$$

4.6. Sizing, a complex problem

Sizing an exchanger means determining its dimensions. For plate-fin heat exchangers, for example, these dimensions include the number of modules on each side, the number of channels per module, the height of a module, the flow lengths, the dead height and the compactness of each circuit. Similarly, for a shell-and-tube heat exchanger, the sizing means the determination of the diameter of the tubes, their arrangement in the shell, the shell's diameter, the length of the tubes, etc.

Sizing also involves setting the operating conditions (flow rates, pressure drops, etc.) that allow the heat transfer sought to be carried out.

It should be noted that at the outset, *only* the thermal schedule of the exchanger (temperatures and flux) is described. As a matter of fact, you will generally be asked to heat (or cool) a given flow rate of a fluid from temperature T_1 to temperature T_2, using a second fluid available at a known temperature. The other parameters that need to be set or calculated by the design engineer are presented in section 4.6.1.

4.6.1. *Parameters to be determined by the designer*

Once the type of exchanger (multitube, plate, plate-fin, PCHE, etc.) to be used to carry out the desired thermal schedule has been chosen and the circuits of each of the fluids have been defined, as well as the flow configuration (co-flow, counter-flow or cross-flow), there will remain a multitude of parameters to be determined or calculated by the designer. These parameters are presented in sections 4.6.1.1 to 4.6.1.11 for shell-and-tube heat exchangers. They include tube-and-shell diameters, the number of passes on each side, the number of tubes per pass, etc. Equivalent parameters for compact heat exchangers are presented in the design algorithms detailed in Chapter 6.

4.6.1.1. *Diameter of the tubes*

Whether for finned-tube compact heat exchangers or for shell-and-tube vessels, the diameter of the tubes is an important design parameter that influences the pressure drop across the heat exchanger, as well as the efficiency of heat transfer.

An outer tube diameter of 1" is very common in tubular heat exchangers. The inner diameter depends on the thickness of the tube, which should withstand internal pressure stress while leaving an adequate corrosion allowance. Once the outer parameter is set on pressure-drop and heat-transfer efficiency bases, the thickness of the tube is chosen by the designer according to the standards presented in Appendix 1 as a function of the outer diameter (section A1.12).

4.6.1.2. *Length of the tubes*

Once the diameter of the tubes is fixed, the length of the tubes directly determines the heat transfer area. The designer has complete freedom in determining the length of tubes, but for the sake of manufacturing convenience (minimization of scrap), tube lengths of 8 m, 6 m, 4 m, 3 m, 2 m, 1.5 m or 1 m are preferably used. Indeed, the tubes are commercially available in the form of bars 12 m long.

It should be noted, however, that longer tubes will lead to a smaller shell diameter, but also to a higher pressure drop. Moreover, determination of the length of tubes should be operated while keeping in mind the eventual constraints on the space where the heat exchanger would have to fit.

4.6.1.3. *Number of tube-side passes*

The number of passages for the tubes in the shell, N_p, is another degree of freedom to be set by the designer during sizing. However, it is generally limited by technical considerations regarding the assembly of the various tube bundles in a given shell. The technical upper limit is 16 tube-side passages. However, in practice, the number of tube-side passes is rather limited to eight; but often *one, two or four passages* are sufficient. It should also be emphasized that for multipass heat exchangers, an even number of passes will be retained, given that odd numbers of passes can lead to manufacturing and operating problems.

For a given heat transfer area, the choice of the number of tube-side passes, N_p, is usually made to keep the length of the device within reasonable limits. The most commonly-used numbers range from 1 to 16. But, the most used numbers of passes are 1, 2 and 4.

4.6.1.4. *Number of shell-side passes*

The number of shell-side passes, N_{sh}, is a degree of freedom that will be set by the designer. In general, N_{sh} is determined through an iterative process. In order to keep the geometries easy to execute, we seek the smallest values of N_{sp}. Iterations are thus started with $N_{sp} = 1$.

RULE OF THUMB.–

The number of shell-side passes will be related to the number of tube-side passes by: $N_p = 2 * N_{sp}$.

4.6.1.5. *Arrangement of tubes in the shell*

It should be recalled that the tubes are placed in the shell according to specific *arrays* (square or triangular) and with well-defined *pitches* (see Figure 4.4).

Thus, both the *array type* and the *pitches* are *degrees of freedom* that will be set by the designer during sizing.

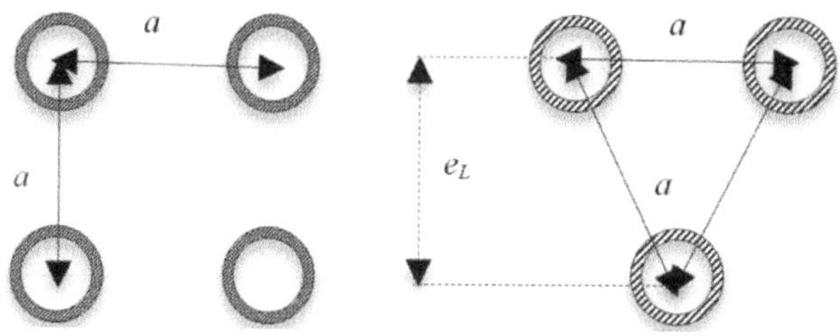

Square array Triangular array

Figure 4.4. *Arrangement of tubes in the shell*

4.6.1.6. *Choosing the array*

The array is generally determined so as to include as many tubes as possible in the shell. This makes it possible to have the largest transfer area in a given shell volume.

Four types of arrays are usually used (see section 2.3.6): *in-line square*, *staggered square*, *in-line triangular* or *staggered triangular*. It should be noted that triangular arrays present a double advantage over square arrays. They make it possible to place *more tubes* in a given shell and induce greater turbulence and therefore a *high heat transfer coefficient*. But these advantages are counterbalanced by the fact that the small size of this configuration makes mechanical cleaning of the tubes practically impossible. Triangular arrays will therefore be limited to cases where the shell-side fluid is clean.

Where the shell-side fluid requires periodic mechanical cleaning, square arrays will be used, with a preference for staggered square arrays, which allow high enough turbulence to be obtained for the heat transfer coefficients to be acceptable.

4.6.1.7. *Choosing the tube pitch*

The pitch of the tubes is defined as the distance between the centers of two adjacent tubes of the same layer. When choosing the pitch, several criteria need to be taken into account. It depends on the array chosen and the minimum values required to ensure the installation of the tubes in the shell and their subsequent cleaning.

Table 4.4 presents the criteria that will be respected when choosing the pitch (TEMA 1999). For a triangular mesh, the minimum pitch will be 1.25 times the outer diameter of a tube, d_o. In contrast, for square arrays, a band of width δ will be planned between adjacent tubes with a view to enabling them to be cleaned (TEMA 1999).

Array type	Pitch
Triangular	$a \geq 1.25\, d_o$
Square	$a \geq \text{Sup}\,\{1.25\, d_o, d_o + \delta\}$ with $\delta \geq 0.25"$ (i.e. 6 mm)

Table 4.4. *Calculation of the pitch*

Table 4.5 gives the most commonly-used array types and pitches.

Square array			Triangular array		
d_o (in inches)	Pitch (in inches)	Pitch (in mm)	d_o (in inches)	Pitch (in inches)	Pitch (in mm)
¾"	1"	24.50	¾"	15/16"	22.97
1"	1" ¼	30.63	¾"	1"	24.50
1" ¼	1" 7/8	38.28	1"	1" ¼	30.63
1" ½	1" 9/16	45.94	1" ¼	1" 9/16	38.28

Table 4.5. *Commonly-used arrays and gaps*

4.6.1.8. *Shell diameter*

For pressures of up to 20 Atm, shells are typically produced from 3/8"-thick standard steel pipes. Diameters of 12 to 24 inches are generally used. The shell diameter determines the shell-side flow, but for a given shell diameter and a fixed tube diameter, higher numbers of tubes lead to greater heat transfer areas. Thus, during sizing, we will try to have a maximum number of tubes in the shell to increase the heat transfer area, as well as turbulence.

It should be noted, however, that this parameter is not quite a degree of freedom since it depends on the number of tubes to be placed in the shell, their outer diameter and the gap between them (see section 2.3.6), etc. It will therefore be chosen not only to ensure a satisfactory flow within the shell, but also so that the tube bundle is well adjusted. Tables A1.11 to A1.19 of Appendix 1 give the shell diameters to be used for a given number of tubes, arranged in square or triangular arrays of different sizes.

The data in Appendix 1 should be used when rapid sizing is sought and when the diameters of the tubes used are given in the tables of that appendix. Otherwise, the relationship between the number of tubes and the shell diameter can also be established by calculation (Kakaç et al. 2002; Murali et al. 2008). In fact, independently of the array used (rectangular or triangular), the number of elementary arrays of cross-section area S that can be placed in a shell, of cross-section Σ, is given by the ratio N_M (see Figure 4.5).

$$N_M = \frac{\Sigma}{S}$$

The number of tubes that can be placed in a shell of diameter D_{sh} is given by:

$$N_t = K_P \frac{\pi D_{sh}^2}{4S}$$

where:

K_P is a filling coefficient, which takes into account the ends in which tubes cannot be easily installed (see Table 4.6).

S is the elementary cross-section area defined by the chosen array. For a square array, $S = a^2$, and for a triangular array: $S = \frac{a^2}{2} cos\left(\frac{\pi}{6}\right)$.

S can therefore be put in the form: $S = K_M\, a^2$, where K_M is a coefficient that depends on the chosen array (see Table 4.7).

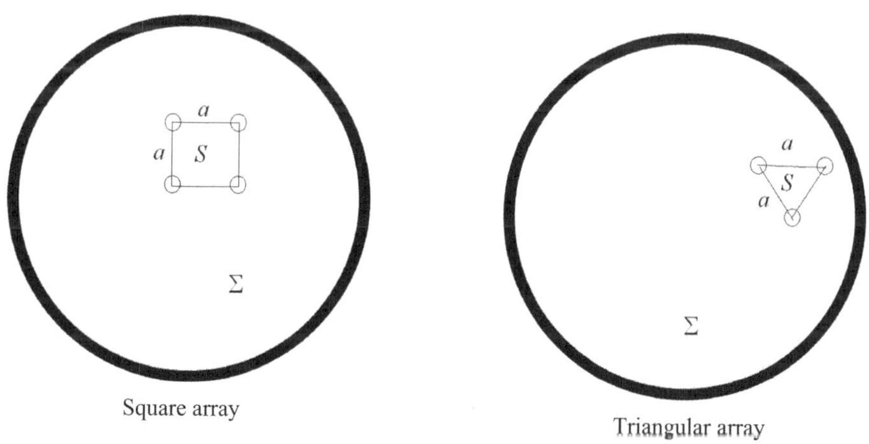

Figure 4.5. *Rectangular and triangular arrays in a shell of cross-section area S*

Number of tube-side passes	K_P
1	0.93
2	0.90
3	0.85
4	0.81
5	0.77
6	0.74

Table 4.6. *Shell filling coefficient*

Array	K_M
Square	1
Triangular	$\frac{1}{2}\cos\left(\frac{\pi}{6}\right)$

Table 4.7. *Array coefficient*

Thus, the number of tubes that can be placed in a shell of diameter D_{sh} can be written as follows[1]:

$$N_t = \frac{\pi\, K_P\, D_{sh}^2}{4\, K_M\, a^2}$$

The shell diameter can therefore be calculated from the following expression:

$$D_{sh} = 2a\sqrt{\frac{N_t\, K_M}{\pi\, K_P}}$$

This type of relationship is preferred when the sizing of the exchanger is performed by numerical methods.

4.6.1.9. *Type of baffles and their spacing, e_c*

It should be recalled that baffles are used to improve heat transfer on the shell side, essentially by increasing flow turbulence.

The spacing, e_c, of adjacent baffles is another degree of freedom to be set by the designer. As we saw in Chapter 2 (section 2.6.3), this parameter influences the value

1 Note that the number of tubes that can be placed in a shell, N_t, is equal to the number of tubes per pass, n_t, multiplied by the number of passes, N_p: $N_t = N_p * n_t$.

of the transverse free area, a_T, and consequently determines the transverse flow rate, G_T (see section 2.6.4.2). It is therefore an important parameter for flow in the shell.

Thus, narrower spacing of the transverse baffles increases the heat transfer coefficient by inducing higher turbulence. But closer baffles (low e_c), generate higher pressure drops in the shell. The choice of e_c will therefore entail finding a *compromise between thermal efficiency and allowable pressure drops.*

RULE OF THUMB.–

The most commonly-used values for e_c lie between 0.2 D_{sh} and D_{sh}.

4.6.1.10. *Free area over the baffles, A_F*

Let us recall that the free area above the baffles is a significant parameter in determining the flow in the shell (see section 2.6.2). Figure 4.6 shows this area in the case of segmental baffles. Segmental baffles are obtained by operating cuts on the discs constituting the baffles. These cuts are generally defined by the cutting ratio, ε, defined by $\varepsilon = \frac{4A_F}{\pi D_{sh}^2}$.

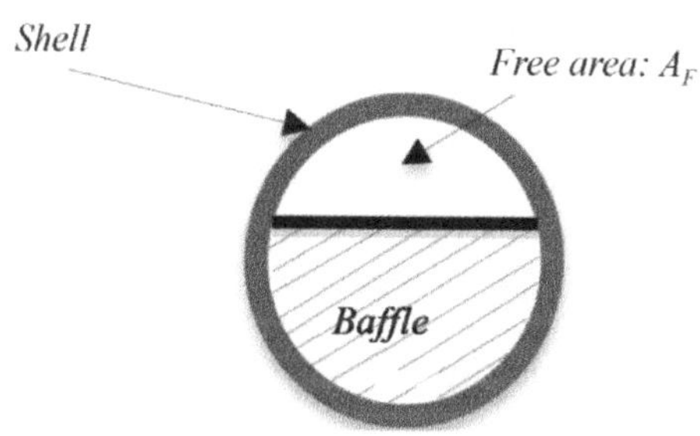

Figure 4.6. *Free area above the baffle. For a color version of this figure, see www.iste.co.uk/benallou/energy6.zip*

RULES OF THUMB.–

– Cutting ratios are usually between 15 and 45 % of the total disk surface.

– Segment cuts of 20 to 25 % generally enable flows in the shell that ensure good energy transfer while maintaining pressure drops within acceptable limits.

Knowledge of the cutting ratio makes it possible to easily determine the free area over the baffles, A_F (see section 2.6.2): $A_F = \frac{\varepsilon \pi D_{sh}^2}{4}$.

4.6.1.11. *Fouling resistance*

Another parameter to be defined when sizing the exchanger concerns fouling. Indeed, as detailed in Appendix 4, most of the process fluids that need to be cooled or heated in the exchanger are charged, to a lesser or greater degree, with particles and some dissolved salts, which tend to deposit on the heat transfer surfaces. Of relatively low heat conductivities, these deposits form crusts, the thickness of which increases with time, thus constituting an additional resistance to conduction.

The exchanger will be designed so as to assure its thermal schedule, even with such fouling.

Thus, when designing an exchanger, an additional resistance to heat transfer, that corresponding to fouling, will be taken into account. This is achieved by including the fouling resistances on the tube side and on the shell side, leading to a change in the expression of the overall heat transfer coefficient (see section 5.3.2.2). This amounts to performing the exchanger calculations while assuming that it is fouled, with an overall heat transfer coefficient, U_F. In fact, this parameter is the basis for starting heat exchanger calculations (see section 6.3).

The values of U_F (overall coefficients for "dirty" heat exchangers) to be used in the design are given in Appendix 1 (see Tables A1.21 to A1.23).

4.6.2. *Origin of the complexity*

As we saw in section 4.6.1, sizing an exchanger involves determining several parameters.

Thus, assuming that all the physical properties (density, viscosity, etc.) are known, sizing a heat exchanger means that the designer will have to set or calculate all of these parameters (see Table 4.8). This is where the difficulty with heat exchanger calculation lies.

Indeed, the parameters that need to be chosen by the designer are related to one another and all have an impact, direct or indirect, on heat-transfer efficiency. For example, the geometrical parameters of the exchanger, in particular the diameter of the tubes (or the gap between the plates) and the number of tubes (or plates), appear to be choosable at will. This is not the case, however, since the tube diameter (or the gap between the plates), for example, influences the Reynolds number, thus the nature of the flow, as well as the convection heat transfer coefficient and the pressure drops. As such, it is clear that caution needs to be taken when making the various choices, bearing in mind that the objective is to design an exchanger that allows the desired thermal flux to be conveyed, with acceptable pressure drops. This is the purpose of Chapter 5.

Parameter type	Designation	Symbol	Way of Determination
Geometric	Array type	Square, triangular	Degree of freedom
	Rotation of the array	30°, 45°	Degree of freedom
	Array pitch	a	Degree of freedom To be set according to the rules of thumb
	Shell diameter	D_{sh}	To be calculated
	Number of baffles	n_c	To be calculated
	Distance between two baffles	e_c	Degree of freedom To be set according to TEMA standards
	Cutting ratio of a baffle	ε	To be set according to the rules of thumb
	Transverse gap	e_T	To be calculated
	Longitudinal gap	e_L	To be calculated
	Outer diameter of the tubes	d_o	Degree of freedom To be set according to tube standards
	Inner diameter of the tubes	d_i	To be determined as a function of d_o
Hydraulic	Longitudinal free area	a_L	To be calculated
	Transverse free area	a_T	To be calculated
	Longitudinal flow rate in the shell	G_L	To be calculated
	Transverse flow rate in the shell	G_L	To be calculated
	Flow velocity in the tubes	v	Degree of freedom To be set according to the rules of thumb

	Shell-side Reynolds number	Re_c	To be calculated
	Tube-side Reynolds number	Re	To be calculated
Thermal	Thermal schedule	ϕ, T_i and t_i	To be calculated from the data
	Convective coefficients	h_i and h_e	To be calculated
	Heat transfer coefficient	U	To be calculated
	Number of tubes	N	To be calculated
	Tube length	L	To be calculated
	Heat transfer area	A	To be calculated
	Number of tube-side passes	N_{pt}	Degree of freedom
	Number of shell-side passes	N_{pc}	Degree of freedom
	Fouling coefficient	h_d	To be estimated
	Fouling resistance	R_d	To be estimated
Mechanical	Shell thickness	δ_c	Set by tube standards
	Tube thickness	δ_t	Set by tube standards

Table 4.8. *The parameters to be set or calculated during sizing*

5

Sizing Methods

5.1. Introduction

In practice, engineers can face two types of design problems, depending on the situation encountered:

– *An evaluation problem:* this is the case where we have an existing heat exchanger that we wish to use to cool (or heat) a fluid that is at a given temperature. In this case, only the inlet temperatures of the fluids are known. The temperatures that could be reached at the outlet are not known, short of proceeding with assembling the installation and measuring them. It will then be necessary to evaluate the available heat exchanger and to determine the temperatures reachable at the device outlet;

– *A sizing problem:* the engineer is asked to construct or order a heat exchanger to implement a thermal schedule. In this case, the flux to be transferred is known, as are the inlet and outlet temperatures of the fluids, and we wish to calculate the heat transfer area of the device that would make it possible to achieve the desired heating or cooling.

Two methods of designing and sizing heat exchangers correspond to these two practical problems:

– a method based on determining efficiency, suitable for the first type of practical problems;

– a method based on the logarithmic mean temperature difference (LMTD), which enables *calculation of the heat transfer area;* this is therefore suitable for the sizing request specified in the second practical problem set out above.

For a color version of all figures in this chapter, see www.iste.co.uk/benallou/energy6.zip

Chapter 5 is devoted to presenting these two methods in detail, based on an analysis of the temperature profiles of the hot and cold fluids in the exchanger and using a coefficient that enables simplification of the calculations of the heat fluxes exchanged: the *overall heat transfer coefficient*.

5.2. Analysis of temperature profiles in an exchanger

Consider a hot fluid and a cold fluid flowing through an exchanger. In steady state, the temperatures of the two fluids are set at one point, but they vary throughout the exchanger. In what way does this variation occur?

To provide a simple visualization, we will assume a *one-pass heat exchanger* operating *co-flow* or *counter-flow*. By convention (see the Rule of thumb below), whatever the flow configuration, *'1' notations are reserved to the side where the hot fluid enters* (see Figure 5.1).

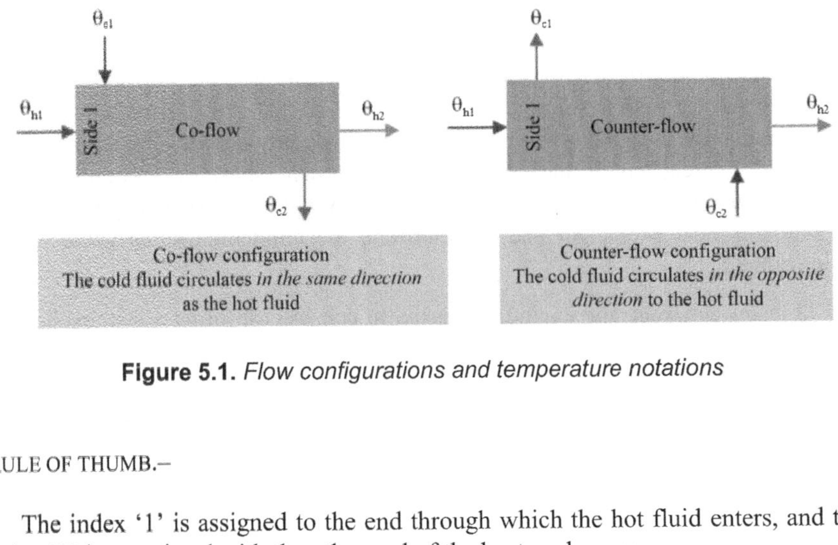

Figure 5.1. *Flow configurations and temperature notations*

RULE OF THUMB.–

The index '1' is assigned to the end through which the hot fluid enters, and the index '2' is associated with the other end of the heat exchanger.

Energy transfer will of course depend on the configuration considered. Similarly, variations in fluid temperatures, hot and cold, depend on the flow configuration. Figure 5.2 gives a symbolic representation of the temperature profiles of the two fluids for co-flow and counter-flow exchangers.

Depending on whether the exchanger is of parallel flow (co-flow) or counter-flow, the curves representing variations in θ_h and θ_c, the temperatures of the hot and cold fluids, respectively, are different (see Figure 5.3).

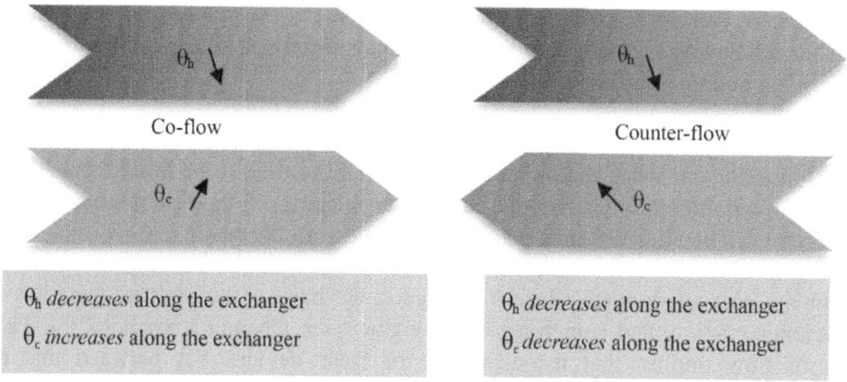

Figure 5.2. *Temperature variations according to the flow mode*

We can see that in co-flow exchangers, the lowest temperature of the hot fluid in the device, θ_{h2}, is necessarily higher than the highest temperature of the cold fluid, θ_{c2}. Whereas in counter-flow exchangers, the outlet temperature of the cold fluid, θ_{c2}, may exceed the lowest temperature of the hot fluid, θ_{h2}.

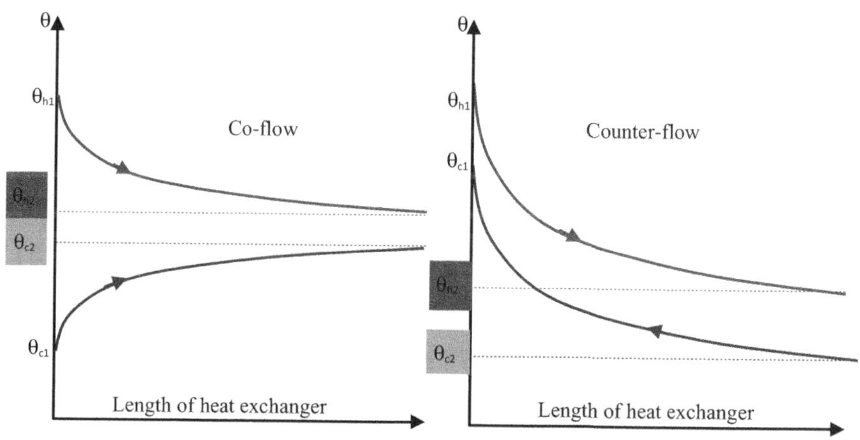

Figure 5.3. *Temperature profiles as a function of the flow configuration*

We can also see that in principle, counter-flow presents better arrangements for heat transfer, since the temperature differences between the two fluids are large throughout the device (see Figure 5.3). This will lead to *significant driving potential differences*.

We can therefore intuitively grant the counter-flow configuration a better transfer efficiency, knowing that the locally transferred heat flux is proportional to the driving potential difference.

Nevertheless, the use of the co-flow configuration may prove useful in some cases; in particular when the outlet temperature of the fluid to be heated should not exceed a certain limit. This is the case in the heating of food or pharmaceutical fluids that can be altered from a determined temperature threshold.

In the case of multipass heat exchangers, we have seen that the operation is sometimes co-flow and sometimes counter-flow (see Chapter 2, section 2.5). This type of flow implies that the efficiency of these devices lies between that of exchangers operating co-flow and those in counter-flow.

5.3. Overall heat transfer coefficient

Consider an exchanger (tubular or compact) in which two fluids are circulated. In steady state, the flowrates of the fluids are constant and the temperature at a given point in the device is constant.

Let us define a volume element of the device, comprised between two cross-section areas perpendicular to the walls of the exchanger (see Figure 5.4), located at distances x and $x+dx$ from the end through which the hot fluid enters (side 1). In this differential volume, the heat transfer area is noted dA_i for the transfer between the interior fluid and the wall, and dA_o for the transfer between the outer fluid and the wall.

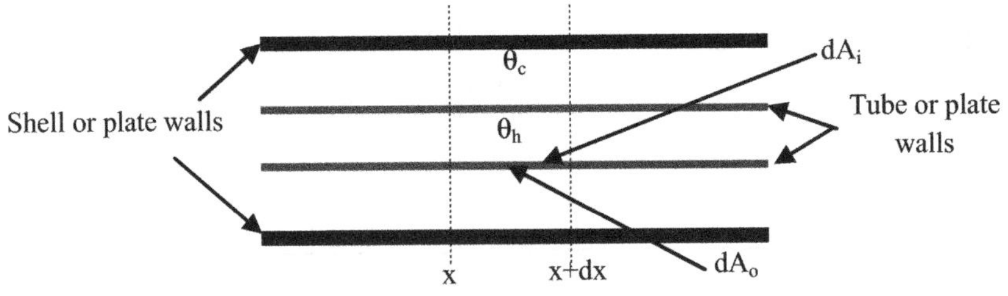

Figure 5.4. *Heat balance on an exchanger*

Note that for a plate heat exchanger, $dA_i = dA_o$, whereas for a tubular heat exchanger, $dA_i = \pi d_i dx$ and $dA_o = \pi d_o dx$, where d_i and d_o are the inner and outer diameters of the tubes, respectively.

Let θ_h and θ_c be the average temperatures of the hot and cold fluids, respectively, in this volume element.

5.3.1. *Overall coefficient for a clean exchanger*

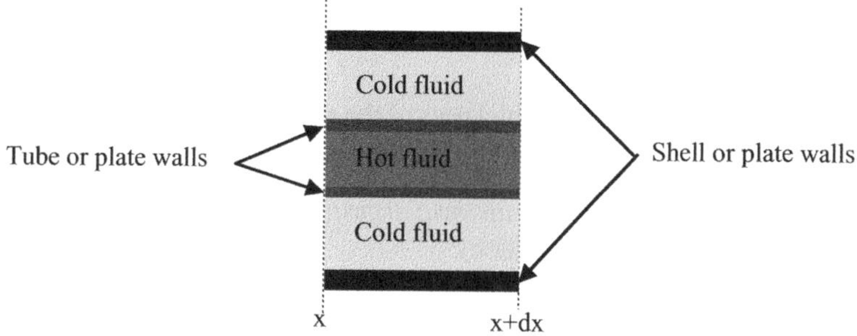

Figure 5.5. *Thermal resistances in a clean exchanger*

Although it represents an ideal situation that does not correspond to the reality of operation in an industrial environment, the case of an exchanger with clean surfaces is of interest in the process of analyzing thermal resistances between the hot fluid and the cold fluid.

In this case, heat transfer between these two fluids involves the following three thermal resistances (see Figure 5.5):

– Resistance to convection between the hot fluid and the inner surface, dA_i. This is given by (see Volume 2, section 2.10.1): $R_{hi} = \frac{1}{h_i dA_i}$; h_i being the convection heat transfer coefficient between the hot fluid and the inner surface;

– Resistance to conduction in the tube (or plate) wall, characterized by thermal conductivity, λ_t. This depends on the wall geometry (Benallou 2018, section 2.10.3 and section 2.13):

- $R_{\lambda_t} = \frac{e}{\lambda_t dA}$ for flat walls ($dA_i = dA_o = dA$),

- $R_{\lambda_t} = \frac{Ln\left(\frac{r+e}{r}\right)}{2\pi\lambda_t L}$ for tubular walls of radius r (L being the length of the tubes),

Resistance to convection between the outer surface, dA_o, and the cold fluid, $R_{ho} = \frac{1}{h_o dA_o}$; h_o being the convection heat transfer coefficient between the cold fluid and the outer surface.

5.3.1.1. *Case of a plate heat exchanger*

In the case of a plate heat exchanger (flat walls), we have $dA_i = dA_o = dA$. The elementary heat flux, $d\phi$, exchanged between the two fluids is then given by (Benallou 2018, section 2.10.4):

$$d\phi = \frac{(\theta_h - \theta_c)}{R_{hi} + R_{\lambda_t} + R_{ho}}$$

or $d\phi = \frac{(\theta_h - \theta_c)dA}{\frac{1}{h_i} + \frac{e}{\lambda_t} + \frac{1}{h_o}}$; where e is the thickness of the plates.

DEFINITION.– We define the *overall heat transfer coefficient*, U, of the volume element of a flat-wall heat exchanger by:

$$\frac{1}{U} = \frac{1}{h_i} + \frac{e}{\lambda_t} + \frac{1}{h_o}$$

U is expressed in kcal/hm^2°C or W/m^{2}K.

Using the definition of the overall heat transfer coefficient, the expression giving the exchanged elementary flux can take the form:

$$d\phi = U\, dA\, \Delta\theta$$

5.3.1.2. *Case of a tubular heat exchanger*

In the case of a tubular heat exchanger, $dA_i = 2\pi r_i dx$ and $dA_o = 2\pi(r_i + e)dx$; where r_i is the (inner) radius of the tubes, and e is their thickness.

The elementary flux exchanged is then given by (Benallou 2018, section 2.13):

$$d\phi = \frac{\theta_h - \theta_c}{\frac{1}{h_i dA_i} + \frac{Ln\left(\frac{r_i + e}{r_i}\right)}{2\pi\lambda_t L} + \frac{1}{h_e dA_o}}$$

DEFINITION.– We define the *overall heat transfer coefficient*, U, of the volume element of a tubular heat exchanger by:

$$\frac{1}{UdA} = \frac{1}{h_i dA_i} + \frac{Ln\left(\frac{r_i + e}{r_i}\right)}{2\pi\lambda_t L} + \frac{1}{h_o dA_o}$$

where dA is the reference area element.

In this case too, the flux expression can take the form:

$$d\phi = U\, dA\, \Delta\theta$$

NOTE.– The reference area appearing in the definition of the overall heat transfer coefficient for tubular heat exchangers is chosen by the user: $dA = dA_i$ or $dA = dA_o$. Note that the value of the overall heat transfer coefficient, U, will then depend on this choice:

– If we choose the inner area of the tubes as the reference area:

$$dA = dA_i \text{ and } UdA = U_i\, dA_i.$$

This will give a value, U_i, normalized with respect to the internal heat transfer area;

– If we choose the outer area of the tubes as the reference area:

$$dA = dA_o \text{ and } UdA = U_o\, dA_o$$

This will give a value, U_o, normalized with respect to the outer transfer area.

Thus, for a tubular heat exchanger, the *overall heat transfer coefficient* is only meaningful in relation to the *corresponding reference area*.

RULE OF THUMB.–

For tubular heat exchangers, the overall heat transfer coefficient is usually normalized with respect to the *outer area* of the tubes.

It will therefore be necessary to retain, in all the developments that follow in this work, *the outer area* as the reference area for U.

5.3.2. *Overall coefficient for a fouled exchanger*

We have already pointed out that, in practice, the walls of heat exchangers that are in contact with the fluids are subjected to deposits of suspended matter, scale or oxides, which cause fouling of these surfaces (see Appendix 4). This involves the development, over time, of a thin layer of solid on the walls, introducing an *additional resistance* to heat transfer, R_d.

Thus, in the case of a fouled exchanger, heat transfer between the hot fluid and the cold fluid involves the following five thermal resistances (see Figure 5.6):

– Internal convective resistance; $R_{hi} = \frac{1}{h_i dA_i}$; h_i being the convection heat transfer coefficient between the hot fluid and the inner surface (see Appendix 4, section A4.4);

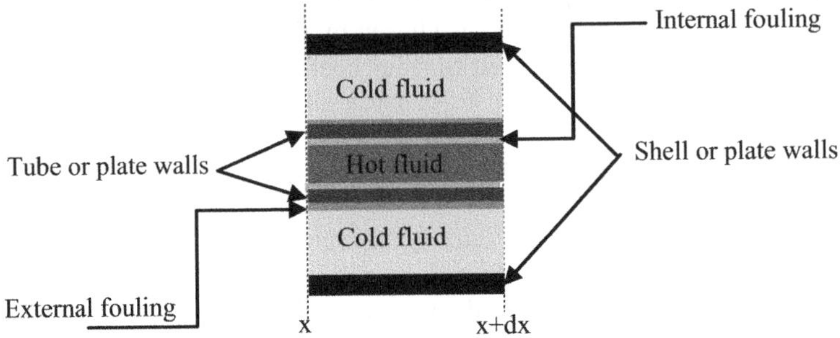

Figure 5.6. *Thermal resistances in a fouled exchanger*

– Resistance to conduction in the solid film deposited by fouling inside the wall, $R_{di} = \frac{1}{h_{di} dA_i}$; h_{di} being the internal surface fouling coefficient;

– Resistance to conduction in the wall separating the two fluids:

$R_{\lambda_t} = \frac{e}{\lambda_t dA}$ for flat walls ($dA_i = dA_o = dA$);

– $R_{\lambda_t} = \frac{Ln\left(\frac{r_i+e}{r_i}\right)}{2\pi\lambda_t L}$ for tubular walls of radius r_i (L being the length of the tubes);

– Resistance due to fouling on the outside of the wall, $R_{do} = \frac{1}{h_{do} dA_o}$;

– Resistance to convection with outer surface: $R_{ho} = \frac{1}{h_o dA_o}$.

5.3.2.1. *Case of a heat exchanger with fouled plates*

In a way similar to that of section 5.3.1.1, it can be shown that the elementary flux is given by:

$$d\phi = \frac{(\theta_h - \theta_c)dA}{\frac{1}{h_i} + \frac{1}{h_{d_i}} + \frac{e}{\lambda_t} + \frac{1}{h_{d_o}} + \frac{1}{h_o}}$$

where: e is the thickness of the plates.

This expression can be written in the form:

$$d\phi = U_F\, dA\, \Delta\theta$$

where: U_F is *the overall heat transfer coefficient for the fouled exchanger.* It is given by:

$$\frac{1}{U_F} = \frac{1}{h_i} + \frac{1}{h_{d_i}} + \frac{e}{\lambda_t} + \frac{1}{h_{d_o}} + \frac{1}{h_o} \text{ i.e.: } U_F = \frac{1}{\frac{1}{h_i}+\frac{1}{h_{d_i}}+\frac{e}{\lambda_t}+\frac{1}{h_{d_o}}+\frac{1}{h_o}}$$

5.3.2.2. *Case of a fouled tubular heat exchanger*

Similarly to section 5.3.1.2, we obtain the following expression for $d\phi$:

$$d\phi = U_{Fo}\, dA_o\, \Delta\theta$$

where U_{Fo} is the overall heat transfer coefficient for the fouled exchanger. It is given by:

$$\frac{1}{U_{Fo} dA_o} = \frac{1}{h_i dA_i} + \frac{1}{h_{d_i} dA_i} + \frac{Ln\left(\frac{r_i+e}{r_i}\right)}{2\pi\lambda_t L} + \frac{1}{h_{d_o} dA_o} + \frac{1}{h_o dA_o}$$

Thus:

$$U_{Fo} = \frac{1}{\frac{1}{h_i}\frac{dA_o}{dA_i}+\frac{1}{h_{d_i}}\frac{dA_o}{dA_i}+\frac{Ln\left(\frac{r_i+e}{r_i}\right)}{2\pi\lambda_t L}dA_o+\frac{1}{h_{d_o}}+\frac{1}{h_o}}$$

5.3.3. *Overall coefficient for finned heat exchangers*

Let us recall that for compact heat exchangers where one of the fluids is a gas, *fins* are often added to the gas-side surfaces, thus increasing the outer transfer area (see sections 1.5.1 to 1.5.3).

In certain applications, especially in the case of shell-and-tube heat exchangers, fins can be arranged on the two (inner and outer) surfaces. In section 5.3.3.1 we give the heat transfer coefficient for the situation where both surfaces are fitted with fins, although this is quite rare. It is obvious that in the case where the fins are placed only on the outer surfaces, the terms corresponding to the internal fins are overlooked in the expression of the heat transfer coefficient.

5.3.3.1. *Case of fouled finned-tube heat exchangers*

The installation of fins means that convections (internal and external) occur across the totality of the surfaces, including those of the fins (Shah et al. 2003). As a result, the convective thermal resistances, R_{hi} for internal convection and R_{ho} for outer convection, become $R_{h_i} = \frac{1}{\eta_i h_i dA_i}$ and $R_{h_o} = \frac{1}{\eta_o h_o dA_o}$; where η_i and η_o are the fin efficiencies, but A_i and A_o denote the *total* inner and outer surfaces, including the surfaces of the fins.

It follows that, for a fouled finned-tube heat exchanger, the overall heat transfer coefficient is given by:

$$\frac{1}{U_{Fo} dA_o} = \frac{1}{\eta_i h_i dA_i} + \frac{1}{\eta_i h_{d_i} dA_i} + \frac{Ln\left(\frac{r_i+e}{r_i}\right)}{2\pi\lambda_t L} + \frac{1}{\eta_o h_{d_o} dA_o} + \frac{1}{\eta_o h_o dA_o}$$

Hence:

$$U_o = \frac{1}{\frac{1}{\eta_i h_i}\frac{dA_o}{dA_i} + \frac{1}{\eta_i h_{d_i}}\frac{dA_o}{dA_i} + \frac{Ln\left(\frac{r_i+e}{r_i}\right)}{2\pi\lambda_t L} dA_o + \frac{1}{\eta_o h_{d_o}} + \frac{1}{\eta_o h_o}}$$

Note that this expression is simplified in the case where the fins are placed only on the outside:

$$U_{Fo} = \frac{1}{\frac{1}{h_i}\frac{dA_o}{dA_i} + \frac{1}{h_{d_i}}\frac{dA_o}{dA_i} + \frac{Ln\left(\frac{r_i+e}{r_i}\right)}{2\pi\lambda_t L} dA_o + \frac{1}{\eta_o h_{d_o}} + \frac{1}{\eta_o h_o}}$$

The efficiencies of the fins depend, in particular, on their dimensions (Kern and Kraus 1972; Icropera and Dewitt 1996; Shah and Sekulic 2003). They are given by:

$$\eta_i = \frac{tanh(m_i L)}{m_i L} \text{ and } \eta_o = \frac{tanh(m_o L)}{m_o L}$$

where m_i and m_o are parameters that depend on the geometry of the fins. They are given by Table 5.1, where:

– e_f is the thickness of the fins;

– d_f is the diameter of the annular or helical fins (see Figure 3.1);

– λ_f is the heat conductivity of the fin construction material;

– L is the length of the fin.

Note, moreover, that the fouling takes place on the totality of the surfaces, including that of the fins. The resistances due to fouling are thus affected by the efficiencies of the fins. They become, for the inner and outer surfaces, respectively: $R_{d_i} = \frac{1}{\eta_i h_{di} dA_i}$ and $R_{d_o} = \frac{1}{\eta_o h_{d_o} dA_o}$.

Fin type	m_i	m_o
Rectangular	$m_i = \sqrt{\frac{2h_i}{e_f \lambda_f}}$	$m_o = \sqrt{\frac{2h_f}{e_f \lambda_f}}$
Annular or helical	$m_i = \sqrt{\frac{4h_i}{d_f \lambda_f}}$	$m_o = \sqrt{\frac{4h_o}{d_f \lambda_f}}$

Table 5.1. *m_i and m_o for the calculation of efficiencies*

5.3.3.2. *Case of fouled plate-fin heat exchangers*

Note that here too, the presence of the fins will introduce their efficiencies at the resistances relating to the internal and external convections ($R_{h_i} = \frac{1}{\eta_i h_i dA_i}$ and $R_{h_o} = \frac{1}{\eta_o h_o dA_o}$) and the fouling resistances ($R_{d_i} = \frac{1}{\eta_i h_{di} dA_i}$ and $R_{d_o} = \frac{1}{\eta_o h_{d_o} dA_o}$).

Thus, for a flat-wall heat exchanger, equipped with fins and fouled, the overall heat transfer coefficient becomes:

$$\frac{1}{U_F} = \frac{1}{\eta_i h_i} + \frac{1}{\eta_i h_{d_i}} + \frac{e}{\lambda_t} + \frac{1}{\eta_o h_o} + \frac{1}{\eta_o h_{do}}$$

$$\text{or: } U_F = \frac{1}{\frac{1}{\eta_i h_i} + \frac{1}{\eta_i h_{d_i}} + \frac{e}{\lambda_t} + \frac{1}{\eta_o h_o} + \frac{1}{\eta_o h_{d_o}}}.$$

5.4. Illustration: overall heat transfer coefficient for a fouled exchanger

A stainless-steel plate heat exchanger is used to recover heat from a process liquid and transfer it to a flowrate, W_2, of a heat-transfer liquid.

The process liquid is available at a flowrate W_1, and temperature T_1.

By defining the unit thermal resistances as being the resistances obtained for a surface, dA, of 1 m²:

1) Calculate the unit convective resistances (internal and external).

2) Calculate the unit fouling resistances (internal and external).

3) Calculate the conduction resistance in the wall.

4) Compare these resistances; can one or more of them be overlooked?

5) Under these conditions, what is the value of the overall heat transfer coefficient, U_F, for the fouled heat exchanger?

6) What would the overall heat transfer coefficient, U, have been for the clean exchanger?

Given data:

– heat conductivity of stainless steel: $\lambda_{ss} = 20$ W/m°C

– plate wall thickness: $e = 1$ mm

– $h_i = 2{,}465$ W/m^2°C $\quad h_o = 3{,}800$ W/m^2°C

– $h_{di} = 3{,}200$ W/m^2°C $\quad h_{do} = 4{,}000$ W/m^2°C

Solution

1) Unit convective resistances

The convective resistances are: $R_i = \frac{1}{h_i dA}$ and $R_o = \frac{1}{h_o dA}$

NUMERICAL APPLICATION

$h_i = 2{,}465$ W/m^2°C $\quad h_o = 3{,}800$ W/m^2°C $\quad dA = 1$ m^2

Hence: $R_i = \frac{1}{2{,}465\text{x}1} = 4.06\ 10^{-4}$°C/W and $R_o = \frac{1}{3{,}800\text{x}1} = 2.63\ 10^{-4}$°C/W

2) Unit fouling resistances

The fouling resistances are: $R_{di} = \frac{1}{h_{di} dA_i}$ and $R_{do} = \frac{1}{h_{do} dA}$.

NUMERICAL APPLICATION

$h_{di} = 3{,}200$ W/m^2°C $\quad h_{do} = 4{,}000$ W/m^2°C $\quad dA = 1$ m^2

Hence: $R_{di} = \frac{1}{3{,}200\text{x}1} = 3.13\ 10^{-4}$°C/W and $R_{do} = \frac{1}{4{,}000\text{x}1} = 2.50\ 10^{-4}$°C/W

3) Conduction resistance in the wall

This is given by: $R_{\lambda_{ss}} = \frac{e}{\lambda_{ss} dA}$

NUMERICAL APPLICATION

$\lambda_{ss} = 20$ W/m°C $\quad e = 1$ mm $\quad dA = 1$ m^2

Hence: $R_{\lambda_{ss}} = \frac{0.001}{20x1} = 0.50\ 10^{-4}$°C/W.

4) Comparison of resistances

Table 5.2 summarizes the different resistances.

Resistance	**Value** (in °C/W)
Internal convection	$4.06\ 10^{-4}$
External convection	$2.63\ 10^{-4}$
Internal fouling	$3.13\ 10^{-4}$
External fouling	$2.50\ 10^{-4}$
Conduction	$0.5\ 10^{-4}$

Table 5.2. *Comparison of different resistances to transfer*

All unit resistances are of the same order of magnitude. *Therefore, none of these can be overlooked.*

5) Calculation of U_F

For a fouled plate heat exchanger, U_s is given by (see section 5.3.2.1):

$$U_F = \frac{1}{\frac{1}{h_i} + \frac{1}{h_{d_i}} + \frac{e}{\lambda_{ss}} + \frac{1}{h_{d_o}} + \frac{1}{h_o}}$$

NUMERICAL APPLICATION

$\frac{1}{h_i} = 4.06\ 10^{-4}\ \text{m}^2 \frac{°\text{C}}{\text{W}}$ $\quad \frac{1}{h_{d_i}} = 3.13\ 10^{-4}\ \text{m}^2 \frac{°\text{C}}{\text{W}}$ $\quad \frac{e}{\lambda_{ss}} = 0.50\ 10^{-4}\ \text{m}^2 \frac{°\text{C}}{\text{W}}$

$\frac{1}{h_{d_o}} = 2.5\ 10^{-4}\ \text{m}^2 \frac{°\text{C}}{\text{W}}$ $\quad \frac{1}{h_o} = 2.63\ 10^{-4}\ \text{m}^2 \frac{°\text{C}}{\text{W}}$

Hence: $U_F = \frac{10^4}{4.06+3.13+0.5+2.5+2.63} = 780\ \frac{\text{W}}{\text{m}^2 °C}$.

6) Overall heat transfer coefficient for the clean exchanger, U

For the clean exchanger (see section 5.3.1): $U = \frac{1}{\frac{1}{h_i}+\frac{e}{\lambda_{SS}}+\frac{1}{h_o}}$

NUMERICAL APPLICATION

$\frac{1}{h_i} = 4.06\ 10^{-4}\ \mathrm{m^2°C/W}$ $\frac{e}{\lambda_{SS}} = 0.50\ 10^{-4}\ \mathrm{m^2°C/W}$ $\frac{1}{h_o} = 2.63\ 10^{-4}\ \mathrm{m^2°C/W}$

Hence: $U = \frac{10^4}{4.06+0.5+2.63} = 1{,}390\ \mathrm{W/m^2°C}$

5.5. Calculation of the heat flux transferred

Consider an exchanger in which a mass flowrate, W_h, of a hot fluid (specific heat, C_{ph}) is used to heat a cold fluid (mass flowrate, W_c, and specific heat, C_{pc}).

5.5.1. *Heat balances*

Whatever the nature of the heat exchanger (compact, plate or multitube), the heat flux can be determined by an overall balance on the fluids, hot or cold.

– Overall balance on the hot fluid:

$$\phi = - W_h C_{ph} (\theta_{h2} - \theta_{h1})$$

– Overall balance on the cold fluid:

$$\phi = \varepsilon W_c C_{pc} (\theta_{c2} - \theta_{c1})$$

where $\varepsilon = \pm 1$ depending on whether the device operates co-flow or counter-flow.

Hence: $\left(\frac{1}{W_h C_{ph}} + \frac{1}{\varepsilon W_c C_{pc}}\right)\phi = \left(\theta_{h_1} - \theta_{c_1}\right) - \left(\theta_{h_2} - \theta_{c_2}\right)$

or: $\psi\,\phi = -\left(\Delta\theta_2 - \Delta\theta_1\right)$

where: $\Delta\theta = \theta_h - \theta_c$

and: $\psi = \left(\frac{1}{W_h C_{ph}} + \frac{1}{\varepsilon W_c C_{pc}}\right)$.

This expression allows the flux to be presented as follows:

$$\phi = -\frac{(\Delta\theta_2 - \Delta\theta_1)}{\psi}$$

5.5.2. *Relationship between flux and transfer area*

As we saw in section 5.3, regardless of the heat exchanger type (compact and finned, flat-plate or tubular) and regardless of its state (clean or fouled), the expression of the elementary flux remains the same:

$$d\phi = U\, dA\, \Delta\theta$$

Moreover, in steady state and within reasonable temperature ranges, W_h, W_c, C_{ph} and C_{pc} are constant. It follows that ψ is constant. Differentiating the expression of ϕ obtained in section 5.5.1, the flux differential can also take the form: $d\phi = -\frac{d(\Delta\theta)}{\psi}$.

Hence: $U\, dA\, \Delta\theta = -\frac{d(\Delta\theta)}{\psi}$ Or: $-\frac{d(\Delta\theta)}{\Delta\theta} = \psi\mathrm{U\ dA}$.

The flux (or, as applicable, the transfer area) is calculated by integrating this differential expression. But the coefficient, U, that appears here is a local coefficient that reflects the situation between x and $x+dx$. It depends on the convection heat transfer coefficients (h_i and h_o), fouling coefficients (h_{di} and h_{do}), and physical properties of the fluids and walls.

When the temperature variations (of the hot fluid and the cold fluid) along the heat exchanger are not large enough to significantly impact the physical properties, the various parameters that determine the value of U can be considered constant. Also, *U can, in turn, be considered constant.* This corresponds to the *majority of cases encountered in practice.*

Hence, integrating the equation between the two ends of the exchanger gives:

$$Ln\frac{\Delta\theta_1}{\Delta\theta_2} = \psi\mathrm{U\ A}$$

Yet: $\psi = -\frac{(\Delta\theta_2 - \Delta\theta_1)}{\phi}$.

Therefore: $Ln\frac{\Delta\theta_1}{\Delta\theta_2} = \frac{(\Delta\theta_1 - \Delta\theta_2)}{\phi}U\ A$; or: $\phi = \frac{(\Delta\theta_1 - \Delta\theta_2)}{\mathrm{Ln}\frac{\Delta\theta_1}{\Delta\theta_2}}U\ A$.

Which is often written in the form: $\phi = UA\ (\Delta\theta_{mL})$,

with: $\Delta\theta_{mL} = \frac{(\Delta\theta_1 - \Delta\theta_2)}{Ln\frac{\Delta\theta_1}{\Delta\theta_2}}$; $\Delta\theta_1 = \theta_{h_1} - \theta_{c_1}$ and $\Delta\theta_2 = \theta_{h_2} - \theta_{c_2}$; *the index '1' being assigned to the end of the exchanger corresponding to the hot fluid inlet.*

REMEMBER. Thus, the flux can be calculated using one of the following equations:

– from the hot fluid data: $\phi = -\ W_h\ C_{ph}\ (\theta_{h2} - \theta_{h1})$;

– from the cold fluid data: $\phi = \varepsilon W_c\ C_{pc}\ (\theta_{c2} - \theta_{c1})$;

– from the transfer area and the overall transfer coefficient: $\phi = UA\ (\Delta\theta_{mL})$.

The choice of either expression depends on the data available.

5.6. Method based on the calculation of efficiency: NTU method

In some situations, we are asked to determine the performance that can be expected of a given exchanger. This is often the case when we wish to evaluate the operation of an exchanger that is already installed and operational: is the exchanger operating as expected or not?

In such a case, the heat transfer area is known since the geometry of the device is set. In addition, since the fluids and operating conditions are known, the overall heat transfer coefficient can be determined. Although the effective temperatures can be measured in such a situation, we do not know what the outlet temperatures and the fluxes to be exchanged should be in the case of normal operation of the heat exchanger. It is then more appropriate to carry out an evaluation of the efficiency of the exchanger considered (Kays and London 1984).

5.6.1. *Definition of efficiency*

DEFINITION.– The efficiency η of a heat exchanger is defined as the *ratio between the heat flux actually transferred and the maximum possible flux, ϕ_{max};* that is, the thermodynamic limit of the flux.

$$\eta = \frac{\phi}{\phi_{\max}}$$

The maximum possible flux, ϕ_{max}, depends on the heat capacities of the two fluids. Indeed:

– If $W_c C_{pc} < W_h C_{ph}$: in such a case, the outlet temperature of the cold fluid, θ_{co}, may reach the inlet temperature of the hot fluid, θ_{hi}. The maximum flux that could be exchanged is then given by: $\phi_{max} = W_c C_{pc}(\theta_{hi} - \theta_{ci})$; θ_{ci} being the inlet temperature of the cold fluid.

– If $W_h C_{ph} < W_c C_{pc}$: in this situation, the outlet temperature of the hot fluid, θ_{ho}, may reach the inlet temperature, θ_{ci}. The maximum flux that could be exchanged is then given by: $\phi_{max} = W_h C_{ph}(\theta_{hi} - \theta_{ci})$.

In both cases, this flux can be written as follows:

$$\phi_{\max} = (WC_p)_{\min}(\theta_{hi} - \theta_{ci})$$

where $(WC_p)_{\min}$ is the smallest of the amounts $W_h C_{ph}$ and $W_c C_{pc}$.

Thus, the efficiency of an exchanger is given by:

$$\eta = \frac{W_h C_{p_h}(\theta_{hi}-\theta_{ho})}{(WC_p)_{\min}(\theta_{hi}-\theta_{ci})} = \frac{W_c C_{p_c}(\theta_{co}-\theta_{ci})}{(WC_p)_{\min}(\theta_{hi}-\theta_{ci})}$$

The heat flux actually exchanged can then be related to $(WC_p)_{\min}$ and η as follows:

$$\phi = \eta\phi_{\max} = \eta(WC_p)_{\min}(\theta_{hi} - \theta_{ci})$$

5.6.2. *Efficiency of a co-flow heat exchanger*

For a co-flow (see Figure 5.7), θ_{hi}, θ_{ho}, θ_{ci} and θ_{co} are noted: θ_{h1}, θ_{h2}, θ_{c1} and θ_{c2}, respectively. Likewise, we have: $\Delta\theta_1 = \theta_{h1} - \theta_{c1}$ and $\Delta\theta_2 = \theta_{h2} - \theta_{c2}$.

The heat balance can be written: $\phi = \dfrac{\theta_{h_1}-\theta_{h_2}}{\frac{1}{W_h C_{p_h}}} = \dfrac{\theta_{c2}-\theta_{c1}}{\frac{1}{W_c C_{p_c}}} = \dfrac{\Delta\theta_1-\Delta\theta_2}{\left(\frac{1}{W_h C_{p_h}}+\frac{1}{W_c C_{p_c}}\right)}$

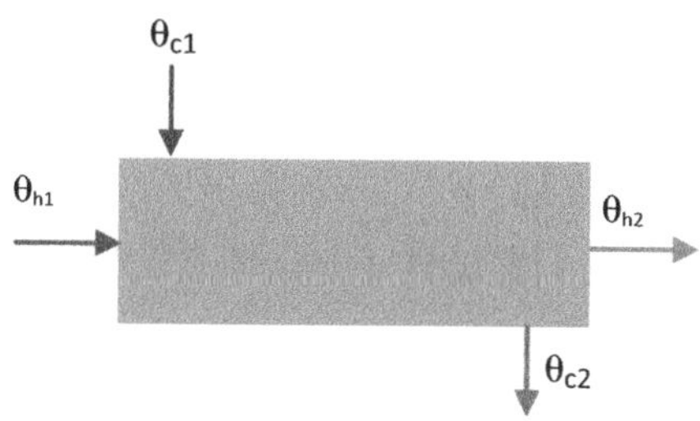

Figure 5.7. *Co-flow*

Hence: $\Delta\theta_2 = \Delta\theta_1 - \phi\left(\frac{1}{W_h C_{p_h}} + \frac{1}{W_c C_{p_c}}\right)$.

Moreover, $\Delta\theta_1$ can be expressed as a function of efficiency: $\Delta\theta_1 = \frac{\phi}{\eta(WC_p)_{\min}}$ where $(WC_p)_{\min}$ is the smallest of the amounts $W_h C_{ph}$ and $W_c C_{pc}$.

NOTE.– In a heat exchanger, it is the fluid whose specific heat flowrate, WC_p, is minimal that *governs* the heat transfer.

$\Delta\theta_2$ can then be written as follows: $\Delta\theta_2 = \phi\left[\frac{1}{\eta(WC_p)_{\min}} - \left(\frac{1}{W_h C_{p_h}} + \frac{1}{W_c C_{p_c}}\right)\right]$.

When U is constant along the device, the equation that links the flux to the transfer area is: $\phi = U\,A\,\frac{(\Delta\theta_1 - \Delta\theta_2)}{Ln\frac{\Delta\theta_1}{\Delta\theta_2}}$.

By replacing $\Delta\theta_1$ and $\Delta\theta_2$ with their expressions, we obtain:

$$\phi = U\,A\,\frac{\phi\left(\frac{1}{W_h C_{p_h}} + \frac{1}{\varepsilon W_c C_{p_c}}\right)}{Ln\left[\frac{\frac{\phi}{\eta(WC)_{\min}}}{\phi\left[\frac{1}{\eta(WC_p)_{\min}} - \left(\frac{1}{W_{ch} C_{p_{ch}}} + \frac{1}{\varepsilon W_{fc} C_{p_{fc}}}\right)\right]}\right]}$$

i.e.: $Ln\left[1 - \eta(WC_p)_{\min}\left(\frac{1}{W_h C_{p_h}} + \frac{1}{\varepsilon W_c C_{p_c}}\right)\right] = -U\,A\left(\frac{1}{W_h C_{p_h}} + \frac{1}{\varepsilon W_c C_{p_c}}\right)$.

or: $\eta = \frac{1 - e^{-U\,A\left(\frac{1}{W_h C_{p_h}} + \frac{1}{\varepsilon W_c C_{p_c}}\right)}}{(WC_p)_{\min}\left(\frac{1}{W_h C_{p_h}} + \frac{1}{\varepsilon W_c C_{p_c}}\right)}$

We can then easily see that, both for $(WC_p)_{\min} = W_h C_{ph}$ and for $(WC_p)_{\min} = W_c C_{pc}$, the expression of η can take the form:

$$\eta = \frac{1 - e^{-\frac{U\,A(1+R)}{(WC_p)_{\min}}}}{1+R}$$

where: $R = \frac{(WC_p)_{\min}}{(WC_p)_{\max}}$; $(WC_p)_{\max}$ being the largest of the amounts $W_h C_{ph}$ and $W_c C_{pc}$.

5.6.3. *Number of transfer units (NTU)*

Consider a heat exchanger having a transfer area A, operating with global transfer coefficient U, to heat a mass flowrate W_c with a hot fluid W_h.

DEFINITION.– The ratio $\frac{U\,A}{(wc_p)_{min}}$ is called the number of transfer units of the exchanger considered (NTU):

$$\text{NTU} = \frac{U\,A}{(wc_p)_{min}}$$

Yet: $U\,A = \frac{\phi}{\Delta\theta_{ML}}$.

We can thus write: $\text{NTU} = \frac{\phi}{\Delta\theta_{ML}(wc_p)_{\text{min}}}$.

A way of visualizing this concept would be to imagine that a heat transfer unit corresponds to the elementary part of a heat exchanger that generates, for the fluid with the lowest sensible heat, $(WC_p)_{\text{min}}$, a temperature change equal to the average temperature difference between the fluids.

Using this concept, the efficiency can then be expressed as a function of the NTU:

$$\eta = \frac{1-e^{-(1+R)\text{NTU}}}{1+R} \text{ where } R = \frac{(wc_p)_{\text{min}}}{(wc_p)_{\text{max}}}.$$

5.6.4. *Illustration: outlet temperatures of a heat exchanger*

An aqueous solution leaves the reactor used for its manufacture with a flowrate W_h and a temperature θ_{h1}. We wish to cool this aqueous solution using well water, available at θ_{c1} with a flowrate, W_c. Your department manager informs you that this task will have to be carried out using a heat exchanger available at the factory. The latter is configured to operate co-flow, it has a transfer area, A, and makes it possible to have an overall heat transfer coefficient, U, for the fluids and the flowrates considered.

1) Which fluid governs the heat transfer?

2) Calculate the maximum exchangeable heat flux.

3) Calculate the efficiency of this exchanger.

4) Infer therefrom the temperature of the hot fluid at the outlet.

Given data:

aqueous solution: $W_h = 0.18$ T/hr $C_{ph} = 1.3$ kcal/kg°C $\theta_{h1} = 95$°C

well water: $W_c = 0.54$ T/hr $C_{pc} = 1.0$ kcal/kg°C $\theta_{c1} = 20$°C

existing exchanger: $A = 55.45$ m^2 $U = 245$ W/m^2°C

Solution

1) The fluid that governs the exchange

This is the fluid with the lowest sensible-heat flowrate. We then calculate the sensible-heat flowrates on both sides:

– Hot side: $W_h\ C_{ph} = 180 \times 1.3 = 234$ kcal/h°C.

– Cold side: $W_c\ C_{pc} = 540 \times 1.0 = 540$ kcal/h°C.

Therefore: $(WC_p)_{\min} = W_h C_{ph} = 234$ *kcal/h°C*. It is therefore the hot side that governs the exchange.

2) Maximum flux

The maximum flux is given by: $\phi_{\text{Max}} = (WC_p)_{\min}(\theta_{h1} - \theta_{c1})$.

NUMERICAL APPLICATION

$(WC_p)_{\min} = 234$ kcal/h°C.

$\theta_{h1} = 95$°C $\theta_{c1} = 20$°C

Hence: $\phi_{Max} = 234\ (95 - 20) = 17{,}550$ kcal/hr.

3) Efficiency of the existing exchanger

The efficiency of the exchanger is given by: $\eta = \frac{1-e^{-(1+R)NTU}}{1+R}$.

with: $R = \frac{(wc_p)_{min}}{(wc_p)_{max}}$ and NTU $= \frac{U\ A}{(wc_p)_{min}}$.

NUMERICAL APPLICATION

$W_h\ C_{ph} = 234$ kcal/h°C $W_c\ C_{pc} = 540$ kcal/h°C

$A = 55.45\ m^2$ $\qquad U = 425\ W/m^2°C = 211\ kcal/hm^2°C$

In this case : $R = \frac{W_h C_{p_h}}{W_c C_{p_c}} = \frac{234}{540} = 0.433$ and $\text{NTU} = \frac{U\,A}{(WC_p)_{min}} = \frac{211*55.45}{234} = 50.$

Hence: $\eta = \frac{1-e^{-50(1+0.433)}}{1+0.433} = 69.76\%$

4) Outlet temperature of the hot fluid

The outlet temperature of the hot fluid, θ_{h2}, is related to the heat flux exchanged by: $\phi = W_h C_{p_h}(\theta_{h1} - \theta_{h2})$.

Thus, in order to determine the outlet temperature, the flux, ϕ, which can be exchanged in this device, needs to be known. The latter is calculated from the efficiency and the maximum flux: $\phi = \eta\ \phi_{Max}$.

Hence: $\theta_{h_2} = \theta_{h_1} - \frac{\eta\phi_{Max}}{W_h C_{p_h}}$.

NUMERICAL APPLICATION

$W_h\ C_{ph} = 234$ kcal/h°C $\qquad \theta_{h1} = 95°C \qquad \eta = 69.76\% \qquad \phi_{Max} = 17{,}550$ kcal/hr

Hence: $\theta_{h_2} = 95 - \frac{0.6976*17{,}550}{234}$.

i.e.: $\theta_{h2} = 43°C$.

5.6.5. *Efficiency of other heat exchanger types*

Regardless of the type of exchanger (multipass, cross-flow, or plate heat exchangers), efficiency η is a function of the ratio $R = \frac{(WC_p)_{min}}{(WC_p)_{max}}$ and the number of transfer units, $\text{NTU} = \frac{U\,A}{(WC_p)_{min}}$. Table 5.3 gives, as a function of the NTU, the expressions of efficiencies of all types of exchangers (one-pass and multipass), operating in co-flow, in counter-flow or in cross-flow, as well as for condensers and for boilers.

Note that, by definition, the value of the efficiency varies from 0 to 1. This property is found in the expressions in Table 5.3 and the resulting graphical representations (Figures 5.7 to 5.11).

Type of heat exchanger	Efficiency	$\lim\limits_{R\to 0} \eta$	$\lim\limits_{NTU\to\infty} \eta$
Co-flow 	$\eta = \frac{1-e^{-(1+R)NTU}}{1+R}$	$1-e^{-NTU}$	$\frac{1}{1+R}$
Counter-flow 	$\eta = \frac{1-e^{-(1-R)NTU}}{1-Re^{-(1-R)NTU}}$	$1-e^{-NTU}$	1
Multitube with one shell-side pass and 2 or more tube-side passes: 1-2, 1-4, 1-6, etc.	$\eta = \frac{2}{1+R+\sqrt{1+R^2}\frac{1+e^{-NTU\sqrt{1+R^2}}}{1-e^{-NTU\sqrt{1+R^2}}}}$	$1-e^{-NTU}$	$\frac{2}{(1+R)+\sqrt{1+R^2}}$
Cross-flows with both fluids channeled 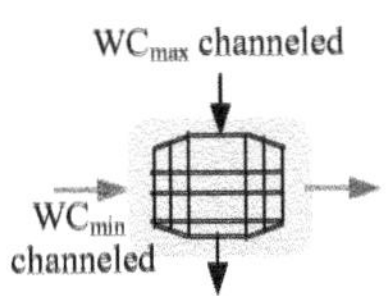	$\eta = 1-e^{\left\{\frac{NTU^{0.22}}{R}\left[e^{(-R\,NTU^{0.78})}-1\right]\right\}}$	$1-e^{-NTU}$	1
Cross-flows with WC_{max} channeled 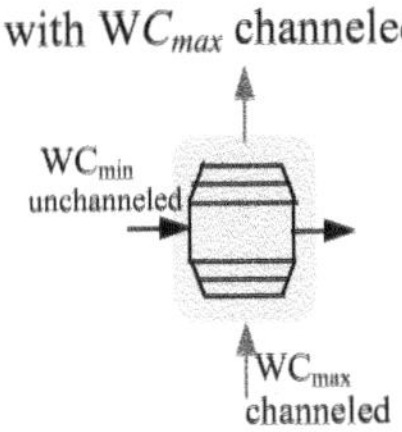	$\eta = 1-e^{\left(-\frac{1-e^{-R(NTU)}}{R}\right)}$	$1-e^{-NTU}$	$1-e^{-\frac{1}{R}}$
Cross-flows with C_{min} channeled 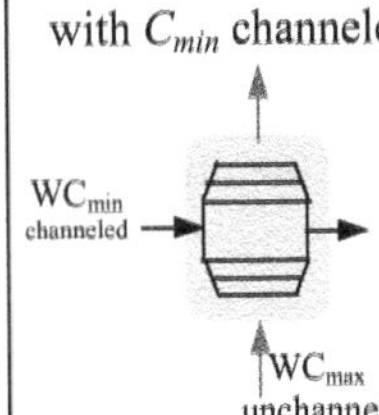	$\eta = \frac{1-e^{[1-R(1-e^{-NTU})]}}{R}$	$1-e^{-NTU}$	$\frac{1}{R}(1-e^{-R})$

In the case where $R = 1$, the expression giving η for the exchanger operating counter-flow becomes indeterminate, but noting the fact that $\lim_{x\to 1}\frac{f(x)}{g(x)} = \lim\limits_{x\to 1}\frac{f'(x)}{g'(x)}$ (L'Hospital's rule), we obtain: $\eta = \frac{NTU}{1+NTU}$.

Table 5.3. *Efficiencies of the different types of exchangers as a function of the NTU*

Likewise, by definition, the value of the heat capacity ratio, R, is between 0 and 1. Thus, for a given value of the NTU, the efficiency reaches a maximum for $R = 0$ and a minimum for $R = 1$.

Figures 5.8 to 5.11 show graphical representations of the expressions presented in Table 5.3.

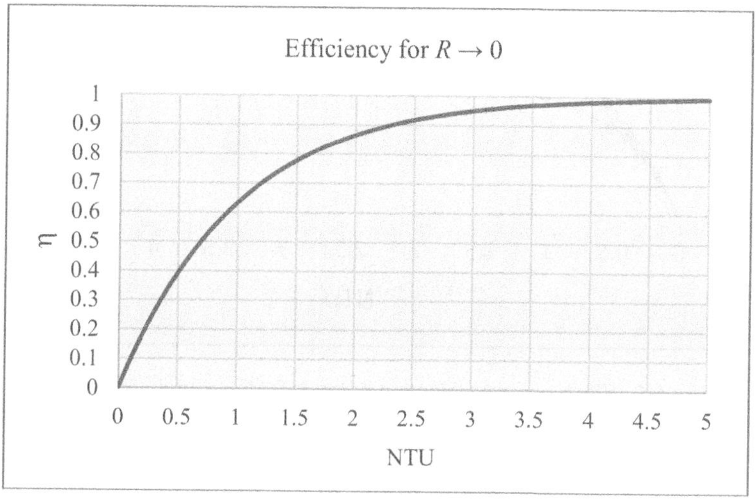

Figure 5.8. *Efficiency as a function of the NTU for condensers and boilers*

The case $R \to 0$ (Figure 5.8) corresponds to $WC_{max} \to \infty$, which is encountered during a phase change process (*condensation* or *boiling*). In these cases, the efficiency reaches the maximum $1 - e^{-NTU}$, regardless of the type of heat exchanger.

Figure 5.9 shows the variation in efficiency as a function of the NTU for a co-flow heat exchanger and for different values of R, ranging from 0 to 1, whereas Figure 5.10 gives the same information for a counter-flow heat exchanger. These two figures clearly illustrate the fact that for a given NTU, the effectiveness of a counter-flow is always greater than that of a co-flow.

Moreover, Figure 5.11 represents variation in efficiency as a function of the NTU for a cross-flow heat exchanger where the two fluids are channeled.

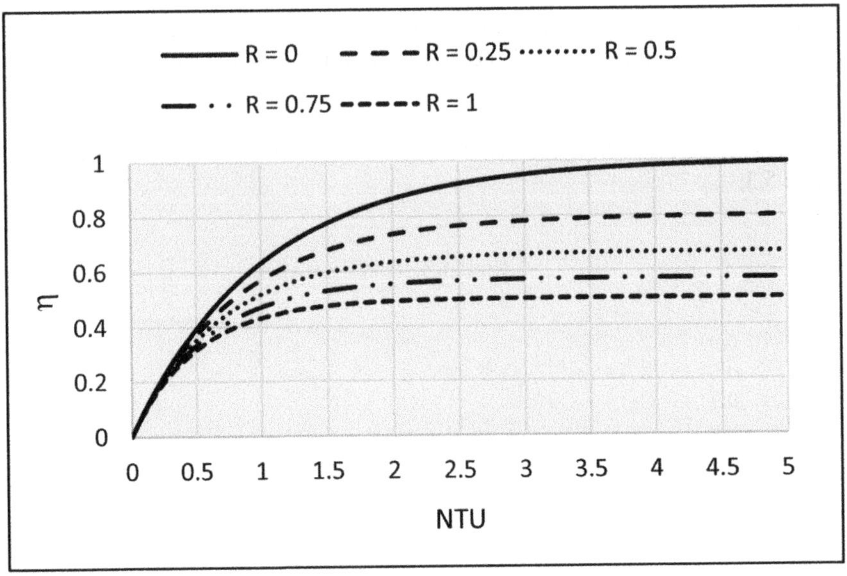

Figure 5.9. *Efficiency as a function of the NTU for co-flow exchangers*

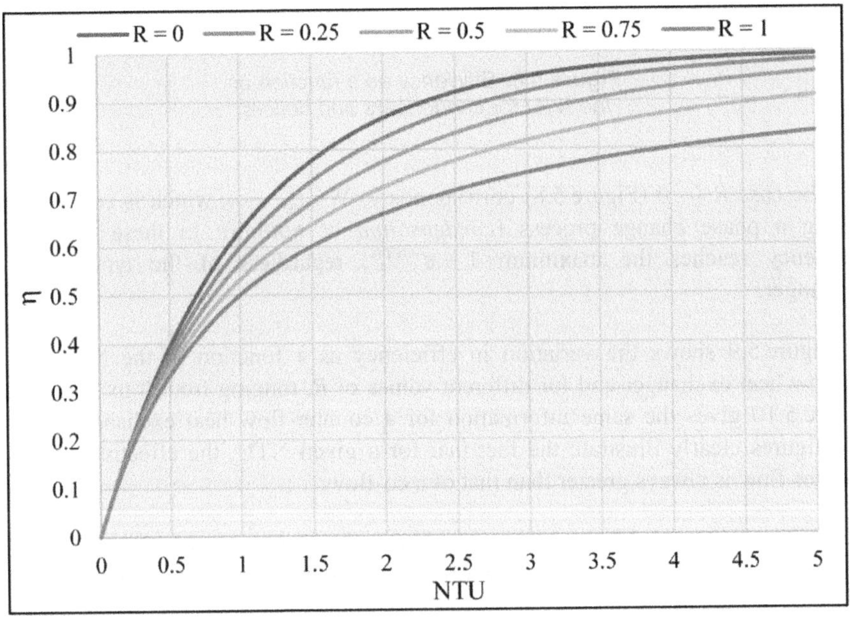

Figure 5.10. *η as a function of the NTU for counter-flow exchangers*

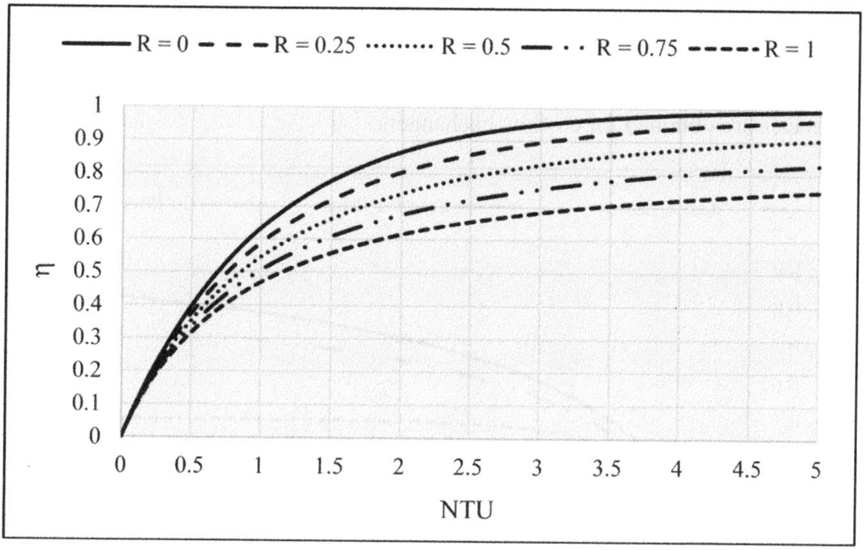

Figure 5.11. *Efficiency as a function of the NTU for cross-flow exchangers where the two fluids are channeled*

Analysis of the curves shows that η increases rapidly for small NTU values, up to around 1.5 to 2. This variation becomes slower for larger NTU values. We deduce therefrom the Rule of thumb below.

RULE OF THUMB.–

In practice, we will not seek a heat exchanger that has a large NTU (generally greater than 3) and is therefore large in size, since a significantly increased NTU results in only slightly greater efficiency. As a result, we will not seek a very high efficiency to the detriment of the cost of the exchanger. Rather, a compromise between efficiency and cost will be sought.

It should also be noted that, for NTU values below about 0.3, the efficiency of a heat exchanger becomes virtually independent of the heat capacity ratio, R.

Figures 5.8 to 5.11 also show that the efficiency is lowest in the limit case of $R = \frac{(wc_p)_{min}}{(wc_p)_{max}} = 1$. Such a case is encountered in practice when the heat capacities of the two fluids are equal.

Moreover, Figure 5.12 shows a comparison between the efficiencies of the different types of exchangers for the case $R = 1$: counter-flow, co-flow and cross-flows with channeled fluids. It shows that, for a given NTU and heat capacity ratio $R = \frac{(wc_p)_{min}}{(wc_p)_{max}}$, the counter-flow has the highest efficiency and that the lowest efficiencies are obtained for co-flow exchangers.

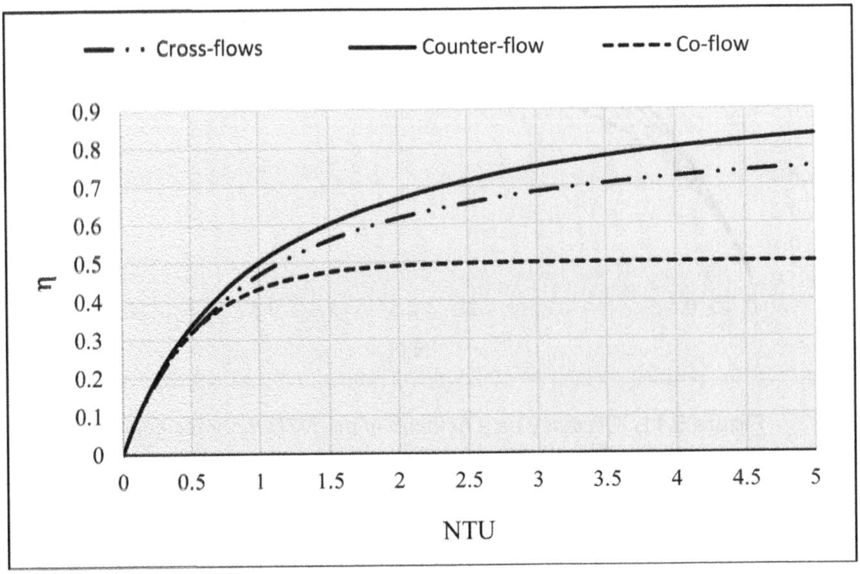

Figure 5.12. *Comparison of the efficiencies of the different exchanger types*

5.6.6. *Practical calculations of efficiency*

Efficiency can be determined from the graphs or from the expressions in Table 5.3. But, for repetitive or iterative calculations, it is of course more convenient to use expressions rather than performing graphical readings.

5.6.7. *Illustration: heat recovery from industrial waste heat*

A manufacturing process generates aqueous waste with a flowrate W_1 and temperature θ_{h1}. Prior to its purification with a view to its recycling within the process, part of the energy contained in this effluent is extracted to heat a flowrate, W_2, of water, available at θ_{c2}, which needs to supply the production line.

For this task, we decide to use a single-pass counter-flow heat exchanger; the hot fluid circulates in the tubes and the water to be heated flows in the shell. The cost of this exchanger is proportional to its number of transfer units (NTU). It has a lifetime of 20 years and its annual maintenance costs are also proportional to its NTU.

Moreover, the amount of energy recovered using this exchanger is related to its efficiency, and therefore to the NTU of the exchanger.

1) Calculate the maximum flux of recoverable energy, ϕ_{Max}.

2) What is the maximum temperature that W_2 could then reach at the outlet?

3) Give, as a function of the NTU and the ratio $= \frac{(wc_p)_{min}}{(wc_p)_{max}}$, the expression of the amount of energy that will be recoverable annually: Q(NTU).

4) Determine the NTU that optimizes the function "Annual energy saving value", V(NTU), and deduce therefrom $V(\text{NTU}_{\text{Max}})$.

5) What, then, is the outlet temperature of the heated water in this case?

6) What is the heat transfer area of the exchanger thus determined?

Given data:

The function "Annual energy saving value" is defined by:

$$V(\text{NTU}) = \pi Q(\text{NTU}) - \frac{P(\text{NTU})}{N} - M(\text{NTU})$$

where π is the energy price in €/kWh, P(NTU) is the cost of the exchanger (in €), N is the exchanger's lifetime in years, M(NTU) is the annual maintenance cost, in €/year.

$W_1 = 1.5$ kg/sec $\quad \theta_{h1} = 95°C \quad Cp_1 = 4.51$ kJ/kg°C $\quad \pi = 0.14$ €/kWh
$W_2 = 2.1$ kg/sec $\quad \theta_{c2} = 18°C \quad Cp_2 = 4.18$ kJ/kg $\quad N = 6$ years
$P(\text{NTU}) = \alpha\,\text{NTU} + \beta \quad \alpha = €308{,}312 \quad \beta = €159{,}800$
$M(\text{NTU}) = \gamma\,\text{NTU} + \delta \quad \gamma = €28{,}500 \quad \delta = €34{,}800$
$U = 25$ W/m^2°C
factory operating time: $\tau = 8{,}500$ hr/year

Solution

1) Calculation of the maximum flux of recoverable energy

$$\phi_{Max} = (WC_p)_{min}(\theta_{h1} - \theta_{c2})$$

NUMERICAL APPLICATION

$W_1 = 1.5$ kg/sec $\quad Cp_1 = 4.51$ kJ/kg°C $\quad \theta_{h1} = 95°C$

$W_2 = 2.1$ kg/sec $\quad Cp_2 = 4.18$ kJ/kg°C $\quad \theta_{c2} = 18°C$

$W_1\ Cp_1 = 1.5 \times 4.51 = 6.77$ kW/°C $\quad W_2\ Cp_2 = 2.1 \times 4.18 = 8.78$ kW/°C

Hence: $(WC_p)_{min} = W_1\ Cp_1 = 6.77$ kW/°C; $\phi_{\text{Max}} = 6.765 \times (95 - 18) = 520.91$ kW.

2) Maximum temperature that W_2 could reach at the outlet

A balance on fluid W_2 gives:

$$\phi_{\text{Max}} = W_2\ Cp_2\ (\theta_{c1\text{Max}} - \theta_{c2})$$

Hence: $\theta_{c1Max} = \theta_{c2} + \frac{\phi_{\text{Max}}}{W_2 Cp_2}$.

NUMERICAL APPLICATION

$W_2\ Cp_2 = 8.778$ kW/°C $\quad \phi_{Max} = 520.905$ kW $\quad \theta_{c2} = 18°C$

$$\theta_{c1Max} = 18 + \frac{520.905}{8.778}$$

I.e.: $\theta_{c1Max} = 77.34°C$.

3) Expression of the amount of energy recoverable annually

The flux effectively exchanged is given by: $\phi = \eta(\text{NTU})\ \phi_{Max}$.

The amount of energy recoverable annually is thus: $Q = \eta(\text{NTU})\ \phi_{Max}\ \tau$.

For a counter-flow exchanger, the efficiency, η, is given by Table 5.3:

$$\eta = \frac{1-e^{-(1-R)\text{NTU}}}{1-Re^{-(1-R)\text{NTU}}}$$

Hence: $Q(\text{NTU}) = \frac{1-e^{-(1-R)\text{NTU}}}{1-Re^{-(1-R)\text{NTU}}}\phi_{\text{Max}}\tau$.

4) Optimization of V(NTU)

$$V(\text{NTU}) = \pi Q(\text{NTU}) - \frac{P(\text{NTU})}{N} - M(\text{NTU})$$

$$V(\mathrm{NTU}) = \pi\frac{1-e^{-(1-R)\mathrm{NTU}}}{1-Re^{-(1-R)\mathrm{NTU}}}\phi_{\mathrm{Max}}\tau - \frac{\alpha\mathrm{NTU}+\beta}{N} - \gamma\mathrm{NTU} - \delta$$

The function V(NTU) has been implemented in a spreadsheet. Figure 5.13 shows the evolution in annual energy savings as a function of the NTU. An optimum clearly appears around 2.

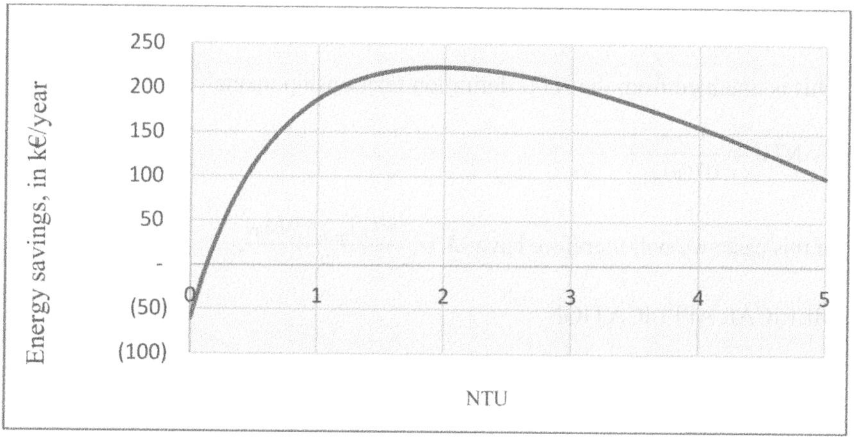

Figure 5.13. *Value of annual energy savings as a function of the NTU*

$\mathrm{NTU_{Max}} = 2$ and $V(\mathrm{NTU_{Max}}) =$ €223,448/year

5) Outlet temperature of the heated water

The water outlet temperature is obtained from the expression of the exchanged flux, namely:

$\phi = W_2\, CP_2\,(\theta_{c1} - \theta_{c2})$ with $\phi = \eta(\mathrm{NTU_{Max}})\,\phi_{\mathrm{Max}}$

Hence: $\eta(\mathrm{NTU_{Max}})\,\phi_{\mathrm{Max}} = W_2\, Cp_2\,(\theta_{c1} - \theta_{c2})$.

Or: $\theta_{c1} = \theta_{c2} + \frac{\phi_{\mathrm{Max}}}{W_2Cp_2}\eta(\mathrm{NTU_{Max}})$

i.e.: $\theta_{c1} = \theta_{c2} + \left(\frac{\phi_{\mathrm{Max}}}{W_2Cp_2}\right)\left(\frac{1-e^{-(1-R)\mathrm{NTU_{Max}}}}{1-Re^{-(1-R)\mathrm{NTU_{Max}}}}\right)$.

NUMERICAL APPLICATION

$W_2\, Cp_2 = 8.778$ kW/°C $\qquad \phi_{\mathrm{Max}} = 520.905$ kW $\qquad \theta_{c2} = 18$°C

$R = 0.771 \qquad \text{NTU}_{\text{Max}} = 2$

$$\theta_{c1} = 18 + \left(\frac{520.905}{8.778}\right)\left(\frac{1-e^{-2(1-0.771)}}{1-0.771e^{-2(1-0.771)}}\right)$$

i.e.: $\theta_{cl} = 60.57°\text{C}$.

6) Transfer area

This is obtained from the NTU definition relationship, namely:

$$\text{NTU} = \frac{U\,A}{(wC_p)_{\text{min}}}$$

In this case, we will therefore have: $A = \dfrac{(wC_p)_{\text{min}}\text{NTU}_{\text{Max}}}{U}$.

NUMERICAL APPLICATION

$(WC_p)_{min} = 6.765$ W/°C $\qquad \text{NTU}_{\text{Max}} = 2 \qquad U = 25$ W/m^2°C

$$A = \frac{6.765\text{x}2}{25}$$

i.e.: $A = 541.2$ m^2.

5.7. Method based on logarithmic mean temperature difference

The LMTD (Logarithmic Mean Temperature Difference) method makes it possible to directly calculate the transfer area required to carry out a given thermal schedule. It is based on *knowledge of the inlet and outlet temperatures of the two fluids* and the *overall heat transfer coefficient, U*. Of course, the LMTD method depends on the chosen flow configuration: one-pass, multipass, cross-flow, etc.

5.7.1. *Case of a one-pass heat exchanger*

For a one-pass heat exchanger, the flux equation established in section 5.5.2 gives:

$$\phi = U\,A\,\Delta\theta_{mL}$$

where:

U is the overall heat transfer coefficient, and

$$\Delta\theta_{mL} = \frac{(\Delta\theta_1 - \Delta\theta_2)}{Ln\frac{\Delta\theta_1}{\Delta\theta_2}}, \text{ with } \Delta\theta_1 = \theta_{h_1} - \theta_{c_1} \text{ and } \Delta\theta_2 = \theta_{h_2} - \theta_{c_2}$$

This equation can be used to calculate the transfer area : $A = \frac{\phi}{U\,\Delta\theta_{mL}}$.

Despite its apparent simplicity, this equation masks several complexities. Indeed, its implementation to calculate the transfer area assumes knowledge of the overall heat transfer coefficient, U. However, as we detailed in section 5.3, this coefficient is related to the technology used (multitube or compact) and to the geometrical choices adopted by the designer (tube arrangement, diameter, etc.). Thus, in the case of multitube heat exchangers, for example, the overall coefficient, and consequently the exchange area, are related to the number of tubes, their diameter, their length, the diameter of the shell and the number of tube passages in the latter, etc.

Similarly, for plate heat exchangers, the transfer area depends on the number of plates, on the gap between the plates and on the unit surface area of a plate. By contrast, for finned compact heat exchangers, the transfer area will depend, in addition to the lateral surface of the tubes, on the shape and the contact area of the fins.

The procedure for calculating the transfer area, based on this equation, is presented in detail in Chapter 6. It is iterative in nature, given that the various calculations require definition of the tube and the shell diameters. The latter, in turn, condition the flow (laminar or turbulent) and, as a consequence, the heat transfer coefficients.

NOTE.– Except in the case of certain condensers and boilers where the temperature of one of the fluids remains constant along the exchanger, *the logarithmic mean temperature of an exchanger operating counter-flow is always higher than that of a co-flow exchanger*, regardless of the temperatures of the fluids, cold and hot.

Thus, for a given thermal load ϕ, the counter-flow will lead to a smaller transfer area.

5.7.2. *Illustration: comparison of flow configurations*

A one-pass multitube exchanger shall be constructed using N tubes of outer diameter d_o, arranged in a 1-inch square array, in a shell of diameter D_{sh}.

This exchanger shall be used to cool a mass flowrate, W_h, of an oil from temperature θ_{h1} to temperature θ_{h2}. The cooling fluid used is sea water, available at $\theta_{c_{in}} = 20°C$.

The latter circulates in the tubes, while the oil circulates in the shell. The exchanger shall be designed to operate counter-flow or co-flow, but in both configurations, the overall heat transfer coefficient based on the external surface of the tubes is the same: *U*.

1) What should the diameter of the shell be?

2) Calculate the heat flux to be exchanged.

3) What should the flowrate of the sea water be so that its outlet temperature does not exceed $\theta_{c_{max}}$?

4) When operating at this flowrate, what would the length of the device be in each of the configurations: co-flow and counter-flow?

Given data:

$N = 61$ $\quad d_o = 0.75$ inches $\quad U = 600$ W/m²°C.

oil: $W_h = 11$ T/hr $\quad C_{ph} = 2{,}900$ J/kg°C $\quad \theta_{h1} = 77°C$ $\quad \theta_{h2} = 35°C$

sea water: $C_{pc} = 3.96$ kJ/kg°C. $\quad \theta_{c_{max}} = 30°C$

Solution

1) *Shell diameter*

We have tubes with an outer diameter of ¾", to be arranged in a 1-inch square array. We then use Table A1.18 (see Appendix 1) to determine the shell diameter which would fit the N tubes.

For one pass of 61 tubes, a shell diameter of $D_{sh} = 11$ inches is recommended.

2) *Calculating the flux*

The heat flux can be obtained from an overall energy balance on the hot fluid:

$$\phi = W_h\, C_{ph}\, (\theta_{h1} - \theta_{h2})$$

NUMERICAL APPLICATION.

$W_h = 3.06$ kg/sec $\theta_{h1} = 77°C$ $\theta_{h2} = 35°C$ $C_{ph} = 2{,}900$ J/kg°C

$\phi = 372$ kW

3) *Minimum seawater flowrate*

This flowrate is obtained for $\theta_{c_{out}} = \theta_{c_{max}}$

In this case, the seawater balance equation is:

$$\phi = W_{c_{min}} C_{pc}\left(\theta_{c_{max}} - \theta_{c_{in}}\right)$$

Hence: $W_{c_{min}} = \frac{\phi}{C_{pc}\left(\theta_{c_{max}} - \theta_{c_{in}}\right)}$

NUMERICAL APPLICATION.

$C_{pc} = 3{,}960$ J/kg°C $\theta_{c_{in}} = 20°C$ $\theta_{c_{max}} = 30°C$ $\phi = 372{,}167$ W

$W_{c_{min}} = \frac{372{,}167}{3{,}960(30-20)}$ $W_{c_{min}} = 9.4$ kg/sec

4) *Length of the device*

When operating at $W_{c_{min}}$, the seawater outlet temperature will be $\theta_{c_{max}}$. With the exchanger comprising N tubes, its length is linked to its exchange area by the relationship: $A = N\pi d_e L$.

Hence: $L = \frac{A}{N\pi d_e}$.

The exchange area is calculated from the equation: $A = \frac{\phi}{U\,\Delta\theta_{mL}}$.

$$\Delta\theta_{mL} = \frac{(\Delta\theta_1 - \Delta\theta_2)}{Ln\frac{\Delta\theta_1}{\Delta\theta_2}}, \text{ with } \Delta\theta_1 = \theta_{h_1} - \theta_{c_1} \text{ and } \Delta\theta_2 = \theta_{h_2} - \theta_{c_2}$$

θ_{c1} and θ_{c2} will depend on the flow configuration (see Figure 5.14).

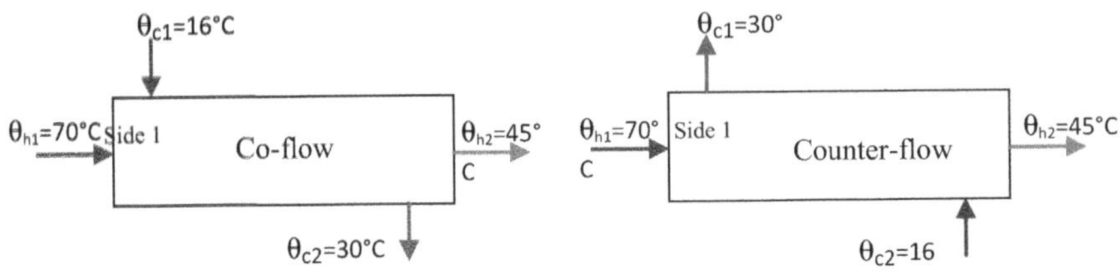

Figure 5.14. *Values of θ_{c1} and θ_{c2} according to the flow configurations*

Therefore, $\Delta\theta_1$ and $\Delta\theta_2$ and $\Delta\theta_{mL}$ depend on the flow configuration. Table 5.4 gives $\Delta\theta_1$ and $\Delta\theta_2$, as well as $\Delta\theta_{mL}$, as applicable.

Flow configuration	Co-flow	Counter-flow
θ_{c1}	20	30
θ_{c2}	30	20
$\Delta\theta_1 = \theta_{h1} - \theta_{c1}$	57	47
$\Delta\theta_2 = \theta_{h2} - \theta_{c2}$	5	15
$\Delta\theta_{mL}$	21	28
A	29	22
L	8	6

Table 5.4. *Expressions of $\Delta\theta mL$ as a function of the flow configuration*

We can thus see that the counter-flow is more efficient, since to transfer the same flux, ϕ, it would take less heat transfer area, and therefore a shorter length than for the co-flow.

The illustration in section 5.7.2 shows that equations giving the transfer area are very simple to use to calculate the exchange area of a device when *the exchanged flux and the inlet and outlet temperatures of the fluids are known.*

5.7.3. *Case of multipass and cross-flow heat exchangers*

In this section, we focus on calculating the transfer area for multipass and cross-flow heat exchangers.

What differentiates these exchangers from one-pass devices is their flow mode. Indeed, in multipass heat exchangers, fluid circulation *is sometimes co-flow, sometimes counter-flow* (see sections 2.5 and 2.6). One would then expect the efficiency of multipass (or cross-flow) heat exchangers to be somewhere between that of the single-pass counter-flow and that of the single-pass co-flow exchangers.

In addition, in multipass heat exchangers, the fluid temperature may not be uniform in a given cross-section area of the exchanger. Similarly, in cross-flow heat exchangers, fluids flow perpendicular to each other, causing temperature variations in all directions.

Thus, the calculation method presented in section 5.7.1 is no longer valid. It needs to be adapted by introducing the effective temperature difference, $\Delta\theta_{eff}$, defined by:

$$\Delta\theta_{eff} = Y\,\Delta\theta_{mL}$$

where:

– Y is called the correction factor (or coefficient) for the multipass (or cross-flow) device compared to the one-pass counter-flow serving as a reference;

– $\Delta\theta_{mL}$ is the logarithmic mean of the temperature differences obtained for the reference counter-flow exchanger, i.e., which has the same inlet and outlet temperatures: $\Delta\theta_{mL} = \frac{(\Delta\theta_1 - \Delta\theta_2)}{Ln\frac{\Delta\theta_1}{\Delta\theta_2}}$.

Under these conditions, the relationship between the transfer area and the exchanged flux becomes: $\phi = U\,A\,Y\,\Delta\theta_m$.

The correction coefficient, Y, *depends on the number of passes*. It is expressed as a function of parameters X and Z, defined as follows in terms of inlet hot and cold temperatures (θ_{hi}, θ_{ci}) and outlet hot and cold temperatures (θ_{ho}, θ_{co}):

$$X = \frac{\theta_{co} - \theta_{ci}}{\theta_{hi} - \theta_{ci}} \text{ and } Z = \frac{\theta_{hi} - \theta_{ho}}{\theta_{co} - \theta_{ci}} = \frac{w_c C_c}{w_h C_h}$$

Parameter X is a measure of the *ratio of heat actually transferred to heat that would be transferred if the outlet temperature of the hot fluid were equal to the inlet temperature of the cold fluid*. Therefore, X is a kind of *thermal efficiency of the heat exchanger*; it measures the difference with respect to a vessel of infinite length, for which the maximum energy exchange between the cold fluid and the hot fluid would be obtained.

NOTE.– The exchange is maximum ($X = 1$) in the case where the outlet temperature of the cold fluid is equal to the inlet temperature of the hot fluid. But this represents an ideal scenario that cannot be achieved in practice.

Parameter Z defines the ratio between the heat capacities, W_cC_c of the cold fluid and W_hC_h of the hot fluid. It is called the *heat capacity rate*.

5.7.4. *Determination of the correction factors*

Correction factors, Y, are often easily determined using charts available in the literature (Nagle 1933; Underwood 1934; Bowman 1936). An example of these charts is given in Figure 5.15 for a 1-2 heat exchanger.

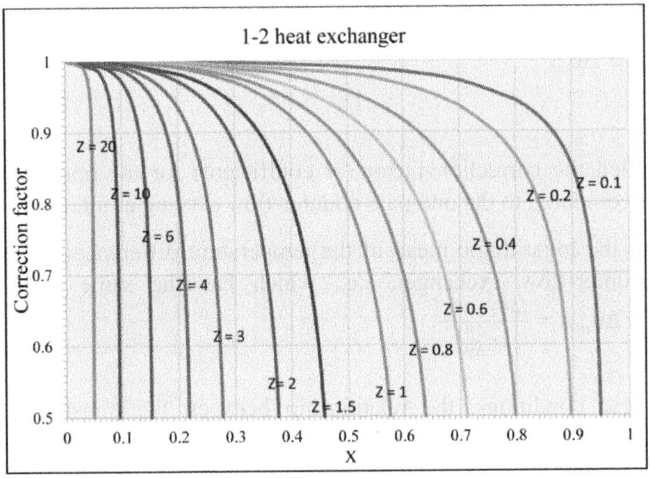

Figure 5.15. *Correction factor, Y, for a 1-2 heat exchanger*

Using charts of course avoids having to perform calculations that are sometimes long and complex. Indeed, this makes it possible to obtain, simply by reading, rapid estimates of the correction coefficients. But, as we will see in Chapter 6, the sizing of exchangers is an *iterative procedure* that will not be satisfied by a single estimate of the correction coefficient, Y.

Indeed, the procedure requires the determination of Y for several configurations. This aspect means that the use of charts limits the sizing procedure, unless the curves are digitized to allow numerical use. But even with chart digitization, repeated calculations can lead to critical deviations due to inaccuracies in reading and extrapolation.

Thus, digital computation techniques should favor the use of analytical expressions of Y (Monteiro-Guimarães et al. 2015), better suited to iterative calculations used in heat exchanger sizing.

5.7.4.1. *Expressions of correction factors for tube-side multipass heat exchangers and shell-side one-pass heat exchangers*

Table 5.5 gives the analytical expressions of the correction factors for shell-side, one-pass heat exchangers.

Type of heat exchanger	Y
1-2 heat exchanger	For $Z \neq 1$: $Y_{1.2} = \dfrac{\frac{\sqrt{1+Z^2}}{(Z-1)}\log_{10}\left(\frac{1-X}{1-XZ}\right)}{\log_{10}\left[\frac{\frac{2}{X}-1-Z+\sqrt{1+Z^2}}{\frac{2}{X}-1-Z-\sqrt{1+Z^2}}\right]}$ For $Z = 1$: $Y_{1.2} = \dfrac{\frac{X\sqrt{1+Z^2}}{2.3\,(1-X)}}{\log_{10}\left[\frac{\frac{2}{X}-1-Z+\sqrt{1+Z^2}}{\frac{2}{X}-1-Z-\sqrt{1+Z^2}}\right]}$
1-4 heat exchanger	Y is calculated as a function of X, Z and the number of transfer units (NTUs) as follows: For $Z \# 1$: $Y = \dfrac{1}{\mathrm{NTU}\,(1-Z)} Ln\left[\frac{1-XZ}{1-X}\right]$ For $Z = 1$: $Y = \dfrac{X}{(1-X)\mathrm{NTU}}$ where NTU is the solution to the equation: $X = \dfrac{4}{2(Z+1)+\sqrt{Z^2+4}\coth\left(\frac{\mathrm{NTU}}{4}\sqrt{Z^2+4}\right)+Z\tanh\left(\frac{\mathrm{NTU}}{4}Z\right)}$
1-6[+] heat exchanger	For $Z \# 1$: $Y_{1.6} = \dfrac{\frac{\sqrt{1+Z^2}}{(Z-1)}\log_{10}\left(\frac{1-X}{1-XZ}\right)}{\log_{10}\left[\frac{\frac{2}{X}-1-Z+\sqrt{1+Z^2}}{\frac{2}{X}-1-Z-\sqrt{1+Z^2}}\right]}$ For $Z = 1$: $Y_{1.6} = \dfrac{\frac{X\sqrt{1+Z^2}}{2.3\,(1-X)}}{\log_{10}\left[\frac{\frac{2}{X}-1-Z+\sqrt{1+Z^2}}{\frac{2}{X}-1-Z-\sqrt{1+Z^2}}\right]}$

Table 5.5. *Shell-side correction coefficient, Y, for one-pass heat exchangers*

Note that the expressions giving the correction factor for 1-4 exchangers are relatively complex and require an iterative procedure to solve the intermediate NTU equation. Nevertheless, the results obtained are so close to those of the 1-2 heat exchanger that the expression relating to the latter can be used as a first approximation of correction factor for 1-4 exchangers (Bowman et al. 1940).

The same remark applies to exchangers with one shell-side pass and more than 4 tube-side passes. It should be pointed out that technical complications encountered in the manufacturing of multiple tube passes are not fairly rewarded in terms of efficiency gain, since the correction factor remains substantially close to (and lower than) that of the 1-2 heat exchanger. Consequently, the rules of thumb hereafter are to be observed during the exchanger calculations.

RULES OF THUMB.–

In practice, the expression (or chart) relating to the 1-2 heat exchanger can also be used for determining the correction factor of 1-4, 1-6 (and more) heat exchangers.

The expression of the correction factor specific to the 1-4 heat exchanger is to be used only if the desired gain in accuracy is justified.

We will always adopt flow configurations that will yield correction factors greater than 0.8.

It should be noted, however, that in many practical applications, one-pass shell-side heat exchangers do not always allow correction factors greater than 0.8, thus leading to low efficiencies. Configurations giving higher values of Y are then sought by using two exchangers in series or a single exchanger with multiple shell-side passes; the 2-4 heat exchanger is often preferred in such situations.

RULES OF THUMB.–

In practice, for a single-tube exchanger operating counter-flow, $Y = 1$.

In the case of a multitube with a single shell and an even number of tube-side passes (1-2k heat exchanger):

– Y is between 0.8 and 1.0.

– Y will be as close to 1.0 as the temperature of one of the fluxes remains constant (condensation and boiling).

– Y will be close to 0.8 when the outlet temperatures of the two fluids are equal.

5.7.4.2. *Correction factors for multiple-pass shell-side heat exchangers*

Table 5.6 presents expressions for the correction factors, Y, for multiple-pass shell-side heat exchangers. In general, an exchanger with N passes on the shell side will have a number of tube-side passes equal to $2N$.

Type of heat exchanger	Y
2-4 heat exchanger	For $Z \# 1$: $Y_{2.4} = \dfrac{\frac{\sqrt{1+Z^2}}{2(Z-1)}\log_{10}\left(\frac{1-X}{1-XZ}\right)}{\log_{10}\left[\frac{\frac{2}{X}-1-Z+\frac{2}{X}\sqrt{(1-X)(1-XZ)}+\sqrt{1+Z^2}}{\frac{2}{X}-1-Z+\frac{2}{X}\sqrt{(1-X)(1-XZ)}-\sqrt{1+Z^2}}\right]}$ For $Z = 1$: $Y_{2.4} = \dfrac{\frac{X\sqrt{1+Z^2}}{4.6\,(1-X)}}{\log_{10}\left[\frac{\frac{2}{X}-1-Z+\frac{2}{X}\sqrt{(1-X)(1-XZ)}+\sqrt{1+Z^2}}{\frac{2}{X}-1-Z+\frac{2}{X}\sqrt{(1-X)(1-XZ)}-\sqrt{1+Z^2}}\right]}$
N-2N exchanger	For $Z \# 1$: $Y_{N,2N} = \dfrac{\frac{\sqrt{1+Z^2}}{(Z-1)}\log_{10}\left(\frac{1-X_{N,2N}}{1-X_{1.2}Z}\right)}{\log_{10}\left[\frac{\frac{2}{X_{N,2N}}-1-Z+\sqrt{1+Z^2}}{\frac{2}{X_{N,2N}}-1-Z-\sqrt{1+Z^2}}\right]}$ with: $X_{N,2N} = \dfrac{1-\left(\frac{1-ZX_{1.2}}{1-X_{1.2}}\right)^N}{Z-\left(\frac{1-ZX_{1.2}}{1-X_{1.2}}\right)^N}$ For $Z = 1$: $Y_{N,2N} = \dfrac{\frac{X_{N,2N}\sqrt{1+Z^2}}{2.3\,(1-X_{N,2N})}}{\log_{10}\left[\frac{\frac{2}{X_{N,2N}}-1-Z+\sqrt{1+Z^2}}{\frac{2}{X_{N,2N}}-1-Z-\sqrt{1+Z^2}}\right]}$ with: $X_{N,2N} = \dfrac{NX_{1.2}}{NX_{1.2}-X_{1.2}+1}$

Table 5.6. *Correction coefficient, Y, for multiple-pass shell-side heat exchangers*

Figure 5.16 shows the correction factors for a 2-4 heat exchanger for several values of parameter Z.

NOTE.–

If the temperature change of one of the fluids is negligible ($\theta_{c1} - \theta_{c2} \approx 0$ or $\theta_{h1} - \theta_{h2} \approx 0$), then $X = 0$ or $Z = 0$.

From the shapes of the charts representing the correction coefficient, we can deduce that for such cases $Y = 1$. This means that the behavior of the exchanger considered is similar to that of the counter-flow heat exchanger. This case (constant temperature) is encountered when one of the fluids undergoes a phase change. This is the case for condensers and boilers.

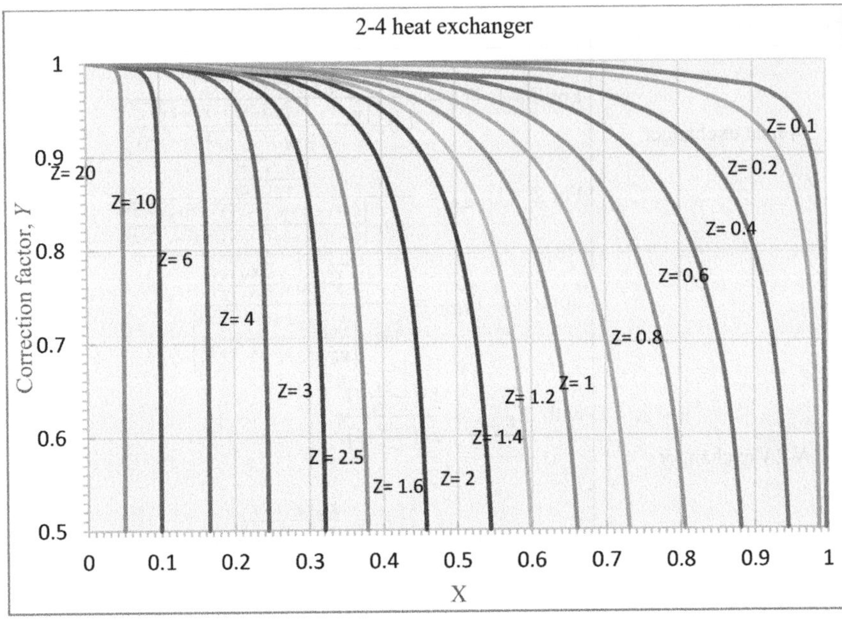

Figure 5.16. *Correction factor, Y, for a 2-4 heat exchanger*

5.7.4.3. *Correction factors for cross-flow heat exchangers*

The correction factors, *Y*, for cross-flow heat exchangers are calculated as a function of parameters *X* and *Z* defined in section 5.7.3, namely:

$$X = \frac{\theta_{co} - \theta_{ci}}{\theta_{hi} - \theta_{ci}} \text{ and } Z = \frac{\theta_{hi} - \theta_{ho}}{\theta_{co} - \theta_{ci}} = \frac{W_c C_c}{W_h C_h}$$

Y is calculated as a function of *X*, *Z* and the number of transfer units (NTU) as follows:

– for $Z \# 1$: $Y = \frac{1}{\text{NTU}\,(1-Z)} Ln\left[\frac{1-XZ}{1-X}\right]$;

– for $Z = 1$: $Y = \frac{X}{(1-X)\text{NTU}}$.

NTU is then calculated, either directly from the relationship NTU = *f*(*X*) given in Table 5.7, or, where appropriate, by solving the equation *X* = *f*(NTU), which appears in the same table.

Type of heat exchanger	Equation giving NTU as a function of X
Cross-flows 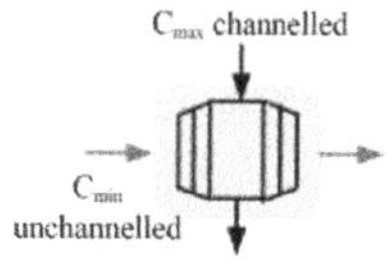	$\text{NTU} = -\frac{1}{Z}\{Ln[1 + ZLn(1 - X)]\}$
Cross-flows 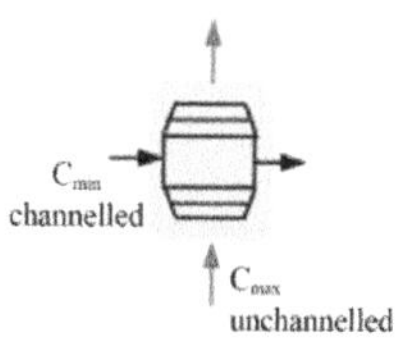	$\text{NTU} = Ln\left[\dfrac{1}{1 + \frac{1}{Z}Ln(1 - ZX)}\right]$
Cross-flows With both sides channeled 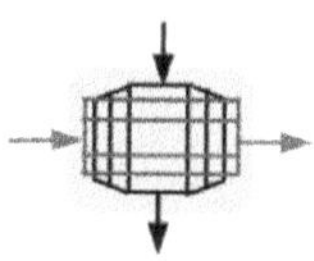	– For $Z < 1$: $X = 1 - \exp\left[\frac{\exp(-Z\ \text{NTU}^{0.78}) - 1}{Z\ \text{NTU}^{-0.22}}\right]$ – For $Z > 1$: Reverse the hot and cold sides or replace X with $X' = XZ$ and Z with $Z' = 1/Z$ in the formulas

Table 5.7. *Calculation of the NTU for cross-flow heat exchangers*

This type of calculation makes it possible to generate correction factor charts. An example of such charts is shown in Figure 5.17 for a cross-flow heat exchanger where fluid C_{min} is channeled, while fluid C_{max} is not channeled.

NOTE.–

Examination of the shapes of the various curves constituting correction factor charts shows that, whatever the configuration considered (parallel-flow or cross-flow) and whatever the number of passes considered, we have:

$$\lim_{X\to 0} Y = 1 \text{ and } \lim_{Z\to 0} Y = 1$$

This means that the behavior of any exchanger configuration tends toward that of a counter-flow condenser or boiler.

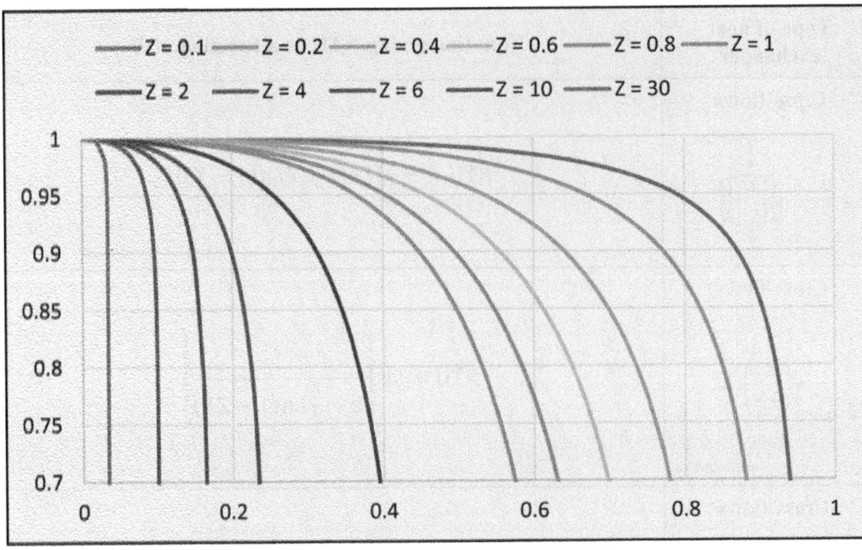

Figure 5.17. *Correction factor, Y, for a cross-flow heat exchanger with C_{min} channeled and C_{max} unchanneled*

5.7.4.4. *Illustration: generation of charts from the expressions of the correction factor and the NTU*

From the expressions of the NTU given in Table 5.6, you are asked to generate a chart giving the correction factor for a cross-flow heat exchanger where fluid C_{min} is not channeled, while fluid C_{max} is channeled.

Solution

For this type of exchanger, the NTU is given by (see Table 5.6):

$$\text{NTU} = -\frac{1}{Z}\{Ln[1 + ZLn(1 - X)]\}$$

The correction factor can then be calculated from the expressions given in section 2.7.4.3, namely:

– for $Z \# 1$: $Y = \frac{1}{\text{NTU}\,(1-Z)} Ln\left[\frac{1-XZ}{1-X}\right]$;

– for $Z = 1$: $Y = \frac{X}{(1-X)\text{NTU}}$.

Spreadsheet implementation of these expressions, for values of Z ranging from 0.1 to 30, has enabled the development of the chart presented in Figure 5.18.

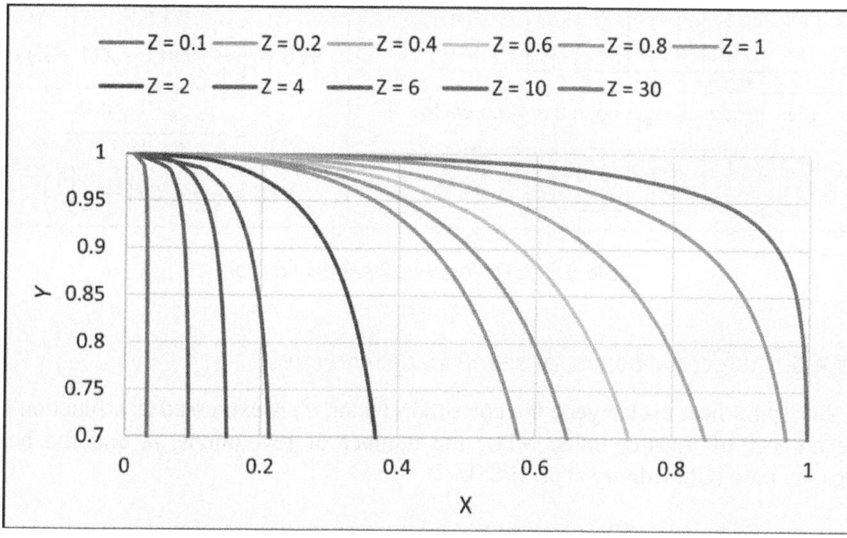

Figure 5.18. *Correction factor, Y, for a cross-flow heat exchanger with C_{min} unchanneled and C_{max} channeled*

5.7.4.5. *Correction factors for plate heat exchangers*

As with cross-flow heat exchangers, correction factors, Y, for plate heat exchangers are also calculated as a function of parameters X and Z defined in section 5.7.3, namely:

$$X = \frac{\theta_{co} - \theta_{ci}}{\theta_{hi} - \theta_{ci}} \text{ and } Z = \frac{\theta_{hi} - \theta_{ho}}{\theta_{co} - \theta_{ci}} = \frac{W_c C_c}{W_h C_h}$$

– for $Z \# 1$: $Y = \frac{1}{\text{NTU}\,(1-Z)} Ln\left[\frac{1-XZ}{1-X}\right]$;

– for $Z = 1$: $Y = \frac{X}{(1-X)\text{NTU}}$.

where the NTU is given in Table 5.8 for both flow configurations.

Type of plate heat exchanger	Equation linking NTU and X
Plate heat exchanger operating co-flow 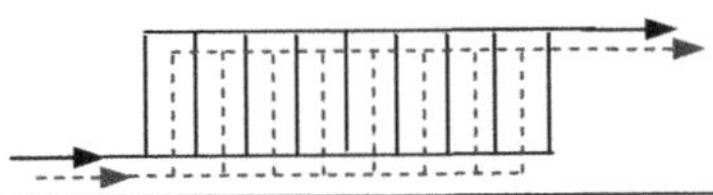	$X = \frac{1 - e^{-(1+Z)\text{NTU}}}{1+Z}$ $\text{NTU} = \frac{-1}{1+Z} Ln\big(1 - X(1+Z)\big)$
Plate heat exchanger operating counter-flow 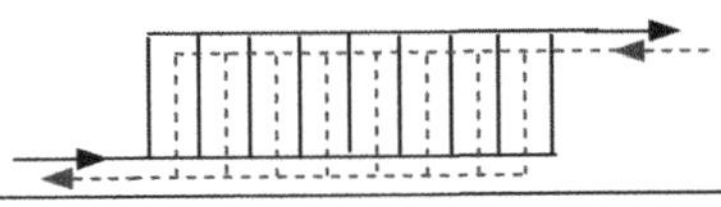	$X = \frac{1 - e^{-(1-Z)\text{NTU}}}{1 - Ze^{-(1-Z)\text{NTU}}}$ $\text{NTU} = \frac{1}{1-Z} Ln\left(\frac{1-ZX}{1-X}\right)$

Table 5.8. *NTU for plate heat exchangers*

5.7.4.6. *Correction factors for spiral heat exchangers*

For spiral heat exchangers, the correction factor, Y, is expressed as a function of the number of transfer units, NTU, the number of revolutions, n, and the heat capacity ratio (Chowdhury et al. 1985):

$$Y = \frac{n}{\text{NTU}} \tanh \frac{\text{NTU}}{n}$$

6

Sizing Algorithms for Heat Exchangers

6.1. Introduction

As we have seen, the multiplicity of parameters to be defined during heat-exchanger design makes their sizing a complex operation. The sizing should in fact lead to the determination of all the parameters of the device needed to carry out a given thermal schedule (thermal sizing), without generating excessive pressure drops (hydrodynamic sizing). In addition, designers usually find themselves in a situation where thermal and hydrodynamic calculations need to be conducted with a large degree of freedom: the geometric parameters are generally not set initially. As an example, for a tubular heat exchanger, the designer needs to choose the diameter of the tubes, shell diameter, arrangement of the tubes in the shell, etc. Similarly, in the case of a plate heat exchanger, it is necessary to set the plate type, and therefore the plates' width, their length and the gap between them.

In reality, these design parameter choices need to be reasoned and coordinated so as to lead to an optimal exchanger configuration (Ravelo-Mendivelso et al. 2023). Indeed, we know that the geometry determines the nature of the flow and that the latter conditions the heat transfer coefficients, which in turn define the overall heat transfer coefficient, U. We also know (see Chapter 5) that in order to carry out a given thermal schedule (Q) in a heat exchanger, the latter needs to have a well-determined transfer area, A[1].

For a color version of all figures in this chapter, see www.iste.co.uk/benallou/energy6.zip

1 The heat transfer area, A, shall be such that $Q = UAY\Delta\theta_{ML}$.

It should also be emphasized that the chosen tube diameters (or gaps between plate sheets), as well as the position of any baffles, need to allow a fluid flow that ensures pressure drops below the allowable limits.

Threrefore, the sizing of an exchanger is often a complicated multiple-criteria decision process, including *a thermal calculation* and *a hydrodynamic calculation*, which are *interdependent* (Ravelo-Mendivelso et al. 2023). The optimal configuration to be retained will be that which maximizes heat transfer, while minimizing pressure drops (El Gharbi et al. 2015).

6.2. Principle of heat exchanger sizing

Figure 6.1 sets out, in general, the tasks necessary for sizing heat exchangers. The first part of this figure shows that when the physical data and thermal schedule are defined, thermal calculations are performed for a given set of geometric parameters (diameters, number of tubes, length, etc.). Indeed, when the geometry is set, it is possible to compute Reynolds numbers, heat transfer coefficients and heat transfer areas, which all satisfy the thermal schedule desired.

The second part of Figure 6.1 shows that, once the thermal schedule is satisfied, the designer should verify if pressure drop conditions are respected; i.e., if the hydrodynamics of the vessel are acceptable. Indeed, at this stage, all the geometric and flow parameters are known. It is therefore possible to compute pressure drops for each of the flowrates (hot and cold).

If these pressure drops are below permitted limits, then the vessel is fit and iterations are stopped; the sizing procedure is complete, retaining the last length, diameters, number of tubes, flowrates, heat transfer area, etc. If pressure drops are not both below permitted limits, then the geometric parameters are changed with regards to lowering pressure drops, and iterations are restarted at the level of thermal calculations.

Therefore, heat exchanger sizing is an iterative procedure that, through successive tests, leads to the transfer area; which enables both the thermal schedule to be conducted and the pressure drop constraints to be satisfied.

Sections 6.3 to 6.12 show how the general procedure described above could be broken down into heat exchanger calculation algorithms, which are implemented through illustrations dealing with practical problems.

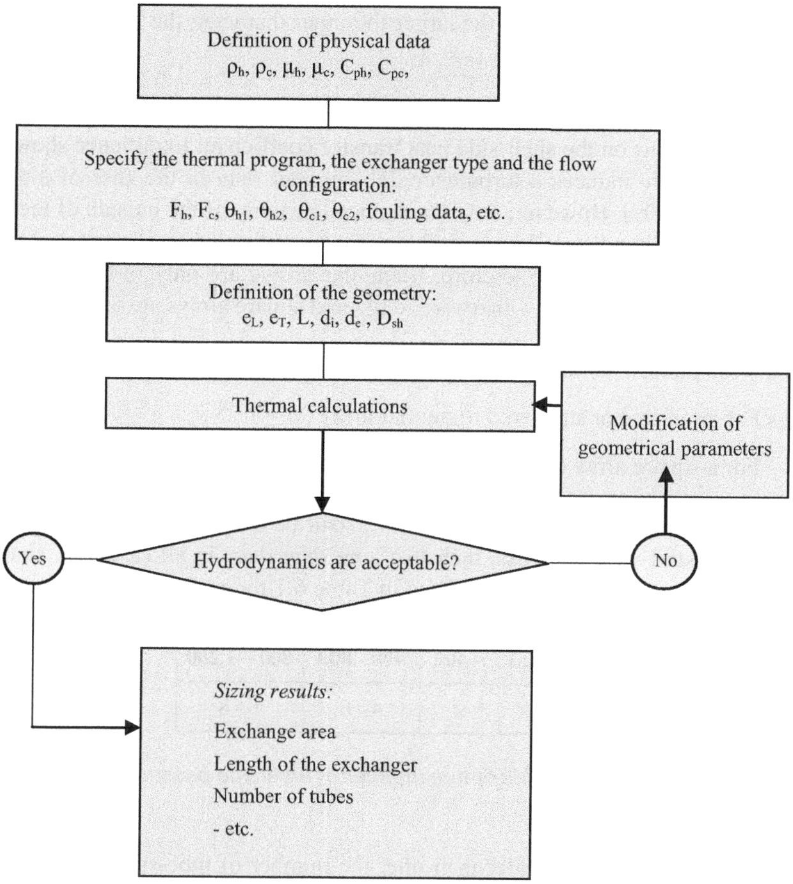

Figure 6.1. *General logic of heat exchanger sizing*

6.3. Calculation algorithm for multitube heat exchangers

The present algorithm breaks down the general procedure presented in Figure 6.1 into 9 elementary steps.

The different choices to be made by the designer at every step are generally guided by the following rules of the art.

RULES OF THUMB.–

R_1.– *The diameter of the tubes* has a direct influence on convection heat transfer coefficients, both inside the tubes, *via the inner and outer tube diameters, d_i* and d_o.

For a given fluid mass flowrate, the larger the inner diameter, the lower the internal turbulence.

R_2.– *The arrangement of the tubes in the shell* has a direct effect on external turbulence and thus on the shell-side heat transfer coefficient. Experience shows that a *triangular array* induces a turbulence 25% greater than in the case of a *square array* (TEMA 1999). However, easy mechanical cleaning of the outside of the tubes requires the tube bundles to be accessible. This is performed with staggered square arrays (see section A1.13). Therefore, triangular arrays are only to be retained for clean fluids on the shell side. Otherwise, staggered square arrays are to be favored.

R_3.– The pitch, a, of the array is chosen so as to have:

- For an in-line or staggered triangular array , $a \geq 1.25\ d_o$;
- For a square array (in-line or staggered): $a \geq \text{Sup}\ \{1.25\ d_o;\ d_o + 6\ \text{mm}\}$.

R_4.– The number of tube-side passes, N_p, shall be less than eight. During design calculations, start with one pass, then *two, and then four*. In all cases, N_p shall not exceed a maximum value, $N_{p\ Max}$, defined in Table 6.1 for each shell diameter.

D_{sh} **(mm)**	**< 400**	**400 - 800**	**800 – 1,200**
N_p *Max*	2	4 - 6	6 - 8

Table 6.1. *Maximum number of tube-side passes*

R_5.– If it needs to be greater than one, the number of tube-side passes shall be even.

R_6.– *Fixed-head* heat exchangers are generally used when the shell-side fluids are clean. In contrast, *floating-head* heat exchangers are used for dirty fluids on both the shell and tube sides.

R_7.– U-shaped heat exchangers are to be retained when the tube-side fluids are clean.

R_8.– *The spacing of the transverse baffles, e_c,* also affects turbulence on the shell side. The smaller the value of e_c, the more baffles there are and the more the flow on the shell side changes directions. This results in greater turbulence, and therefore in better heat transfer between the tubes and the outer fluid (Brownell et al. 1959).

However, smaller values of e_c lead to larger pressure drops. The TEMA recommendation is for e_c to be *greater than 1/5th of the shell diameter and never less than 2 inches*. The calculations will always start with $e_c \cong D_{sh}$, then if more turbulence is sought on the shell side, the value of e_c is decreased, while ensuring that:

$$\frac{D_{sh}}{5} \leq e_c \leq D_{sh}$$

R_9.– If the fluid circulating in the tubes needs to enter and exit the same side of the exchanger, the number of tube-side passes shall be even.

Keeping in mind these rules of thumb, the 9 steps are deployed as follows:

Step 1. Using the problem data and physical properties

1.1- Define the circulation mode: parallel flow, counter-flow or cross-flow.

1.2- Calculate the flux, ϕ, to be exchanged.

Step 2. Knowing ϕ

2.1- Choose a value of the "dirty" (fouled exchanger) heat transfer coefficient (U_F) according to the experimental orders of magnitude presented in Tables A1.21 to A1.23.

2.2- Calculate the approximate heat transfer area, A_0: $A_0 = \frac{\phi}{U_F \, \Delta\theta_{ML}}$.

Step 3. Knowing A_0

3.1- Choose the outer diameter of the tubes, d_o, from the standards (Table A1.10).

3.2- Choose the SCH of the tube[2], hence the tube thickness and its inner diameter, d_i.

3.3- Calculate the number of tubes per pass, n, to ensure a suitable velocity of the fluid in the tubes: 0.2 to 0.5 m/s for a viscous fluid; 1 to 2 m/s for a low-viscosity liquid; 10 to 20 m/s for a gas. If F_t is the mass flowrate of the fluid circulating in

2 SCH5 to SCH10 for ordinary pressures, and up to SCH160 for very high pressures.

tubes of inner diameter d_i, the flow velocity, v, is given by[3] $v = \frac{4F_t}{\rho_c n \pi d_i^2}$ where ρ_t is the fluid density. The number of tubes per pass is then given by $n = \frac{4F_t}{\rho_t v \pi d_i^2}$.

Take the integer, n_t, immediately greater than or equal to the result, n, found.

Step 4. Determine number of passes (tube and shell sides) to yield acceptable Y

This calculation is performed in successive iterations in order to yield an acceptable correction factor[4], Y, that is to say, greater than 0.8.

4.1- Calculate X and Z: $X = \frac{\theta_{c1} - \theta_{c2}}{\theta_{h1} - \theta_{c2}}$ and $Z = \frac{\theta_{h1} - \theta_{h2}}{\theta_{c1} - \theta_{c2}}$.

4.2- We initially set the number of shell-side passes to 1: $N_{sh} = 1$.

4.3- The number of tube-side passes[5] is then $N_p = 2 * N_{sh}$.

4.4- Calculate Y using the expressions given in section 5.7.4.

4.5- If $Y < 0.8$, repeat the iteration in step 4.3 after increasing the value of N_{sh}.

Step 5. Hydraulic calculations on the tube side

5.1- Calculate the exact fluid velocity: $v = \frac{4F_t}{\rho_t n_t \pi d_i^2}$.

5.2- Calculate the length of the device[6] that ensures an exchange area close to A_0.

5.3- Calculate *the pressure drop* (ΔP_t) *in the tubes*[7] (using the exact flow velocity) and compare this value with *the allowable; (i.e. admissable) pressure drop* (ΔP_t^{Adm}).

3 For n tubes per pass, the flowrate through each tube is $f_t = \frac{F_t}{n}$. The flow velocity in each tube is then $v = \frac{f_t}{\rho_t \pi \frac{d_i^2}{4}} = \frac{4F_t}{\rho_t n \pi d_i^2}$.

4 See section 5.7.3.

5 If the heat exchanger needs to have more than one pass, we will ensure that N_p is an even number less than or equal to 8.

6 $L = \frac{A_0}{N_p n_t \pi d_o}$. Since tubes are generally commercially available in lengths of 12 m, we will round the result to one of the following lengths: 6 m, 4 m, 3 m, 2 m, 1.5 m or 1 m. Note that it will also be necessary to take into consideration possible constraints on the space available to install the heat exchanger, or if applicable, transportation constraints.

5.4- If $\Delta P_t \leq \Delta P_t^{Adm}$, proceed to step 6.

5.5- If $\Delta P_t > \Delta P_t^{Adm}$, repeat calculations starting at the beginning of step 3, modifying either the diameter of the tubes[8], or the fluid velocity.

Step 6. Designing the shell

At this step, the length (L) of the device, the diameters of the tubes (d_i and d_o), the number of tubes per pass (n_t) and the number of tube-side passes (N_p) are known.

6.1- Calculate the total number of tubes to be inserted into a cross-section area of the shell: $N_t = N_p\, n_t$.

6.2- For the arrangement of the tubes in the shell, choose the type of array[9]: square or triangular.

6.3- Choose the pitch:

- For a square array , take $a = \text{Sup}\ \{1.25\ d_e, d_o + 6\ \text{mm}\}$;
- For a triangular array, take $a = 1.25\ d_e$.

6.4- Calculate the transverse and longitudinal spacings of the tubes, e_T and e_L (see section 2.4.6.4):

- For a staggered square array of pitch a: $e_T = a\sqrt{2}$ and $e_L = a\,\frac{\sqrt{2}}{2}$;
- For a staggered triangular array of pitch a: $e_T = a\sqrt{3}$ and $e_L = \frac{a}{2}$.

6.5- Determine the shell diameter, D_{sh}, from the standards given in Tables A1.11 to A1.19 or by calculation[10].

---

7 Tube-side pressure drop is given by (see Appendix 3): $\Delta P_t = \frac{\rho_t v^2}{2}\left(N_p \frac{L}{d_i} f_M + 4\,N_p + 1.5\right)$ where f_M is the Moody friction factor (see Volume 3, section 2.4.4.2.1), calculated as follows: for smooth tubes (Moody 1944): $f_M = \frac{1}{[1.82\log_{10}(Re) - 1.64]^2}$; $Re = \frac{d_i v \rho_t}{\mu_t}$. And for tubes presenting roughness ε (Knudsen and Katz 1958): $f_M = \frac{1}{[2.28 - 1.74\,Ln(\varepsilon)]^2}$.

8 Taking a larger diameter.

9 If we are not sure that the fluid circulating in the shell is clean (which is usually the case), take a staggered square array. Where we are sure the shell-side fluid is clean, we can take a staggered triangular array, which will give greater compactness.

10 The diameter of the shell, D_{sh}, can be calculated (see section 4.6.1.8) as a function of the total number of tubes to be placed in the shell, N_t, and the pitch, a: $D_{sh} = 2a\sqrt{\frac{N_t}{\pi}\frac{K_M}{K_P}}$; K_M and K_P being the parameters given by Tables 4.5 and 4.6, respectively.

6.6- Calculate the number of longitudinal baffles: $N_{LB} = N_{sh}- 1$. N_{sh} being the number of shell-side passes.

6.7- Define the spacing, e_c, of the transverse baffles and the ratio, ε, of the free area (see sections 2.6.1 and 2.6.2).

6.8- Calculate the number of transverse baffles[11]: $N_{TB} = Ent\left(\frac{L}{e_c}\right)$.

Step 7. Hydraulic calculations of the shell

7.1- Calculate the transverse (a_T) and longitudinal (a_L) free areas (see sections 2.6.5 and 2.6.6): $a_T = \frac{e_c D_{sh}\left(\sqrt{e_T e_L}-d_o\right)}{\sqrt{e_T e_L}}$ and $a_L = \frac{\varepsilon\pi}{4}(D_{sh}^2 - n_t d_o^2)$.

7.2- F_{sh} being the mass flowrate of the fluid circulating in the shell, calculate the longitudinal (G_L), transverse (G_T), average (G_M), and maximum (G_{Max}) unit mass flowrates (see sections 2.6.7.1 to 2.6.7.4): $G_L = \frac{F_{sh}}{a_L}$; $G_T = \frac{F_{sh}}{a_T}$; $G_M = \sqrt{G_L G_T}$; $G_{Max} = \frac{v_{sh}\rho_{sh}e_T}{e_T-d_o}$[12].

7.3- We calculate the pressure drop in the shell (ΔP_{sh})[13].

7.3.1- If $\Delta P_{sh} \leq \Delta P_{sh}^{Adm}$, we proceed to step 8.

7.3.2- If $\Delta P_{sh} > \Delta P_{sh}^{Adm}$, we return to step 6.5. We take a larger diameter for the shell and reset the gaps, e_T and e_L, between the tubes accordingly.

---

11 e_c shall be close to the shell diameter: $e_c \cong D_{sh}$. ε shall be from 25% to 30%.

12 ρ_{sh} is the density of the fluid circulating in the shell and v_{sh} is the flow velocity in the shell, far from the baffles. It is given by: $v_{sh} = \frac{4F_{sh}}{\rho_{sh}\pi D_{sh}^2}$.

13 For a shell without phase change, provided with N_{TB} transverse baffles, ΔP_{sh} is given by (see section A3.2.2): $\Delta P_{sh} = (N_{TB} - 1)\Delta P_T + N_{TB}\Delta P_L$; $\Delta P_T = 1.5\frac{D_{sh}-\delta}{\sqrt{e_T e_L}}\left[\frac{\mu_f}{G_T\sqrt{e_T e_L} - d_e}\right]^{0.2}\frac{G_T^2}{\rho_{sh}}$; $\Delta P_L = \frac{G_L^2}{\rho_{sh}}$. δ being the height above the baffle: $\delta = R_c\left[1 - \cos\left(\frac{\theta}{2}\right)\right]$; θ is the solution of $\theta - \sin\theta = \frac{8a}{D_{sh}^2}$ (see illustration in section 2.6.3) and $a = \frac{\varepsilon\pi D_{sh}^2}{4}$. For a shell with phase change, with N_{TB} baffles (section A3.2.3.1): $\Delta P_{sh} = \frac{1}{2}\frac{G_{mv}^2 D_{sh}(N_{TB}+1)}{D_{eq}\ \rho_v} f$ where D_{eq} is the equivalent diameter of the shell, G_{mv} is the average unit flow rate of vapor, f is the friction factor: $f = \frac{0.36}{Re^{0.925}}$ for $10 < Re = \frac{D_{eq}\ G_{mv}}{\mu_v} < 250$ and $f = \frac{1.19\,10^{-2}}{Re^{0.187}}$ for $250 < Re < 10^6$.

Step 8. Thermal check

At this level, we have an exchanger of surface A^* (which may well be different from A_0). We now need to check that this exchanger will ensure the transfer of the flux, ϕ.

8.1- We calculate the exchange areas: $A_e^* = n_t \pi d_e L$; $A_i^* = n_t \pi d_i L$ and $A_m^* = \frac{A_i^* + A_e^*}{2}$.

8.2- We calculate h_i: see Appendix 2, section A2.2.1.

8.3- We calculate h_e: see Appendix 2, section A2.2.2.

8.4- We calculate the heat transfer coefficient for the clean heat exchanger:

$$\frac{1}{U_e^* A_e^*} = \frac{1}{h_i A_i^*} + \frac{d_o - d_i}{\lambda_t A_m^*} + \frac{1}{h_e A_e^*}$$

λ_t being the heat conductivity of the tubes.

8.5- We re-evaluate the fouled heat transfer coefficient, U_F^* :

$$\phi = U_F A_0 \, \Delta\theta_{ML} = Y \, U_F^* \, A_e^* \, \Delta\theta_{ML} \Rightarrow \frac{1}{U_F^*} = \frac{Y \, A_e^*}{U_F \, A_0}$$

Step 9. Checking the margin allowed for fouling

9.1- From given data or from estimations given in Appendix 4, calculate $R_{fi} + R_{fe}$.

9.2- Operate the following tests on $\frac{1}{U_F^*} - \frac{1}{U_e^*}$:

- If $\frac{1}{U_F^*} - \frac{1}{U_e^*} \cong R_{fi} + R_{fe}$, then $\frac{1}{U_F^*} - \frac{1}{U_e^*}$ is of the same order of magnitude as the sum of the fouling resistances. This means that the margin allowed by the overall coefficient of the clean heat exchanger will enable it to assure the thermal schedule, even in a fouled state. Therefore, *the device is correctly sized.*

- If $\frac{1}{U_F^*} - \frac{1}{U_e^*} < 0$; this means that, to realize the desired thermal program, the heat transfer resistance under fouling conditions would need to be less than the one for the clean exchanger. This is not possible. *The device is thus undersized.* We will then choose a more suitable value for U_F and repeat the calculations, starting from step 2.

- If $\frac{1}{U_F^*} - \frac{1}{U_e^*} \ll R_{fi} + R_{fe}$, then $\frac{1}{U_F^*} - \frac{1}{U_e^*}$ is too small compared to $R_{fi} + R_{fe}$. The exchanger will therefore not be able to assure the thermal schedule under the expected fouling conditions. Therefore, *the device is undersized.* We will then choose a more suitable value for U_F and repeat the calculations, starting from step 2.

- If $\frac{1}{U_F^*} - \frac{1}{U_e^*} \gg R_{fi} + R_{fe}$, then $\frac{1}{U_F^*} - \frac{1}{U_e^*}$ is too large compared to $R_{fi} + R_{fe}$. This means that the device is oversized. We will, here again, repeat the calculation at step 2, using a more suitable value for U_F.

6.4. Illustration: implementation of the shell-and-tube algorithm

At the end of a refining operation, we seek to heat a flow rate, D_1, of kerosene from temperature t_2 to temperature t_1, by conducting heat recovery from a flowrate of diesel prior to its storage. For this, we decide to use a counter-flow multitube heat exchanger. The diesel is available at T_1.

Implement the different steps of the algorithm to size the heat exchanger, which will make it possible to conduct the desired thermal schedule. If it is necessary to use transverse baffles, those that are available are segmental and have a 30% cutting ratio ε.

Given data:

Kerosene:

- F_k = 70 tonnes/hr t_1 = 50°C t_2 = 23°C λ_k = 0.1437 W/m°C
- C_{pk} = 0.48 kcal/kg°C ρ_k = 800 kg.m^{-3} μ_k =1.6 cp
- Allowable pressure drop ΔP_k^{Adm} = 10 psi
- Fouling resistance R_k = 0.001 m^2°C/W

Diesel:

- F_d = 33 tonnes/hr T_1 = 93°C ρ_d = 685 kg.m^{-3} λ_d = 0.1298W/m°C
- μ_d= 0.2 cp C_{pd}= 0.57 kcal/kg C.
- Allowable pressure drop ΔP_d^{Adm} = 0.5 Atm
- Fouling resistance R_d = 0.0005 m^2°C/W

Construction material of the exchanger tubes: λ_t = 109 W/m°C

Solution

Step 1. Based on the problem data and the physical properties define the circulation mode, then calculate the flux, ϕ, to be exchanged

– *Circulation mode: counter-flow*:

- The index '1' is reserved for the hot fluid inlet;

- The hot fluid (diesel) will circulate inside the tubes.

– *Calculation of the flux, ϕ, to be exchanged*: the thermal load is obtained by a heat balance on kerosene:

$$\phi = F_k C_{pk}(t_1 - t_2)$$

NUMERICAL APPLICATION.–

$F_k = 19.44$ kg/sec $\quad t_1 = 50°C \quad t_2 = 23°C \quad C_{pk} = 2{,}006.4$ J/kg°C

$$\phi = 19.44 \times 2{,}006.4(50 - 23) \qquad \phi = 1.05 \text{ MW}$$

– *Calculating the diesel outlet temperature, T_2*: T_2 is obtained from the following expression of the heat balance:

$$F_k C_{pk}(t_1 - t_2) = F_d C_{pd}(T_1 - T_2)$$

Hence: $T_2 = T_1 - \frac{F_k C_{pk}}{F_d C_{pd}}(t_1 - t_2)$.

NUMERICAL APPLICATION.–

$F_k = 19.44$ kg/sec $\quad t_1 = 50°C \quad t_2 = 23°C \quad C_{pk} = 2{,}006.4$ J/kg°C

$F_d = 9.17$ kg/sec $\quad t_1 = 93°C \quad \phi = 1.05$ MW $\quad C_{pd} = 2{,}382.6$ J/kg°C

$$T_2 = 93 - \frac{19.44 \text{ x } 2{,}006.4}{9.17 \text{ x } 2{,}382.6}(50 - 23) \qquad T_2 = 44.77°C$$

Step 2. Knowing ϕ, we perform a first iteration on U_F

– Choosing U_F: from Table A1.21 (two low-viscosity liquids): U_F = 700 to 1,800 kcal/hr m^2°C.

We thus take: U_F = 1,250 kcal/hr m^2°C; i.e.: U_F = 1,451.4 W/m^2°C.

– Calculating the approximate exchange area, A_0:

$$A_0 = \frac{\phi}{U_F\,\Delta\theta_{ML}}, \text{ with } \Delta\theta_{mL} = \frac{(\Delta\theta_1 - \Delta\theta_2)}{Ln\frac{\Delta\theta_1}{\Delta\theta_2}}$$

$$\Delta\theta_1 = T_1 - t_1 \text{ and } \Delta\theta_2 = T_2 - t_2$$

NUMERICAL APPLICATION.–

ϕ = 1.05 MW U_F = 1,451.4 W/m^2°C T_1 = 93°C T_2 = 44.77°C.

t_1 = 50°C t_2 = 23°C

We thus have: $\Delta\theta_1 = 93 - 50$; $\Delta\theta_1 =$ 43°C; $\Delta\theta_2 = 44.77 - 23$; $\Delta\theta_2 =$ 21.77°C

Hence: $\Delta\theta_{ML} =$ 31.19°C and $A_0 = \frac{1.05\ 10^6}{1{,}451.39 \times 31.19}$; i.e.: A_0 = 23.27 m^2.

Step 3. Knowing A_0, we calculate the number of tubes per pass

3.1- We choose the outer diameter of the tubes, d_o, from the standards (Table A1.10): kerosene and diesel being rather clean, we start with a small nominal diameter, DN 15, NPS ½". We then have: d_o = 21.3 mm.

3.2- We choose SCH10, hence the inner diameter, d_i = 17.08 mm (see Table A1.10).

3.3- We set v_d: Diesel is a low-viscosity liquid (v_d = 1 to 2 m/sec).

We will take: v_d = 1.7 m/sec.

3.4- Calculating the number of tubes per pass to ensure v_d.

The number of tubes per pass is: $n = \frac{4F_d}{\rho_d v_d \pi d_i^2}$.

NUMERICAL APPLICATION.–

F_d = 9.17 kg/sec ρ_d = 685 kg/m^3 v_d = 1.7 m/sec d_i = 1.708 10^{-2}m.

$n = \frac{4*9.17}{685*1.5*\pi*0.01708^2}$ n = 34.36; we take n_t = 35 tubes per pass.

Step 4. Calculating the number of tube-side (N_p) and shell-side (N_{sh}) passes, to ensure an acceptable value of Y

4.1- Calculating X and Z.

The correction factor, Y, depends on $Z = \frac{T_1 - T_2}{t_1 - t_2}$ and $X = \frac{t_1 - t_2}{T_1 - t_2}$ (see Chapter 5, section 5.7.4).

NUMERICAL APPLICATION.–

$T_1 = 93°C \quad T_2 = 44.77°C \quad t_1 = 50°C \quad t_2 = 23°C$

$X = \frac{50 - 23}{93 - 23}$; $X = 0.386$

$Z = \frac{93 - 44.77}{50 - 23}$; Z = 1.786 Z is therefore different from 1.

4.2 and 4.3- *First iteration on N_p and N_{sh}: Let's take $N_{sh} = 1$ and $N_p = 2$.*

4.4- $Z \neq 1$, $Y_{1\text{-}2}$ is then given by:

$$Y_{1-2} = \frac{\frac{\sqrt{1+Z^2}}{(Z-1)} log_{10}\left(\frac{1-X}{1-XZ}\right)}{log_{10}\left[\frac{\frac{2}{X} - 1 - Z + \sqrt{1+Z^2}}{\frac{2}{X} - 1 - Z - \sqrt{1+Z^2}}\right]}.$$

NUMERICAL APPLICATION.–

$X = 0.386 \qquad Z = 1.786$

$$Y_{1-2} = \frac{\frac{\sqrt{1+1.786^2}}{(Z-1)} \log_{10}\left(\frac{1-0.386}{1-0.386*1.786}\right)}{\log_{10}\left[\frac{\frac{2}{0.386} - 1 - 1.786 + \sqrt{1+1.786^2}}{\frac{2}{0.386} - 1 - 1.786 - \sqrt{1+1.786^2}}\right]};$$ i.e.: $Y_{1\text{-}2} = 0.699$.

4.5- Test on Y: *$Y < 0.8$, so we return to step 4.2.*

We return to point 4.2; we change *N_{sh} and N_p and recalculate Y.*

4.2 and 4.3- *Second iteration on N_p and N_{sh}: Let's take $N_{sh} = 2$ and $N_p = 4$.*

Z being different from 1, the correction factor, $Y_{2\text{-}4}$, is given by (see Chapter 5, section 5.7.4):

$$Y_{2-4} = \frac{\frac{\sqrt{1+Z^2}}{2(Z-1)} log_{10}\left(\frac{1-X}{1-XZ}\right)}{log_{10}\left[\frac{\frac{2}{X} - 1 - Z + \frac{2}{X}\sqrt{(1-X)(1-XZ)} + \sqrt{1+Z^2}}{\frac{2}{X} - 1 - Z + \frac{2}{X}\sqrt{(1-X)(1-XZ)} - \sqrt{1+Z^2}}\right]}.$$

NUMERICAL APPLICATION.–

$X = 0.386 \qquad Z = 1.786$

$$\frac{1-X}{1-XZ} = 1.975; \sqrt{1+Z^2} = 2.047; \frac{\sqrt{1+Z^2}}{Z-1} = 2.604$$

$$\frac{2}{X} - 1 - Z + \frac{2}{X}\sqrt{(1-X)(1-XZ)} + \sqrt{1+Z^2} = 6.712$$

$$\frac{2}{X} - 1 - Z + \frac{2}{X}\sqrt{(1-X)(1-XZ)} - \sqrt{1+Z^2} = 2.618$$

$$Y_{2-4} = \frac{\frac{2.604}{2}\log_{10}(1.975)}{\log_{10}\left[\frac{6.712}{2.618}\right]}; \text{ i.e. } Y_{2\text{-}4} = 0.941.$$

4.5- Test on Y: $Y > 0.8$

Thus $N_p = 4$ *and* $N_{sh} = 2$

The retained device will therefore be a 2-4 heat exchanger.

We thus proceed to step 5.

Step 5. Hydraulic calculations on tube side

5.1- Exact fluid velocity: $v_d = \frac{4F_d}{\rho_d n_t \pi d_i^2}$.

NUMERICAL APPLICATION.–

$F_d = 9.17$ kg/sec $\qquad \rho_d = 685$ kg/m^3 $\qquad n_t = 35$ tubes $\qquad d_i = 1.708\ 10^{-2}$m.

$$v_d = \frac{4*9.17}{685*35*\pi*0.01708^2}\ ; \text{ i.e.: } v_d = 1.67 \text{ m/sec}$$

v_d is thus within the acceptable interval (1 - 2 m/sec).

5.2- Calculating the length of the exchanger:

$$L = \frac{A_0}{N_p n_t \pi d_e}$$

NUMERICAL APPLICATION.–

$A_0 = 23.27$ m^2 $\qquad N_p = 4$ $\qquad n_t = 35$ tubes $\qquad d_o = 2.13\ 10^{-2}$m.

Hence: $L = 2.48$ m; we take $L = 3$ m.

5.3- Pressure drops in the tubes, ΔP_t, see section A3.3.1.2:

$$\Delta P_t = \frac{\rho_d v_d^2}{2}\left(N_p \frac{L}{d_i} f_M + 4\,N_p + 1.5\right)$$

where: $f_M = \frac{1}{[1.82 log_{10}(Re) - 1.64]^2}$, where $Re = \frac{d_i v_d \rho_d}{\mu_d}$.

NUMERICAL APPLICATION.–

$d_i = 1.708\ 10^{-2}$m $\quad v_g = 1.67$ m/sec $\quad \rho_d = 685$ kg/m^3 $\quad L = 3$ m

$\mu_d = 2\ 10^{-4}$ kg/m/sec $\quad \Delta P_d^{Adm} = 0.5$ Atm $\quad N_p = 4$

$Re = \frac{0.01708*1.67*685}{2\ 10^{-4}}$; i.e.: $Re = 9.76\ 10^4$.

The flow in the tubes is then turbulent.

Hence: $f_M = \frac{1}{[1.82 \log_{10}(9.76\ 10^4) - 1.64]^2}$; i.e.: $f_M = 1.81\ 10^{-2}$.

We thus have: $\Delta P_t = \frac{685*1.67^2}{2}\left(2 * \frac{6}{1.708\ 10^{-2}} * 1.81\ 10^{-2} + 4 * 2 + 1.5\right)$.

I.e.: $\Delta P_t = 1.51\ 10^4$ Pa ; i.e.: $\Delta P_t = 0.15$ Atm.

$\Delta P_t < \Delta P_d^{Adm}$, we thus proceed to step 6.

Step 6. Designing the shell

6.1- Total number of tubes in a cross-section area of the shell: $N_t = N_p\, n_t$.

NUMERICAL APPLICATION.–

$n_t = 35 \quad N_p = 4$

$N_t = 140$ tubes

6.2- Array type.

We choose a staggered square array.

6.3- Pitch.

$a \geq$ Sup $\{1.25\ d_e, d_o + 6$ mm$\}$.

NUMERICAL APPLICATION.–

$d_o = 21.3$ mm

$1.25\ d_o = 26.63$ mm and $d_o + 6$ mm $= 27.3$ mm

Hence: $a > 27.3$ mm.

To be within the standards, we take $a = 30.63$ mm, corresponding to a = 1"1/4.

6.4- Transversal and longitudinal spacings of the tubes: $e_T = a\sqrt{2}$ and $e_L = a\,\frac{\sqrt{2}}{2}$.

NUMERICAL APPLICATION.–

$a = 30.63$ mm

$e_T = 43.32$ mm $\qquad e_L = 21.66$ mm

6.5- Shell diameter.

6.5.1- From the standards in the tables in Appendix 1.

For a square array with a pitch of 1"1/4 and 4 passes, Table A1.12 provides the data closest to our situation. This table is valid for NPS 1"; it gives, for 4 passes and 158 tubes instead of our 140 tubes: D_{sh} = 21"1/4; i.e. $D_{sh} = 0.52$ m.

6.5.2- By calculating: $D_{sh} = 2a\sqrt{\frac{N_t}{\pi}\frac{K_M}{K_P}}$.

NUMERICAL APPLICATION.–

$a = 30.63$ mm $\qquad N_t = 140$ tubes $\qquad K_M = 1 \qquad N_p = 4 \qquad K_p = 0.81$

Hence: $D_{sh} = 2 * 0.03063\sqrt{\frac{140}{\pi} * \frac{1}{0.81}}$; $D_{sh} = 0.45$ m.

Ideally, we should retain $D_{sh} = 0.45$ m for our 140 tubes, but a D_{sh} of 0.45 m does not exist in commercial standards. We therefore retain the closest diameter that is in the standards given by Table A1.12; i.e., $D_{sh} = 0.52$ m corresponding to 21"1/4.

6.6- Number of longitudinal baffles: $N_{LB} = N_{sh} - 1$.

I.e.: $N_{LB} = 1$.

6.7- Arrangement of transverse baffles:

- Baffle spacing: $e_c = D_{sh} = 0.52$ m;

- Given data specifies that $\varepsilon = 30\%$. Table 2.2, (illustration in section 2.6.3) gives the height to be left free over the baffles[14]: $\delta_F = 0.18$ m.

6.8- Number of transverse baffles: $N_{TB} = \text{Ent}\left(\frac{L}{e_c}\right)$; $N_{TB} = 5$.

Step 7. Hydraulic calculations on shell side

7.1- Longitudinal (a_L) and transverse (a_T) free areas (see sections 2.6.5 and 2.6.6):

$$a_T = \frac{e_c D_{sh}\left(\sqrt{e_T e_L} - d_e\right)}{\sqrt{e_T e_L}} \qquad a_L = \frac{\varepsilon\pi}{4}\left(D_{sh}^2 - N_t d_o^2\right)$$

NUMERICAL APPLICATION.–

$e_T = 43.32$ mm $e_L = 21.66$ mm $\varepsilon = 0.30$ $D_{sh} = 0.52$ m

$d_o = 21.3$ mm $e_c = 0.52$ m $N_t = 140$

Hence: $a_L = 0.049$ m^2 and $a_T = 0.083$ m^2.

7.2- Unit mass flow rates (see sections 2.6.7):

- Longitudinal (G_L): $G_L = \frac{F_k}{a_L}$;
- Transverse (G_T): $G_T = \frac{F_k}{a_T}$;
- Average (G_M): $G_M = \sqrt{G_L G_T}$.

NUMERICAL APPLICATION.–

$a_T = 0.049$ m^2 $a_L = 0.083$ m^2 $F_k = 19.44$ kg/sec

Hence: $G_L = 397.65$ kg/m^2/sec $G_T = 235.51$ kg/m^2/s $G_M = 306.02$ kg/m^2/sec

14 δ_F can also be calculated from expressions given in section 2.6.2.

7.3- Pressure drop in the sleeve: ΔP_k:

$$\Delta P_k = (N_c - 1)\Delta P_T + N_c \Delta P_L$$

with: $\Delta P_T = 1.5 \frac{D_{sh}-\delta}{\sqrt{e_T e_L}} \left[\frac{\mu_k}{G_T\sqrt{e_T e_L} - d_o}\right]^{0.2} \frac{G_T^2}{\rho_k}$ and $\Delta P_L = \frac{G_L^2}{\rho_k}$.

NUMERICAL APPLICATION.–

G_L=397.65 kg/m²/sec G_T=235.51 kg/m²/sec G_M=306.02 kg/m²/sec N_c=5

δ_F = 18 cm e_T = 43.32 mm e_L = 21.66 mm μ_k =0.0016 kg m^{-1}s^{-1}

D_{sh} = 0.52 m d_o=21.3 mm ρ_k = 800 kg.m^{-3} ΔP_k^{Adm} = 10 psi

Hence: ΔP_T=217 Pa; ΔP_L =198 Pa.

$\Delta P_k = (5 - 1)\Delta P_T + 5\,\Delta P_L$; i.e. ΔP_k = 1,856 Pa, or 0.018 Atm.

ΔP_k^{Adm} = 10 psi = 0.678 Atm

Therefore: $\Delta P_k \leq \Delta P_k^{Adm}$, we thus proceed to step 8.

Step 8. Thermal check

8.1- Calculating transfer areas:

A_e^* = 28.10 m²; A_i^* = 22.54 m²; $A_m^* = \frac{A_i^* + A_e^*}{2}$ = 25.32 m²

8.2- Calculating h_i.

The convective heat transfer coefficient on the tube side is calculated from the correlations given in Table A2.1 (see section A2.2.1).

Reynolds domain	Regime	Recommended correlation
$Re < 2{,}300$	Laminar	$h_i = \frac{48}{11}\frac{\lambda_i}{d_i}$
$Re \geq 2{,}300$	Turbulent	$Nu_i = 0.023\, Re_i^{0.8}\, Pr_i^{\frac{1}{3}}$
$Nu_i = \frac{h_i d_i}{\lambda_i}$;	$Re_i = \frac{d_i v_i \rho_i}{\mu_i}$;	$Pr_i = \frac{Cp_i \mu_i}{\lambda_i}$.

Table 6.2. *Correlations for the calculation of h_i*

NUMERICAL APPLICATION.–

$d_i = 1.708\ 10^{-2}$m $\quad v_d = 1.67$ m/sec $\quad \rho_d = 685$ kg m^{-3}

$\mu_d = 2\ 10^{-4}$ kg m^{-1}s^{-1} $\quad C_{pd}$=2,382.6 J/kg°C. $\quad \lambda_d = 0.1298$ W/m°C.

$Re_i = \frac{1.708\ 10^{-2} * 1.67 * 685}{2\ 10^{-4}}$ $\quad Re_i = 9.76\ 10^4$ $\quad Nu_i = 0.023\ Re_i^{0.8}\ Pr_i^{\frac{1}{3}}$

$Pr_i = \frac{2{,}382.6 * 2\ 10^{-4}}{0.1298}$ $\quad Pr_i = 3.67$ $\quad Nu_i = 348$

Hence: $h_i = \frac{348 * 0.1298}{1.708\ 10^{-2}}$; $h_i = 2{,}645$ W/m^2°C.

8.3- Calculating h_e.

The shell comprising baffles, h_e, is calculated using the Donohue correlation (see section A2.2.2.2.1): $Nu_e = 0.22\ Re_M^{0.6}\ Pr^{\frac{1}{3}}$; with $Re_M = \frac{d_o G_M}{\mu_e}$.

NUMERICAL APPLICATION.–

$d_o = 0.0213$ m $\quad G_M = 306.02$ kg m^{-2}s^{-1} $\quad \rho_k = 800$ kg m^{-3}

$\mu_k = 0.0016$ kg m^{-1}s^{-1} $\quad C_{pk} = 2{,}006.4$ J/kg°C $\quad \lambda_k = 0.1437$ W/m°C

$d_i = 0.01708$ m $\quad \lambda_t = 109$ W/m°C.

$Re_M = \frac{0.0213 * 306.02}{0.0016}$ $\quad Re_M = 4{,}074$

$Pr = \frac{2{,}006.4 * 0.0016}{0.1437}$; $Pr = 22.34$ and $Nu_e = 0.22 \text{ x } 4{,}074^{0.6}\ 22.34^{\frac{1}{3}}$

I.e. $Nu_e = 90.81$. Hence: $h_e = \frac{90.81 * 0.1437}{0.0213}$; i.e. $h_e = 612.56$ W/m^2°C.

8.4- Calculating the overall heat transfer coefficient, U_e^*, for the clean heat exchanger: U_e^* is given by: $\frac{1}{U_e^* A_e^*} = \frac{1}{h_i A_i^*} + \frac{d_o - d_i}{\lambda_t A_m^*} + \frac{1}{h_e A_e^*}$.

NUMERICAL APPLICATION.–

$A_i^* = 22.54$ m^2 $\quad A_e^* = 28.10$ m^2 $\quad A_m^* = 25.32$ m^2 $\quad \lambda_t$ =109 W/m°C

$h_i = 2{,}645$ W/m^2°C $\quad h_e = 612.56$ W/m^2°C $\quad d_i = 0.01708$ m $\quad d_o = 0.0213$ m

We thus have: $\frac{1}{U_e^* A_e^*} = \frac{1}{2{,}645 * 35.63} + \frac{0.02130 - 0.01708}{109 * 25.32} + \frac{1}{612.56 * 28.10}$.

I.e.: $\frac{1}{U_e^* A_e^*} = 7.64\ 10^{-5}$°C/W and $U_e^* = 466$ W/m^2°C.

8.5- Re-evaluating heat transfer coefficient for the fouled exchanger, U_F^*: $\frac{1}{U_F^*} = \frac{Y}{U_F} \frac{A_e^*}{A_0}$.

NUMERICAL APPLICATION.–

$Y_{2\text{-}4} = 0.941$ $U_F = 1{,}451.39$ W/m^2°C $A_0 = 23.27$ m^2 $A_e^* = 28.10$ m^2

Hence: $\frac{1}{U_F^*} = 7.83\ 10^{-4}$ m^2°C/W.

Thus: $U_F^* = 1{,}277$ W/m^2°C.

Step 9. Checking the margin allowed for fouling

9.1- We calculate: $R_{fi} + R_{fe}$: $R_{fi} + R_{fe} = 1.50\ 10^{-3}$ m^2°C/W.

9.2- Tests on $\frac{1}{U_F^*} - \frac{1}{U_e}$:

$\frac{1}{U_F^*} - \frac{1}{U_e} = -1.36\ 10^{-3}$ m^2°C/W

$\frac{1}{U_F^*} - \frac{1}{U_e} < 0$ *The device is thus undersized.* We will then choose a more suitable value for U_F and repeat the calculations, starting from step 2.

Step 2. Second iteration on U_F

– Choosing U_F: we now take $U_F = 300$ kcal/h m^2°C; that is 348.33 W/m^2°C.

– Hence the new value of A_0: $A_0 = \frac{\phi}{U_F\, \Delta\theta_{ML}}$.

NUMERICAL APPLICATION.–

Hence: $A_0 = \frac{1.05\ 10^6}{348.33 \text{x} 31.19}$, i.e.: $A_0 = 96.95$ m^2.

Steps 3 and 4.

The tube diameters remain unchanged: $d_o = 21.3$ mm and $d_i = 17.08$ mm, as well as:

– the flow velocity in the tubes;

– the number of tubes per pass $n_t = 35$ *tubes*;

– the numbers of passes, on the pipe side $N_p = 4$ and on the shell side, $N_{sh} = 2$;

– the correction factor, $Y = 0.941$.

$Y > 0.8$, so we proceed to step 5.

Step 5. Hydraulic calculations on tube side

5.1- The exact velocity of the fluid remains unchanged: v_d =1.67 m/sec.

5.2- Conversely, the length of the heat exchanger becomes (given the new value of A_0):

NUMERICAL APPLICATION.–

$A_0 = 96.95$ m^2 $N_p = 4$ $n_t = 35$ tubes $d_o = 2.13\ 10^{-2}$m.

Hence: $L = 10.35$ m; we take $L = 12\ m$.

5.3- Pressure drop in the tubes, ΔP_t:

Re and f_M remain unchanged. However, with a new length of the exchanger, ΔP_t becomes: $\Delta P_t = 0.33\ Atm$. $\Delta P_t < \Delta P_d^{Adm}$, we thus proceed to step 6.

Step 6. Designing the shell

Apart from the length, the other parameters of the shell remain unchanged, namely:

– array: staggered square;

– pitch: a = 1"1/4;

– transversal and longitudinal spacings of tubes: $e_T = 43.32$ mm; $e_L = 21.66$ mm;

– shell diameter: $D_{sh} = 0.52$ m;

– number of longitudinal baffles: $N_{LB} = 1$;

– gap between the transverse baffles: $e_c = D_{sh} = 0.52$ m, $\varepsilon = 30\%$ and $\delta = 0.18$;

– free areas: longitudinal, $a_L = 0.049$ m² and transverse, $a_T = 0.083$ m²;

– number of transverse baffles : $(N_{TB} = Ent\left(\frac{L}{e_c}\right))$; $N_{TB} = 5$.

Step 7. Hydraulic calculations on shell side

7.1 and 7.2- The unit mass flow rates remain unchanged.

7.3- The transverse and longitudinal pressure drops remain unchanged: ΔP_T=217 Pa and ΔP_L =198 Pa.

Conversely, since N_c has changed, ΔP_k becomes:

ΔP_k = 9,278 Pa, or 0.092 Atm

$$\Delta P_k \leq \Delta P_k^{Adm}$$

We thus proceed to step 8.

Step 8. Thermal check

8.1- The transfer areas become:

$$A_e^* = 112.42\ m^2;\ A_i^* = 90.15\ m^2;\ A_m^* = \frac{A_i^*+A_e^*}{2} = 101.28\ m^2$$

8.2 to 8.4- The tube diameter and the flow rate being unchanged.

NUMERICAL APPLICATION.–

h_i, h_e and U_e^* remain the same: h_i = 2,645 W/m^2°C; h_e = 612.56 W/m^2°C.

U_e^* = 466 W/m^2°C

8.5- Re-evaluating the global heat transfer coefficient for the fouled exchanger, U_S^*: $\frac{1}{U_F^*} = \frac{Y}{U_F}\frac{A_e^*}{A_0}$.

NUMERICAL APPLICATION.–

$Y_{2\text{-}4}$ = 0.941 $\quad U_F$ = 348.33 W/m^2°C

A_0 = 96.95 m^2 A_e^* = 112.42 m^2

Hence: $\frac{1}{U_F^*}$ =3.13 10^{-3} m^2°C/W; thus: U_F^* = 319 W/m^2°C.

Step 9. Checking the margin allowed for fouling

9.1- $R_{fi} + R_{fe}$ = 1.50 10^{-3} m^2°C/W.

9.2- $\frac{1}{U_F^*} - \frac{1}{U_e} = 0.99\ 10^{-3}$ m²°C/W:

The resistances being of the same order of magnitude, we retain this design for the heat exchanger.

Details of the retained device are presented in Table 6.3.

<table>
<tr><td colspan="3">Characteristics of the retained heat exchanger</td><td colspan="2" rowspan="2">Type: Shell -and-tube</td></tr>
<tr><td>Exchange area = 112 m²</td><td colspan="2">$U_e^* = 466$ W/m²°C</td></tr>
<tr><td>Length = 12 m</td><td colspan="2">$U_F = 319$ W/m²°C</td><td colspan="2">Shell-tube passes: 2 - 4</td></tr>
<tr><td colspan="5"></td></tr>
<tr><td rowspan="4">Tube side</td><td>Tubes/pass = 35</td><td>NPS = ½"</td><td>SCH = 10</td><td>Total no. of tubes = 140</td></tr>
<tr><td>Flowrate = 33 T/hr.</td><td>$T_1 = 93$°C</td><td>$T_2 = 44.77$ °C</td><td>$\phi = 1.05$ MW</td></tr>
<tr><td>$d_i = 17.08$ mm</td><td>$d_o = 21.3$ mm</td><td>$v = 1.67$ m/sec</td><td>$Re = 9.76\ 10^4$</td></tr>
<tr><td>Pitch = 1" ¼</td><td>$e_L = 21.6$ mm</td><td>$e_T = 43.3$ mm</td><td>$\Delta P_t = 0.33$ Atm</td></tr>
<tr><td colspan="5"></td></tr>
<tr><td rowspan="4">Shell side</td><td>Flow rate = 70 T/hr.</td><td>$t_1 = 50$°C</td><td>$t_2 = 23$°C</td><td>$Y_{2\text{-}4} = 0.941$</td></tr>
<tr><td>$D_{sh} = 21$" ¼</td><td>$a_L = 0.049$ m²</td><td>$a_T = 0.083$ m²</td><td>$\Delta P_{sh} = 0.09$ Atm</td></tr>
<tr><td colspan="2">No. of longitudinal baffles = 1</td><td colspan="2">No. of transverse baffles = 23</td></tr>
<tr><td>No. of rows= 16</td><td>$e_c = 520$ mm</td><td>$\varepsilon = 30\%$</td><td>$\delta_F = 180$ mm</td></tr>
</table>

Table 6.3. *Details of the retained heat exchanger*

6.5. Calculation algorithm for cross-flow heat exchangers

Cross-flow heat exchangers are often used for liquid/gas heat transfers, where low gas-side heat transfer coefficients are partially compensated for by an extension of the transfer area using fins.

However, it should be noted that addition of fins introduces a significant specific aspect to the calculation of the exchange coefficients and the gas-side pressure drop, given that both of these parameters depend on the geometrical characteristics of the fins used: thickness, height and pitch. The rules of thumb below help guide the choosing of the most important design parameters.

RULES OF THUMB.–

R_1 – Materials to be used:

- *For ventilation and air-conditioning equipment,* opt for copper tubes, individually finned, with fins also made of copper (crimped or welded).

- *For equipment intended for refrigeration,* use copper tubes with fins made of aluminum (individually or collectively crimped) or copper (individually or collectively welded).

- *For embedded heat exchangers,* particularly those intended for vehicles, collectively-finned tubes are used. Tubes and fins are generally made of Cu, Aluminum or stainless steel. The fins can be welded or forced through the tubes and assembled mainly by press fit. The latter process is used for the manufacture of exchangers intended for refrigeration systems.

Lately, continuous-fin heat exchangers have been manufactured in one piece by molding or plastic injection.

R_2 – Fin arrangement.

The distance, s, at the base between two fins[15] ($s = a_a - e_a$), depends on the fouling potential of the gas flowing between the fins. The higher the fouling resistance, the more we must tend towards higher values of s. Table 6.4 gives the recommended minimum values for the distance to be kept between two fins (Reid and Taborek 1994).

Fluid		Characteristic	Minimum s
Air		Pure	1.8 - 3 mm
Combustion gas	Originating from	Natural gas	3 - 4.2 mm
		Low-viscosity oil	4.2 - 6.3 mm
		Heavy oil	5 - 8.5 mm
		Solid fuels	12 - 25 mm

Table 6.4. *Standard pitch of fins*

R_3 – Fin thickness.

Although the thickness of the fins has an impact on their efficiency, this parameter is determined more by considerations of manufacturing feasibility and

15 See section 3.3.1.

material strength during installation and operation. The standard thickness values to be used for different fin types are shown in Table 6.5.

Fin type	Thickness (mm)
Spiral fins	0.9 - 1.3
Straight fins[16]	1.6 - 2.5
Ailerons[17]	0.3 - 0.7

Table 6.5. *Standard thicknesses of fins*[16][17]

R_4 – Tube diameter, pitch and flow velocity on the gas side.

The tubes passing through the fins have diameters between 6 and 13 mm, the pitch, a_a, of the fins is generally between 1 and 8.7 mm; whereas, the gas-side flow velocities are between 1 and 16.5 m/sec (Hugo 2012). For car radiators, this velocity can reach 35 m/sec (i.e. 120 km/hr).

R_5 – Case of embedded exchangers.

In the most general case, the overall dimensions – L_1, L_2 *and* L_3 – of the heat exchanger (see Figure 3.5) are not set. The iterations for the calculation of the exchanger can therefore operate on these dimensions.

By contrast, for embedded exchangers, these dimensions are set by constraints on the space available on the vehicle in question (car, locomotive, aircraft, space shuttle, etc.). In this case, we will check that the maximum exchange area will allow the desired flux to be transferred.

6.5.1. *General algorithm*

In the general case, the design calculations of a cross-flow heat exchanger can be organized into a five-step algorithm.

Step 1. Definition of the device[18] *dimensions*

1.1- Set the values of L_1 (direction of liquid flow), L_2 (direction of gas flow) and L_3.

16 Used for gases from solid fuels.

17 Generally made of Cu and Al, used in ventilation and air conditioning.

18 Generally, these dimensions are limited by space constraints.

1.2- Calculate the flux, ϕ, to be exchanged and, where applicable, the missing temperature.

1.3- Define the interval of U_F (see Tables A1.21 to A1.23) and set a value that is close to the operating conditions.

1.4- Calculate the approximate exchange area, A_{init}: $A_{init} = \frac{\phi}{U_F \, \Delta\theta_{ML}}$.

Step 2. Calculate the number of rows, N_n, the number of tubes per row, n_T

2.1- Choose the outer diameter of the tubes, d_o, from the standards (see Table A1.10).

2.2- Choose the SCH of the tube, hence the tube thickness and its inner diameter, d_i.

2.3- Define the type of array (square or triangular), in-line or staggered.

2.4- Choose the pitch of the tubes[19].

2.5- Calculate the gaps between the tubes, e_T and e_L[20].

2.6- Calculate[21] the number of rows (direction L_2), N_n, and the number of tubes per row (direction L_3), n_T.

2.7- Deduce therefrom (see Table 3.1) the total number of tubes, N_T.

Step 3. Hydraulic checks

3.1- Tube side: Mass flow rate, W_L.

3.1.1- Calculate the flow velocity in the tubes: If W_L is the liquid mass flow rate, the flow velocity, v, is given by[22] $v_L = \frac{4W_L}{\rho_L N_T \pi d_i^2}$.

3.1.2- If this velocity is suitable (0.2 to 0.5 m/sec for a viscous fluid; 1 to 2 m/sec for a low-viscosity fluid), proceed to step 3.1.3.

---

19 For a square array, take $a = \text{Sup}\,\{1.25d_e;\, d_o + 6\text{ mm}\}$, for a triangular array, $a = 1.25\, d_o$.

20 See section 2.3.6.

21 N_n and n_T are calculated from the expressions in Table 3.1.

22 For N_T tubes, the flow rate that passes through each tube is $w_L = \frac{W_L}{N_T}$. The flow velocity in each tube is then $v_L = \frac{w_L}{\rho_L \pi \frac{d_i^2}{4}} = \frac{4W_L}{\rho_L N_T \pi d_i^2}$.

Otherwise, if $d_o > 1/8"$, repeat step 2.1 with a smaller diameter.

3.1.3- Calculate the Reynolds number: $Re_L = \frac{d_i v_L \rho_L}{\mu_L}$.

3.1.4- Calculate the pressure drop (ΔP_L) on the tube side[23] and compare to the allowable pressure drop:

- If $\Delta P_L \leq \Delta P_L^{Adm}$, we proceed to step 3.2;

- *If not, we repeat the calculation at the beginning of step 2, with a larger value of d_i.*

3.2- Fin side: mass flow rate, W_g.

3.2.1- Set the pitch, a_a, and the thickness, e_a, of the fins.

3.2.2- Determine the number of sheets of fins, N_a (see Table 3.1).

3.2.3- Calculate the gas velocity between two fins[24].

3.2.4- If this velocity is greater than 10 m/sec, proceed to step 3.2.5.

Otherwise, repeat step 3.2.1 with another pitch value for the fins.

3.2.5- Calculate the transfer area, A (see Table 3.1).

3.2.6- Determine the free flow area, A_0 (see Table 3.1).

3.2.7- Calculate the hydraulic diameter (see Table 3.1): $D_H = \frac{4 A_0 L_2}{A}$.

3.2.8- Calculate the pressure drops on the fin side (see section A3.6.2.3):

---

23 The pressure drop on the tube side (flow in direction L_1) is given by (see Appendix 3, section A3.2.1.1): $\Delta P_L = \frac{\rho_L v^2}{2}\left(\frac{L_1}{d_i} f + 1.5\right)$ where f is the friction factor (see Appendix 3, section A3.2.1), calculated as follows: for smooth tubes (Moody 1944): $f = \frac{1}{[1.02\log_{10}(Re) - 1.64]^2}$; $Re = \frac{d_i v \rho_c}{\mu_c}$; for tubes presenting roughness ε (Haaland 1983): $\frac{1}{\sqrt{f}} \cong -1.8\log_{10}\left\{\left[\frac{\left(\frac{\varepsilon}{D}\right)}{3.7}\right]^{1.11} + \frac{6.9}{Re}\right\}$.

24 For a gas mass flow rate, W_g, the velocity between two fins is given by: $v_{g_a} = \frac{W_g}{(N_a - 1)\rho_g L_3 a_a}$.

$$\Delta P_g = \frac{\rho_g v_\infty^2}{2}\frac{L_2}{D_H} f;\text{ with } v_\infty = \frac{W_g}{\rho_g L_1 L_3};\ f = 0.0267 Re_{d_o}^{k_1}\left(\frac{e_T}{e_L}\right)^{k_2}\left(\frac{a_a}{d_o}\right)^{k_3}$$

$$k_1 = -0.764 + 0.739\left(\frac{e_T}{e_L}\right) + 0.177\left(\frac{a_a}{d_o}\right) - \frac{0.00758}{N_n};\ k_2 = -15.689 + \frac{64.021}{Ln(Re_{d_o})}$$

$$k_3 = 1.696 - \frac{15.695}{Ln(Re_{d_o})}\text{ and } Re_{d_o} = \frac{d_o v_\infty \rho_g}{\mu_g}$$

3.2.9- If $\Delta P_g \leq \Delta P_g^{Adm}$, proceed to step 4.

Otherwise, repeat step 1.1 with a larger value of L_3.

Step 4. Thermal checks

4.1- Calculate h_i: see Appendix 2, section A2.5.1[25].

4.2- Calculate h_e: see Appendix 2, section A2.5.3.

4.3- Calculate the effective external exchange area[26]: A_e.

4.4- If $A_e \geq A_{init}$, proceed to step 4.5.

Otherwise, take a larger value of U_F, recalculate A_{init} and repeat 4.4.

4.5- Calculate the average exchange area of the tubes: $A_m = \frac{N_T \pi (d_i + d_o) L}{2}$.

4.6- Calculate the overall heat transfer coefficient[27]: $\frac{1}{U_e^* A_e^*} = \frac{1}{h_i A_i^*} + \frac{d_e - d_i}{\lambda_t A_m^*} + \frac{1}{h_e A_e^*}$.

4.7- Calculate X and Z: $X = \frac{\theta_{c1} - \theta_{c2}}{\theta_{h1} - \theta_{c2}}$ and $Z = \frac{\theta_{h1} - \theta_{h2}}{\theta_{c1} - \theta_{c2}}$

4.8- Calculate the correction factor, Y[28].

If $Z \# 1$: $Y = \frac{1}{\text{NTU}(1-Z)} Ln\left[\frac{1-XZ}{1-X}\right]$.

---

25 In a turbulent regime (see section A2.2.1): $Nu_i = 0.023\ Re^{0.8}\ Pr_i^{\frac{1}{3}}$.

26 $A_e = N_T\,\pi d_o (L_1 - N_a e_a) + 2N_a\left(L_2 L_3 - N_T \frac{\pi d_e^2}{4}\right)\eta_e$ with: $\eta_e = \frac{\tanh(m_e L_3)}{m_e L_3}$ and $m_e = \sqrt{\frac{2h_e}{e_a \lambda_a}}$ (see Table 3.1).

27 λ_t is the thermal conductivity of the tubes.

28 The NTU is given by Table 5.7 as a function of the type of cross-flow exchanger.

If Z=1: $Y = \frac{X}{(1-X)\mathrm{NTU}}$.

4.9- Evaluate the "fouled" heat transfer coefficient: U_F^*:

$$\phi = U_F A_{init}\,\Delta\theta_{ML} = Y\,U_F^*\,A_e^*\,\Delta\theta_{ML} \Rightarrow U_F^* = \frac{U_F\,A_{init}}{YA_e^*}$$

Step 5. Checking the margin allowed for fouling

5.1- Calculate $R_{fi} + R_{fe.}$

5.2- Proceed with tests on $\frac{1}{U_F^*} - \frac{1}{U_e^*}$:

If $\frac{1}{U_F^*} - \frac{1}{U_e^*} \cong R_{fi} + R_{fe}$: then the device is correctly sized.

If $\frac{1}{U_F^*} - \frac{1}{U_e} < 0$ *or* $\frac{1}{U_F^*} - \frac{1}{U_e} \ll R_{fi} + R_{fe}$: then the device is undersized. *Thus repeat the calculation at step 1.1, choosing a larger value of L_1.*

If $\frac{1}{U_F^*} - \frac{1}{U_e} \gg R_{fi} + R_{fe}$: then the device is oversized. *Thus repeat the calculation at step 1.1, choosing a smaller value of L_1.*

6.5.2. *Case of embedded heat exchangers*

The geometric constraints on the overall dimensions (L_1, L_2 and L_3) of an embedded heat exchanger mean that it is not possible to vary the exchange area at will. However, for this type of exchanger, the maximum possible dimensions are generally defined as: L_{1Max}; L_{2Max}; L_{3Max}.

The sizing algorithm, in this case, needs to check whether the calculated exchange area will allow the desired energy flux to pass. The design algorithm results from an adaptation of the general algorithm presented in section 6.5.1. It begins with the definition of the liquid flow direction: L_1.

Step 1. From data including L_{1Max}; L_{2Max} and L_{3Max}

1.1- Define L_1 (liquid flow direction)[29]: $L_1 \leq \mathrm{Sup}\{L_{1Max}; L_{2Max}; L_{3Max}\}$.

1.2- Set L_2 (gas flow direction) to[30]: $L_2 \leq \mathrm{Sup}\ \{L_2'; L_3'\}$.

29 The maximum dimensions of the heat exchanger are now L_1; L_2'; L_3'.

1.3- Calculate the flux, ϕ, to be exchanged and, where applicable, the missing temperature.

1.4- Define the interval of U_F (see Table A1.21) and set a value of U_F for initial calculations: $U_{FMin} \leq U_F \leq U_{FMax}$.

1.5- Calculate the approximate exchange area, A_{init}: $A_{init} = \frac{\phi}{U_F \, \Delta\theta_{ML}}$.

Step 2. Knowing L_1, define L_2 and L_3.

2.1- Choose the outer diameter of the tubes, d_o, from the standards (see Table A1.10).

2.2- Choose the SCH of the tube, hence the tube thickness and its inner diameter, d_i.

2.3- Define the type of array (square or triangular), in-line or staggered.

2.4- Choose the pitch of the tubes.

2.5- Calculate the gaps between the tubes, e_T and e_L.

2.6- Calculate the number of rows (direction L_2), N_n, and the number of tubes per row (direction L_3), n_T. and deduce therefrom the total number of tubes, N_T (see Table 3.1).

Step 3. Hydraulic checks

3.1- Tube side: mass flow rate, W_L.

3.11- Calculate the flow velocity in the tubes: If W_L is the mass flow rate of the liquid, the flow velocity, v, is given by $v_L = \frac{4W_L}{\rho_L N_T \pi d_i^2}$.

3.1.2- If this velocity is suitable (0.2 to 0.5 m/sec for a viscous fluid; 1 to 2 m/sec for a low-viscosity fluid), proceed to step 3.1.3.

Otherwise, if d_o > 1/8", repeat step 2.1 with a smaller diameter.

3.1.3- Calculate the Reynolds number: $Re_L = \frac{d_i v_L \rho_L}{\mu_L}$.

30 By setting L_2, L_3 is also set.

3.1.4- Calculate the pressure drop (ΔP_L) on the tube side and compare the allowable pressure drop:

If $\Delta P_L \leq \Delta P_L^{Adm}$, proceed to step 3.2.

If not, repeat the calculation at the beginning of step 2 , with a larger value of d_i.

3.2- Fin side: Mass flow rate, W_g.

3.2.1- Set the pitch, a_a, and the thickness, e_a, of the fins.

3.2.2- Determine the number of sheets of fins, N_a (see Table 3.1).

3.2.3- Calculate the gas velocity between two fins.

3.2.4- If this velocity is greater than 10 m/sec, proceed to step 3.2.5.

Otherwise, repeat step 3.2.1 with another pitch value for the fins.

3.2.5- Calculate the transfer area, A (see Table 3.1).

3.2.6- Determine the free flow area, A_0 (see Table 3.1).

3.2.7- Calculate the hydraulic diameter (see Table 3.1): $D_H = \frac{4\,A_0 L_2}{A}$.

3.2.8- Calculate the pressure drops on the fin side (see section A3.6.2.3):

$$\Delta P_g = \frac{\rho_g v_\infty^2}{2}\frac{L_2}{D_H} f;\ \text{with } v_\infty = \frac{W_g}{\rho_g L_1 L_3};\ f = 0.0267 Re_{d_e}^{k_1}\left(\frac{e_T}{e_L}\right)^{k_2}\left(\frac{a_a}{d_o}\right)^{k_3}$$

$$k_1 = -0.764 + 0.739\left(\frac{e_T}{e_L}\right) + 0.177\left(\frac{a_a}{d_o}\right) - \frac{0.00758}{N_n};\ k_2 = -15.689 + \frac{64.021}{Ln(Re_{d_o})}$$

$$k_3 = 1.696 - \frac{15.695}{Ln(Re_{d_o})}\ \text{and}\ Re_{d_e} = \frac{d_o v_\infty \rho_g}{\mu_g}$$

3.2.9 Compare ΔP_g to the allowable pressure drop:

- If $\Delta P_g \leq \Delta P_g^{Adm}$, proceed to step 4.

- *If the maximum values are reached for* L_1 *and* L_3, *stop the calculations: the available dimensions are insufficient.*

- *Otherwise, repeat step 1.1 with a larger value of* L_3.

Step 4. Thermal checks

4.1- Calculate h_i: see Appendix 2, section A2.5.1.

4.2- Calculate h_e: see Appendix 2, section A2.5.3.

4.3- Calculate the effective external exchange area: A_e.

4.4- If $A_e \geq A_{init}$, proceed to step 4.5.

Otherwise, as long as U_{FMax} *is not reached, take a larger value of* U_F, *recalculate* At_{init} *and repeat 4.4.*

If U_{FMax} is reached, take $U_F = U_{FMax}$, change L_1 (and, where applicable, L_2 or L_3) and return to step 2.

4.5- Calculate the average exchange area of the tubes: $A_m = \frac{N_T \pi (d_i + d_o) L}{2}$.

4.6- Calculate the overall heat transfer coefficient: $\frac{1}{U_e^* A_e^*} = \frac{1}{h_i A_i^*} + \frac{d_o - d_i}{\lambda_t A_m^*} + \frac{1}{h_e A_e^*}$.

4.7- Calculate X and Z:

$$X = \frac{\theta_{c1} - \theta_{c2}}{\theta_{h1} - \theta_{c2}} \text{ and } Z = \frac{\theta_{h1} - \theta_{h2}}{\theta_{c1} - \theta_{c2}} = \frac{W_c C_c}{W_h C_h}$$

4.8- Calculate the correction factor, Y:

- If $Z \# 1$: $Y = \frac{1}{\text{NTU}\,(1-Z)} Ln\left[\frac{1-XZ}{1-X}\right]$;

- If Z=1: $Y = \frac{X}{(1-X)\text{NTU}}$.

4.9- Evaluate the "fouled" heat transfer coefficient, U_F^*:

$$\phi = U_F A_{init}\, \Delta\theta_{ML} = Y\, U_F^*\, A_e^*\, \Delta\theta_{ML} \Rightarrow U_F^* = \frac{U_F\, A_{init}}{Y A_e^*}$$

Step 5. Checking the margin allowed for fouling

5.1- Calculate $R_{fi} + R_{fe}$.

5.2- Tests on $\frac{1}{U_F^*} - \frac{1}{U_e^*}$.

- If $\frac{1}{U_F^*} - \frac{1}{U_e^*} \cong R_{fi} + R_{fe}$: then the device is correctly sized.

- If $\frac{1}{U_F^*} - \frac{1}{U_e} < 0 \ or \ \frac{1}{U_F^*} - \frac{1}{U_e} \ll R_{fi} + R_{fe}$: then the device is undersized. If the maximum possible dimensions have already been reached, stop the calculations: the available space is too small to ensure the requested thermal schedule. Otherwise, repeat the calculation at step 1.1, choosing a larger value of L_1.

- If $\frac{1}{U_F^*} - \frac{1}{U_e} \gg R_{fi} + R_{fe}$: then the device is oversized. *Repeat the calculation at step 1.1, choosing a smaller value of* L_1.

6.6. Illustration: sizing of car radiators

Car radiators come under the category of embedded heat exchangers. As such, their design is subject to space constraints, reflected in maximum dimensions not to be exceeded. In addition, in this case, the arrangement of the radiator with respect to the air flux is set; the dimensions L_1, L_2 *and* L_3 are therefore determined in advance: L_1 *corresponds to the width,* L_2 *the depth, and* L_3 *the height.*

The radiator of a new model of a well-known series of cars shall be sized to provide (thermal) engine cooling. During operation, the engine generates a significant amount of heat: between 1/6 and 1/3 of the power. A significant part of this heat shall be evacuated by the cooling water, which in turn shall be cooled: this is the role of the radiator. Figure 6.2 presents a block diagram of the cooling circuit. The radiator to be sized consists of a bundle of N_n rows of n_T tubes, collectively passing through a series of vertical flat plates acting as fins (see Figure A2.5). It shall be sized for a steady state, when the vehicle is traveling at velocity $\bar{v}$. At this velocity, the water pump delivers a mass flow rate, W_1, and the water arrives at the radiator at temperature T_1 and shall leave it at a temperature less than or equal to T_2. The ambient air is at temperature t_2; it attacks the radiator with a flow rate, W_2.

Size the necessary radiator.

Data:

Water:

Flowrate: $W_l = 0.4$ kg/sec $\rho_l = 980$ kg/m^3 $\lambda_l =$ v0.89 Wm^{-1}°C^{-1}

Radiator inlet temperature: $T_1 = 85$°C

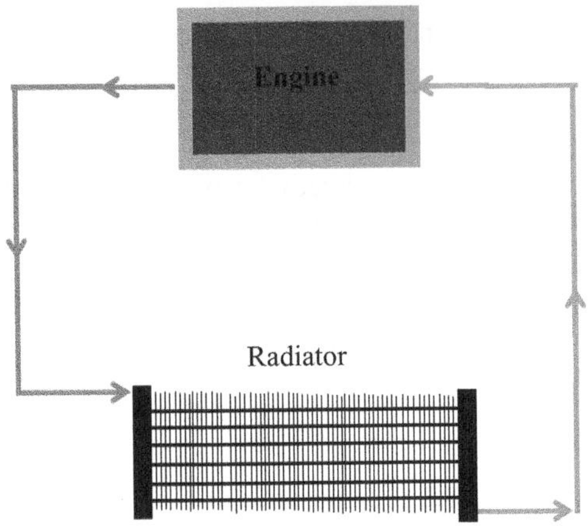

Figure 6.2. *Car engine cooling radiator*

Radiator outlet temperature: $T_2 = 66°C$

$Cp_l = 4{,}190$ J kg^{-1}°C^{-1} $\mu_l = 4\ 10^{-4}$ kgm^{-1}s^{-1}

$\Delta P_l^{Adm} = 0.5$ Atm $R_{fi} = 3\ 10^{-4}$ m²°C/W

Air:

$\bar{v} = 120$ km/hr $t_2 = 25°C$ $\rho_g = 0.86$ kg/m^3

$\lambda_g = 0.04291$ Wm^{-1}°C^{-1} $Cp_g = 1{,}006$ Jkg^{-1}°C^{-1} $\mu_g = 2.9\ 10^{-5}$ kgm^{-1}s^{-1}

$\Delta P_g^{Adm} = 0.5$ Atm $R_{fe} = 5\ 10^{-5}$ m²°C/W

Radiator tubes:

Thickness: $\Delta r > 2$ mm

Arrangement: in-line square array with pitch a

Construction material: tin; $\lambda_p = 60$ W/m°C

Radiator fins:

Construction material: aluminum; $\lambda_a = 210$ W/m°C

Thickness: $e_a = 1$ mm

Constraints for radiator location:

– maximum radiator height: $H_{max} = 0.55$ m;

– maximum width: $L_{max} = 0.95$ m;

– maximum depth: $E_{max} = 15$ cm.

Solution

Step 1. Dimensions of the heat exchanger

1.1- The car radiator is a specific embedded exchanger, in the sense that the liquid and gas flow directions are established beforehand.

Therefore, we need to have: $L_1 \leq L_{max}$, $L_2 \leq E_{max}$ and $L_3 \leq H_{max}$.

We start with $L_1 = 0.75$ m, $L_2 = 0.08$ m and $L_3 = 0.45$ m

1.2- Heat flux to be evacuated.

It is given by: $\phi = W_l\, C_{pl}(T_1 - T_2)$.

NUMERICAL APPLICATION.–

$W_l = 0.4$ kg/sec $\quad Cp_l = 4{,}190$ J/kg°C $\quad T_1 = 85°C \quad T_2 = 66°C$

Hence: $\phi = 31.84$ kW

Air outlet temperature.

The air heat balance gives: $\phi = W_g C_{pg}(t_1 - t_2)$ where $W_g = \bar{v}\, L_1 L_3 \rho$.

Hence: $t_1 = t_2 + \frac{\phi}{W_g Cp_g}$.

NUMERICAL APPLICATION.–

$\bar{v} = 120$ km/hr $\quad L_3 = 0.45$ m $\quad L_1 = 0.75$ m $\quad Cp_g = 1{,}006$ J/kg°C.

$\rho_g = 0.86$ kg/m^3 $\quad t_2 = 25°C \quad \phi = 31.84$ kW

Hence: $W_g = 14.98$ kg/sec and $t_1 = 27.1°C$.

1.3- Choosing U_F.

For an exchanger with fins, operating with a gas and a low-viscosity liquid, Table A1.21 suggests an overall "fouled" coefficient between 50 and 175 W/m²°C.

We start with U_F *=50 W/m²°C.*

1.4- Estimating the approximate exchange area: A_{init}.

$$A_{Init} = \frac{\phi}{U_F\,\Delta\theta_{ML}},\ \text{where } \Delta\theta_{mL} = \frac{(\Delta\theta_1 - \Delta\theta_2)}{Ln\frac{\Delta\theta_1}{\Delta\theta_2}},\ \Delta\theta_1 = T_1 - t_1 \text{ and } \Delta\theta_2 = T_2 - t_2$$

NUMERICAL APPLICATION.–

$T_1 = 85°C$	$T_2 = 66°C$	$t_1 = 27.1°C$	$t_2 = 25°C$
$\phi = 31.84$ kW	$U_F = 50\ W/m^2°C$		

Hence: $\Delta\theta_1$ =57.9°C; $\Delta\theta_2$ =41.0°C and $\Delta\theta_{ML}$ =49.0°C.

Thus: A_{init} = 13.01 m².

Step 2. Knowing L_1, L_2 *and* L_3

2.1- Outer diameter of the tubes, d_o.

We take d_o=10.3 mm (see Table A1.10).

2.2- SCH and inner diameter of the tubes.

Since the thickness needs to be greater than 2 mm, we shall have an SCH of 80 with a thickness, Δr, r = 2.41 mm and an inner diameter, d_i = 5.48 mm.

2.3- Arrangement of the radiator tubes: we choose an in-line square array

2.4- Choosing the tube pitch: for this type of array, $a \geq$ Sup $\{1.25\, d_o,\ d_o + 6\text{ mm}\}$.

NUMERICAL APPLICATION.–

d_o = 10.3 mm

1.25 d_o = 1.29 cm and d_o + 6 mm = 1.63 cm

Hence: $a \geq$ 1.63 cm.

We will take a =1.63 cm = e_L.

2.5- Gaps between the tubes, e_T and e_L[31].

For an in-line square array (see Table 2.1): $e_T = e_L = a$.

2.6- Number of rows, N_n, and number of tubes per row, n_T.

For an in-line square array of pitch a (see Figure 2.16), the number of rows (perpendicular to the flow) is set by the depth of the radiator, L_2; only $\frac{L_2}{e_L}$ rows can be placed (see section 3.3.3 and Table 3.1). We will take $N_n = Ent(\frac{L_2}{e_L})$; where: $e_L = a$.

NUMERICAL APPLICATION.–

$L_2 = 8$ cm $\qquad e_L = 1.63$ cm

Hence: $\frac{L_2}{e_L} = 4.91$. We thus take $N_n = 4$ rows.

Likewise, the number of tubes per row is set by the height of the radiator (direction L_3 in the sense of section 3.3.3): we shall have $n_T \leq \frac{L_3}{e_T}$. We will take $n_T = Ent(\frac{L_3}{e_T})$; with $e_T = a$ (in-line square array).

NUMERICAL APPLICATION.–

$L_3 = 45$ cm $\qquad e_T = a = 1.63$ cm

Hence: $\frac{L_3}{e_T} = 27.61$. We then take $n_T = 27$ tubes per row.

2.7- Total number of tubes, N_T: $N_T = N_n n_T$.

NUMERICAL APPLICATION.–

$N_n = 4$ rows $\qquad n_T = 27$ tubes per row

Hence: $N_t = 108$ tubes.

Step 3. Hydraulic checks

3.1 *Tube side*:

31 See section 2.3.6.

3.1.1- Velocity of the liquid in the tubes: $v_l = \frac{4W_l}{\rho_l N_T \pi d_i^2}$;

3.1.2- v_l is indeed between 0.2 and 0.5 m/sec (see Table 6.6);

3.1.3- Reynolds number: $Re = \frac{d_i v_l \rho_l}{\mu_l}$;

3.1.4- Pressure drop ΔP_l: $\Delta P_l = \frac{\rho v^2}{2}\left(\frac{L_1}{d_i} f + 1.5\right)$ with $f = \frac{1}{[1.82 \log_{10}(Re) - 1.64]^2}$.

Table 6.6 gives the results of the calculations of the various hydraulic parameters. Since ΔP_l is less than ΔP_l^{Adm} (ΔP_l^{Adm} =*0.5 Atm*), we proceed to step 3.2.

v_l (m/sec)	*Re*	*f*	ΔP_l (Atm)
0.160	2151	5.11 10^{-2}	1.29 10^{-3}

Table 6.6. *Calculation of the tube-side pressure drop*

3.2- Fin side:

3.2.1- Pitch of the fins: a_a.

3.2.2- Number of fin-sheets, N_a: $N_a = Ent\left(\frac{L_1}{a_a + e_a} - 1\right)$.

3.2.3- Velocity between fins: $v_{ga} = \frac{W_g}{(N_a - 1)\rho_g L_3 a_a}$.

NUMERICAL APPLICATION.–

$a_a = 1.8$ mm $\quad L_1 = 0.75$ m $\quad e_a = 10^{-3}$ m $\quad W_g = 14.98$ kg/sec

$L_3 = 0.45$ m $\quad d_o = 10.3$ mm $\quad \rho_g = 0.86$ kg/m^3

Hence: $N_a = 338$ fins and $v_{ga} = 52$ m/sec.

3.2.4- Test on v_{ga}: v_{ga} *> 10 m/sec, we proceed to step 3.2.5.*

3.2.5- Transfer area: $A = N_T \pi d_o (L_1 - N_a e_a) + 2N_a\left(L_2 L_3 - N_T \frac{\pi d_o^2}{4}\right)$.

3.2.6- Free flow area: $A_0 = (e_T - d_o) L_1 N_T - N_a e_a L_3$.

3.2.7- Hydraulic diameter: $D_H = \frac{4 A_0 E_{Max}}{A}$.

NUMERICAL APPLICATION.–

$L_1 = 0.75$ m $L_2 = 0.08$ m $L_3 = 0.45$ m $N_T = 108$ tubes

$N_a = 338$ fins $d_o = 10.3$ mm $e_a = 10^{-3}$ m

Hence: $A = 33.24$ m²; $A_0 = 0.43$ m² and $D_H = 5.17$ mm.

3.2.8- Pressure drops (see section A3.6.2.3):

$$\Delta P_g = \frac{\rho_g v_\infty^2}{2}\frac{L_2}{D_H} f \text{ ; with } v_\infty = \frac{W_g}{\rho_g L_1 L_3} \text{ ; } f = 0.0267 Re_{d_o}^{k_1}\left(\frac{e_T}{e_L}\right)^{k_2}\left(\frac{a_a}{d_o}\right)^{k_3}$$

$$k_1 = -0.764 + 0.739\left(\frac{e_T}{e_L}\right) + 0.177\left(\frac{a_a}{d_o}\right) - \frac{0.00758}{N_n}; \; k_2 = -15.689 + \frac{64.021}{Ln(Re_{d_o})}$$

$$k_3 = 1.696 - \frac{15.695}{Ln(Re_{d_e})} \text{ and } Re_{d_o} = \frac{d_o v_\infty \rho_g}{\mu_g}$$

NUMERICAL APPLICATION.–

$\rho_g = 0.86$ kg/m³ $v_\infty = \bar{v} = 33.33$ m/sec $D_H = 5.17$ mm $W_g = 14.98$ kg/sec

$L_1 = 75$ cm $L_3 = 45$ cm $a_a = 1.8$ mm $e_L = e_T = 1.63$ cm

$d_o = 10.3$ mm $N_n = 4$ rows $\mu_g = 2.9\ 10^{-5}$ kgm^{-1}s^{-1}

Hence:

$Re_{do} = 1.018\ 10^4$; $k_1 = 0.253$; $k_2 = -8.75$; $k_3 = -4.74\ 10^{-3}$; $f = 0.276$ and $\Delta P_g = 2.42\ 10^4$ Pa.

I.e.: $\Delta P_g = 0.24$ Atm.

3.2.9- $\Delta P_g < \Delta P_g^{Adm}$ ($\Delta P_g^{Adm} = 0.5\ Atm$), we proceed to step 4.

Step 4. Thermal checks

4.1- Calculation of h_i.

Since the flow inside the tubes is turbulent (Re = 17,287, see step 3.1.3), the Nusselt number is given by (see Appendix 2, section A2.5.1):

$$Nu_i = 0.023\ Re^{0.8}\ Pr_i^{\frac{1}{3}}$$

NUMERICAL APPLICATION.–

$Re = 2{,}151$ $Cp_l = 4{,}190$ J/kg°C. $\mu_l = 4\ 10^{-4}$ kgm^{-1}s^{-1} λ_l=0.89 Wm^{-1}°C^{-1}

Hence: $Pr_l = 1.88$ and $Nu_i = 13.2$; i.e., $h_i = 2{,}139$ W/m²°C.

4.2- Calculating h_e.

The Nusselt number for 4 or more rows is given by (see section A2.5.2.3):

$$Nu_4 = 0.14Re_d^{0.672}\ (Pr)^{\frac{1}{3}}\left(\frac{e_L}{e_T}\right)^{0.502}\left(\frac{s}{d_o}\right)^{0.0312}$$

NUMERICAL APPLICATION.–

Re_{do} =1.018 10^4 Cp_g= 1,006 J/kg°C $\mu_g = 2.9\ 10^{-5}$ kgm^{-1}s^{-1}

$e_L = e_T$= 1.63 cm $\lambda_g = 0.04291$ Wm^{-1}°C^{-1} a_a = 1.8 mm

e_a = 1 mm d_o= 10.3 mm

Hence: $s = a_a - e_a = 0.8$ mm; $Pr_g = 0.68$ and $Nu_4 = 56.1$; i.e., h_e= 234 W/m²°C.

4.3- Calculating the effective exchange area: A_e:

$$A_e = N_T\ \pi d_o(L_1 - N_a e_a) + 2N_a\left(L_2 L_3 - N_T \frac{\pi d_o^2}{4}\right)\eta_e$$

with: $\eta_e = \frac{tanh(m_e L_3)}{m_e L_3}$ and $m_e = \sqrt{\frac{2h_e}{e_a \lambda_a}}$ (see Table 3.1).

NUMERICAL APPLICATION.–

N_T= 108 d_o= 10.3 mm L_1 = 0.75 m L_2 = 0.08 m L_3= 0.45 m

N_a = 338 e_a = 1 mm h_e= 234 W/m²°C λ_a = 210 W/m°C

Hence: m_e= 47.2; $\eta_e = 3.85\ 10^{-2}$ and $A_e = 3.34$ m².

4.4- Test on A_e: $A_e < A_{init}$; *we perform the calculations again with a new value of* U_F. *Table 6.7 presents the results of such calculations.*

U_F	A_{init}	Test	Action
50	13.15	$A_e < A_{init}$	Recalculate A_{init} with another value of U_F
100	6.57	$A_e < A_{init}$	Recalculate A_{init} with another value of U_F
175	3.76	$A_e < A_{init}$	Since U_F is at its maximum value, we set U_F at U_{FMax} = 175 W/m²°C and we repeat step 1 with L_1 = 0.95 m

Table 6.7. *Suitable value of U_F*

The calculations generated with $L_1 = 0.95$ m are presented in Table 6.8.

ϕ (kW)	31.84	N_n	4
W_g (kg/sec)	14.98	n_T	27
t_1 (°C)	27.1	N_T	108
$\Delta\theta_1$ (°C)	57.9	a_a (m)	$1.8\ 10^{-3}$
$\Delta\theta_2$ (°C)	41.0	N_a	338
$\Delta\theta_{ML}$ (°C)	49.0	A_0 (m²)	0.43
d_i (m)	$5.48\ 10^{-3}$	D_H (m)	$5.17\ 10^{-3}$
ΔP_L (Atm)	$1.29\ 10^{-3}$	d_o (m)	$1.03\ 10^{-2}$
h_i (W/m²°C)	2,139	ΔP_g (Atm)	0.24
m_e	47.2	h_e (W/m²°C)	234
UF (W/m²°C)	175	η_e	$3.85\ 10^{-2}$
A_{init} (m²)	3.74	A_e (m²)	3.34

Table 6.8. *Results with L_1 = 95 cm, L_2 = 8 cm and L_3 = 45 cm*

Table 6.8 shows that we still have $A_e < A_{init}$.

We thus repeat step 1 with $L_1 = 0.95$ m and $L_3 = 0.55$ m.

Table 6.9 presents the only results to have changed with respect to Table 6.8; U_F still being set at U_{FMax} = 175 W/m²°C.

n_T	33	N_T	132
A_0 (m²)	$5.67\ 10^{-1}$	D_H (m)	$7.00\ 10^{-3}$
ΔP_L (Atm)	$9.61\ 10^{-4}$	ΔP_g (Atm)	0.176
h_i (W/m²°C)	1,821	UF (W/m²°C)	175
A_{init} (m²)	3.72	A_e (m²)	3.76

Table 6.9. *Results with L_1 = 95 cm, L_2 = 8 cm and L_3 = 55 cm*

We now have $A_e > A_{init}$. We therefore retain $U_F = 175\ W/m^2°C$ and $A_{init} = 3.72\ m^2$ *and we proceed to step 4.5, with* $L_1 = 95\ cm$, $L_2 = 8\ cm$ *and* $L_3 = 55\ cm$.

4.5- Calculating the internal and average exchange areas of the tubes:

$$A_i = N_T \pi d_i L_1 \quad A_m = \frac{N_T \pi (d_i + d_e) L_1}{2}$$

NUMERICAL APPLICATION.–

$N_T = 132$ $d_i = 5.48$ mm $d_o = 10.3$ mm $L_1 = 0.95$ m

Hence: $A_i = 2.16$ m² and $A_m = 3.11$ m².

4.6- Calculating the overall heat transfer coefficient[32]: $\frac{1}{U_e^* A_e^*} = \frac{1}{h_i A_i^*} + \frac{d_o - d_i}{\lambda_t A_m^*} + \frac{1}{h_e A_e^*}$.

NUMERICAL APPLICATION.–

$h_i = 1{,}821$ W/m²°C $A_i = 2.16$ m² $A_e = 3.76$ m² $A_m = 3.11$ m²

$d_i = 5.48$ mm $d_o = 10.3$ mm $h_e = 234$ W/m²°C

Hence: $U_e^* = 188$ W/m²°C.

4.7- Calculating X and Z:

$$X = \frac{\theta_{c1} - \theta_{c2}}{\theta_{h1} - \theta_{c2}} \text{ and } Z = \frac{\theta_{h1} - \theta_{h2}}{\theta_{c1} - \theta_{c2}} = \frac{W_c C_c}{W_h C_h}$$

NUMERICAL APPLICATION.–

$W_l = 0.4$ kg/sec $Cp_l = 4{,}190$ J/kg°C $T_1 = 85$°C $T_2 = 66$°C

$W_g = 14.98$ kg/sec $Cp_g = 1{,}006$ J/kg°C $t_2 = 25$°C $t_1 = 27.1$°C

Hence: $X = 0.03522$ and $Z = 8.991$.

4.8- Calculation of the correction factor, Y[33]:

Z #1, therefore $Y = \frac{1}{(1-Z)\text{NTU}} Ln\left[\frac{1-XZ}{1-X}\right]$.

---

32 λ_t is the thermal conductivity of the tubes.

33 The NTU is given by Table 5.7 as a function of the type of cross-flow exchanger.

where: NTU is the solution to the equation: $X = 1 - \exp\left[\frac{\exp(-Z\,\mathrm{NTU}^{0.78})-1}{Z\,\mathrm{NTU}^{-0.22}}\right]$.

NUMERICAL APPLICATION.–

$X = 0.03522 \qquad Z = 8.991$

The solution to the equation in NTU is obtained by successive iterations on X:

NTU	X	NTU	X	NTU	X
1	0.10525	**0.08**	0.04402	**0.057**	0.03534
0.5	0.09060	**0.06**	0.03659	**0.565**	0.03513
0.1	0.05013	**0.055**	0.03448	**0.056722**	0.035222

Table 6.10. *Resolution of the NTU equation*

We therefore retain NTU = 0.056722. Hence: $Y = 0.76$.

4.9- We re-evaluate the "fouled" heat transfer coefficient: U_F^*:

$$\phi = U_F A_{Init}\,\Delta\theta_{ML} = Y\,U_F^*\,A_e^*\,\Delta\theta_{ML} \Rightarrow U_F^* = \frac{U_F\,A_{Init}}{Y A_e^*}$$

NUMERICAL APPLICATION.–

$U_F = 175$ W/m²°C $\qquad A_{Init} = 3.72$ m² $\qquad Y = 0.76$

Hence: $U_F^* = 227$ W/m²°C.

Step 5. Checking the margin allowed for fouling

NUMERICAL APPLICATION.–

$R_{fi} + R_{fe} = 3.5\ 10^{-4}$ m²°C/W and $\frac{1}{U_F^*} - \frac{1}{U_e^*} = -\,9.33\ 10^{-4}$ m²°C/W

$\frac{1}{U_F^*} - \frac{1}{U_e^*} < 0$

The device is then undersized.

Since L_1 and L_3 are already at their maximum, we repeat the calculation at step 1.1, choosing a larger value of L_2.

We take $L_2 = L_{2Max} = 0.10$ m.

Table 6.11 shows the only results that have changed, compared to Table 6.9.

N_n	6	N_T	198
A_0 (m²)	9.43 10^{-1}	D_H (m)	1.26 10^{-2}
ΔP_L (Atm)	4.67 10^{-4}	ΔP_g (Atm)	9.86 10^{-2}
h_i (W/m²°C)	1,317	UF (W/m²°C)	175
A_e (m²)	4.92	A_{init} (m²)	3.72

Table 6.11. *Results with L_1 = 95 cm, L_2 = 10 cm and L_3 = 55 cm*

$A_e > A_{init}$ we retain U_F = 175 W/m²°C and A_{init} = 3.72 m² and we proceed to step 4.5.

4.5- Calculating the internal and average exchange areas of the tubes:

$$A_i = N_T \pi d_i L_1 \quad A_m = \frac{N_T \pi (d_i + d_o) L_1}{2}$$

NUMERICAL APPLICATION.–

$N_T = 198$ $d_i = 5.48$ mm $d_o = 10.3$ mm $L_1 = 0.95$ m.

Hence: $A_i = 3.24$ m² and $A_m = 4.66$ m².

4.6- Calculating the overall heat transfer coefficient[34]:

$$\frac{1}{U_e^* A_e^*} = \frac{1}{h_i A_i^*} + \frac{d_o - d_i}{\lambda_t A_m^*} + \frac{1}{h_e A_e^*}$$

NUMERICAL APPLICATION.–

$h_i = 1{,}317$ W/m²°C $A_i = 3.24$ m² $A_e = 4.92$ m² $A_m = 4.66$ m²

$d_i = 5.48$ mm $d_o = 10.3$ mm $h_e = 234$ W/m²°C

Hence: $U_e^* = 181$ W/m²°C.

4.7- X and Z do not change: $X = 3.522\ 10^{-2}$ and $Z = 8.991$.

4.8- NTU and Y do not change either: NTU = 0.0567 and $Y = 0.761$.

34 λ_t is the thermal conductivity of the tubes.

4.9- We re-evaluate the "fouled" heat transfer coefficient: U_F^*:

$$\phi = U_F A_{Init}\, \Delta\theta_{ML} = Y\, U_F^*\, A_e^*\, \Delta\theta_{ML} \Rightarrow U_F^* = \frac{U_F A_{Init}}{Y A_e^*}$$

NUMERICAL APPLICATION.–

$U_F = 175$ W/m²°C $\quad A_{Init} = 3.72$ m² $\quad Y = 0.76$

Hence: $U_F^* = 174$ W/m²°C.

Step 5. Checking the margin allowed for fouling

NUMERICAL APPLICATION.–

$$R_{fi} + R_{fe} = 3.5\ 10^{-4}\ \text{m}^2\text{°C/W}\ \frac{1}{U_F^*} - \frac{1}{U_e^*} = 2.42\ 10^{-4}\ \text{m}^2\text{°C/W}$$

The device is therefore suitable. Table 6.12 below summarizes its characteristics.

<table>
<tr><td colspan="3">Characteristics of the retained heat exchanger</td><td colspan="2" rowspan="2">Type: cross-flow, embedded</td></tr>
<tr><td colspan="3">Effective exchange area = 4.92 m²</td></tr>
<tr><td colspan="2">$\eta_e = 3.85\ 10^{-2}$</td><td>$U_e^* = 181$ W/m²°C</td><td>$U_F^* = 174$ W/m²°C</td><td>$\phi = 31.84$ kW</td></tr>
<tr><td colspan="2">Dimensions</td><td>$L_1 = 95$ cm</td><td>$L_2 = 10$ cm</td><td>$L_3 = 55$ cm</td></tr>
<tr><td rowspan="3">Tube side</td><td>Flowrate = 0.4 kg/sec</td><td>$T_1 = 85$°C</td><td>$T_2 = 66$ °C</td><td>Rows: 6</td></tr>
<tr><td>Tubes/row = 33
Total tubes = 198</td><td>NPS = 1/8"
SCH = 80</td><td>$d_i = 5.48$ mm
$d_o = 10.3$mm</td><td>$\Delta P_t = 4.67\ 10^{-4}$ Atm</td></tr>
<tr><td colspan="2">Array: in-line square
Pitch = 1.63 cm</td><td>$e_L = 1.63$ cm
$e_T = 1.63$ cm</td><td>$Re = 1.173\ 10^3$</td></tr>
<tr><td rowspan="2">Fin side</td><td>Flowrate = 14.98 kg/sec</td><td>$t_1 = 27.1$°C</td><td>$t_2 = 25$ °C</td><td>$Y = 0.76$</td></tr>
<tr><td>Fin pitch:
$a_a = 1.8$ mm</td><td>$N_a = 338$</td><td>$D_H = 1.26$ cm</td><td>$Re = 1{,}018\ 10^4$
$\Delta P_c = 0.098$ Atm</td></tr>
</table>

Table 6.12. *Details of the retained heat exchanger*

6.7. Calculation algorithm for plate heat exchangers

The design of this type of exchanger depends closely on the characteristics of the considered plates and, consequently, on the suppliers. Each supplier has a series of plates with specific geometric characteristics (*height, width, spacing, thickness, connection diameter, δ*), which affect both the flow in the exchanger and the heat transfer efficiency.

In practice, plate heat exchanger manufacturers ask to be provided with the desired thermal schedule and offer to conduct the sizing themselves, and, in particular, the choice of which plate to use. Sizing is often conducted based on the experience gathered for each of the manufactured plates, in observance of a set of *rules of thumb* established on the basis of the operation of several installations. However, this type of procedure does not allow the engineer to put their finger on either the determinants of the choice, or the influence of the different parameters in the calculations. Nor does it allow a broad choice, unrestricted by the series of plates existing at a given manufacturer.

Sections 6.8 to 6.12 present a sizing algorithm that is not related to a specific supplier. It is based on a certain number of *rules of thumb* that stem from the recommendations of different manufacturers. It uses a set of "standard" plates that make it possible to perform the various calculations without having to refer to a given manufacturer. The characteristics of these plates are given in Tables A1.24 to A1.27.

RULES OF THUMB.–

R_1 – Both heat transfer and fouling resistance are directly affected by the level of turbulence reached within the plates. The number of plates per pass, as well as the dimensions of the plates, shall consequently be chosen so as to have a Reynolds number located in turbulent regime. *An increased number of plates does not automatically secure a good design!*

R_2 – Similarly, provision for possible *fouling* shall not be made by introducing an additional transfer area. It shall be sought through *an increase in turbulence*, which is reflected in an increase in the shear stress on the walls, and therefore in the friction exerted by the fluid on the surface of the plates. This increase in friction enables any particles to remain suspended in the bulk fluid.

R_3 – We will seek plates that generate significant turbulence in the flow. However, while an increase in turbulence minimizes fouling whilst ensuring good thermal efficiency, it leads to an increase in pressure drops in the plates and in the inlet and outlet connections.

R_4 – The pressure drop in the connections is in the order of 30 to 35% of the total pressure drop across the plates.

R_5 – The plate heat exchanger calculation is based on a compromise between turbulence and pressure drops. We will choose the type of plate that gives the best turbulence, while maintaining pressure drops below the permissible levels.

R_6 – The allowable pressure drops for each circuit are in the order of 30 kPa per NTU (Kuppan 2000).

R_7 – In the scenario where we decide to take an excess surface, this shall not exceed 10% of the surface of the clean heat exchanger. It is also recommended for a filter to be used upstream of the exchanger, with openings of about 75% of the spacing, *e*, between the two sheets forming a plate[35].

R_8 – The thickness, Δx, of the metal sheets forming a plate (see Figure 3.17) is generally around 0.4 mm. The standard plate thicknesses used in the sizing algorithm are given in Tables A1.24-A1.27.

Taking these rules of thumb into account, plate heat exchangers sizing can be organized into the following five steps.

Step 1. Define the circulation mode and calculate the flux

Based on the problem data (fluid flow rate, hot fluid inlet temperature, cold fluid inlet and outlet temperature, allowable pressure drops, allowable fouling, etc.) and the physical properties.

1.1- Define the *circulation mode*: counter-flow is favored.

1.2- *Calculate the flux, ϕ, to be exchanged* and the missing temperature.

Step 2. Knowing ϕ

2.1- Choose a value of the "fouled" heat transfer coefficient (U_F) according to the experimental orders of magnitude presented in Tables A1.21 to A1.23.

2.2- Calculate the approximate exchange area, A_0, from: $A_0 = \frac{\phi}{U_F \, \Delta\theta_{ML}}$.

Step 3. Determination of the types of plates that can be used

3.1- Calculate the numbers of transfer units[36]:

$$\mathrm{NTU}_h = \frac{(\theta_{in} - \theta_{out})_h}{\Delta\theta_{ML}} \text{ and } \mathrm{NTU}_c = \frac{(\theta_{out} - \theta_{in})_c}{\Delta\theta_{ML}}$$

35 See Figure 3.17.

36 See section 3.5.3.

3.2- Calculate the average number of transfer units: $\mathrm{NTU}_{\mathrm{Average}} = \frac{\mathrm{NTU}_h + \mathrm{NTU}_c}{2}$.

3.3- Calculate the number of passes[37]: $N_p = \frac{\mathrm{NTU}_{\mathrm{Average}}}{\mathrm{NTU}_{\mathrm{Max}}}$.

3.4- Eliminate the plates that lead to $N_p > 1$.

Step 4. Hydraulic checks

At this step, one or more plates may be suitable. We will choose between them based on the level of turbulence reached in each plate.

4.1- Knowing the geometric parameters of the retained plates *(height, a, width, b, spacing, e, thickness, Δx, connection diameter, δ)*, for each we calculate:

- the gaps $\mathrm{e}_L = \mathrm{a} - \delta$ and $\mathrm{e}_T = \mathrm{b} - \delta$;
- the flow length[38]: $\mathrm{L} = \sqrt{\mathrm{e}_L^2 + \mathrm{e}_T^2}$;
- the hydraulic diameter[39]: $\mathrm{D_H} = \frac{2\mathrm{e}\sqrt{\mathrm{e}_L^2 + \mathrm{e}_T^2}}{a}$.

4.2- Calculate:

4.2.1- The transfer area of each plate: $A_p = 2(a - \delta)b$.

4.2.2- The number of plates[40] (using A_0): $N = Int(\frac{A_0}{N_p A_p} + 0.5)$.

4.2.3 The Reynolds number for each circuit[41]: $Re_{E_i} = \frac{2}{\mu_i} \frac{W_i}{N_i' b_i} \frac{\sqrt{e_{L_i}^2 + e_{T_i}^2}}{a_i - \delta_i}$; i = 1; 2

4.3- Keep only the plates that give the largest Reynolds numbers. For these plates:

4.3.1- Calculate the pressure drop in the plates, for each circuit[42]:

---

37 $\mathrm{NTU}_{\mathrm{Max}}$ is given in Table A1.24 for each type of plate.

38 See section 3.5.3.5.

39 See section 3.5.2.5: e is the plate's internal spacing, manufacturer datum, e = 1.5 - 5 mm (Kuppan 2000). We often retain: $D_H \cong 2e$.

40 Note that the number of plates after assembly will have to be $N' = N + 2$ because it will be necessary to add the terminal plates not participating in the transfer.

41 Or $Re_{E_i} = \frac{2}{\mu_i} \frac{W_i}{N_i b_i}$, if we retain the approximation $D_H \cong 2e$.

42 See section A3.7.3: $\mathrm{G}_i = \frac{\mathrm{W}_i}{b_i e_i}$.

$$\Delta P_{p_i} = 2N_{p_i}\, C_{f_i} \frac{G_i^2}{\rho_i} \frac{L}{D_H};\ \text{with}\ C_{f_i} = \frac{2.5}{4Re_{E_i}^{0.3}}$$

4.3.2- Calculate the pressure drop in the connections for each circuit:

$$\Delta P_{pc_i} = 0.3 \Delta P_{p_i}$$

4.3.3- Calculate the total pressure drop for each circuit:

$$\Delta P_{pT_i} = \Delta P_{p_i} + \Delta P_{pc_i}$$

4.3.4- Keep only the plates that give pressure drops below the allowable pressure drops.

Step 5. Thermal check

Would the retained plates ensure transfer of the flux, ϕ?

5.1.- Calculate the Nusselt numbers, Nu_i, for each of the circuits ($i = 1; 2$)[43]:

$$Nu_i = 0.374\, Re_{E_i}^{0.668}\, Pr_i^{\frac{1}{3}} \left(\frac{\mu}{\mu_p}\right)_i^{0.15}$$

We deduce therefrom the heat transfer coefficients, h_i ($i = 1; 2$): $h_i = \frac{\lambda_i Nu_i}{D_{H_i}}$.

5.2- Calculate the overall heat transfer coefficient for the clean heat exchanger, for each of the retained plates: $\frac{1}{U^*} = \frac{1}{h_1} + \frac{\Delta x}{\lambda_p} + \frac{1}{h_2}$; λ_p being the heat conductivity of the plate.

5.3- Calculate the thermal capacity ratio: $Z = \frac{\theta_{h1} - \theta_{h2}}{\theta_{c1} - \theta_{c2}} = \frac{W_c C_c}{W_h C_h}$.

5.4- Calculate the correction factor, Y[44]:

- for $Z \neq 1$: $Y = \frac{1}{\text{NTU}\,(1-Z)} Ln\left[\frac{1-XZ}{1-X}\right]$;

43 See section A2.6.3: $Pr_i = \frac{Cp_i\, \mu_i}{\lambda_i}$.

44 See section 5.7.4.5: $X = \frac{\theta_{c1} - \theta_{c2}}{\theta_{h1} - \theta_{h2}}$. Here we use the $NTU_{Average}$ value, defined by: $NTU_{Average} = \frac{NTU_c + NTU_h}{2}$; with: $NTU_c = \frac{(\theta_{out} - \theta_{in})_c}{\Delta\theta_{ML}}$ and $\text{NTU}_h = \frac{(\theta_{in} - \theta_{out})_h}{\Delta\theta_{ML}}$.

- for Z=1: $Y = \frac{X}{(1-X)\text{NTU}}$.

5.5- Calculate, for each of the retained heat exchangers, the "fouled" heat transfer coefficient, U_F^*:

$$\phi = U_F A_0 \, \Delta\theta_{ML} = Y \, U_F^* \, A^* \Delta\theta_{ML} \Rightarrow U_F^* = \frac{U_F \, A_0}{Y A^*}; \; A^* = N A_p$$

5.6- Calculate the sum of the allowable fouling resistances: $R_{d_1} + R_{d_2}$.

5.7- Tests on $\frac{1}{U_F^*} - \frac{1}{U_e}$:

- If $\frac{1}{U_F^*} - \frac{1}{U^*} \cong R_{f_1} + R_{f_2}$: $\frac{1}{U_F^*} - \frac{1}{U^*}$ is of the same order of magnitude as the sum of the fouling resistances, therefore, this exchanger will assure the thermal schedule, even in fouled state. Therefore, *the device is correctly sized.* It is retained among the satisfactory exchangers;

- If $\frac{1}{U_F^*} - \frac{1}{U_e} < 0 \; or \; \frac{1}{U_F^*} - \frac{1}{U_e} \ll R_{fi} + R_{fe}$: the exchanger will not be able to assure the thermal schedule under the expected fouling conditions. Then *the device is undersized.* We eliminate it from the list of retained heat exchangers;

- If $\frac{1}{U_F^*} - \frac{1}{U^*} \gg R_{fi} + R_{fe}$: $\frac{1}{U_F^*} - \frac{1}{U^*}$ is *too large* compared to the sum of the fouling resistances. This means that *the device is oversized.* We eliminate it from the list.

6.8. Illustration: implementation of the plate heat exchanger algorithm

Energy recovery from the cooling of phosphoric acid.

The phosphoric acid manufacturing process implemented delivers a mass flow rate, F_A at temperature T_1.

Prior to its storage, the acid needs to be cooled to temperature, T_2. The energy thus degraded is used to heat the water intended for another point in the plant. Available at the average temperature, t_2, the water needs to be brought to temperature t_1.

You are an engineer recently recruited to the factory's design office. Based on their experience, your department manager recommends that you opt for a plate heat exchanger. What design would you suggest to them?

Available data:

Water: F_W = 150.12 tons/hr t_1 = 45°C t_2 = 25°C

λ_W = 0.62 W/m°C ρ_W = 993 kg.m^{-3} μ_W = 0.75 Cp

C_{pW} = 4,160 J/kg°C

Fouling resistance: R_W= 3 10^{-5} m^2°C/W

Phosphoric acid:

F_A = 180 tons/hr T_1 = 84°C

λ_A = 0.1298 W/m°C ρ_A = 1,780 kg.m^{-3} μ_A = 5Cp C_{pA} = 1,530 J/kg°C

Fouling resistance: R_A = 3 10^{-5} m^2°C/W

Heat conductivity of the plates: λ_p = 0.36 W/m°C

Allowable pressure drops for circuit *i*: $\Delta P_i^{Adm} = 20\ \text{kPa/NTU}$

Solution

Step 1. Define the circulation mode and calculate the flux

1.1- Circulation mode: counter-flow.

1.2- Flux, ϕ, to be exchanged and missing temperature:

$$\phi = F_W C_{pW}(t_1 - t_2) \text{ and } T_2 = T_1 - \frac{F_W C_{pW}}{F_A C_{pA}}(t_1 - t_2)$$

NUMERICAL APPLICATION.–

F_W = 41.70 kg/sec C_{pW} = 4,160 J/kg°C t_1 = 45°C t_2 = 25°C

F_A = 50 kg/sec C_{pA} = 1,530 J/kg°C T_1 = 84°C

ϕ = 3.47 MW T_2 = 38.65°C

Step 2. Knowing ϕ

2.1- Choosing U_F: Table A1.21 (Low-visc liq/ Low-visc liq): U_F between 200 and 550 kcal/hm^2°C ; thus U_F between 230 and 640 W/m^2°C.

We take: U_F = 450 W/m²°C.

2.2- Calculating A_0:

$$A_0 = \frac{\phi}{U_F\,\Delta\theta_{ML}} \text{ with } \Delta\theta_{ML} = \frac{\Delta\theta_1 - \Delta\theta_2}{Ln\left(\frac{\Delta\theta_1}{\Delta\theta_2}\right)}\,;\ \Delta\theta_1 = T_1 - t_1\,;\ \Delta\theta_2 = T_2 - t_2$$

NUMERICAL APPLICATION.–

T_1 = 84°C t_1 = 45°C T_2 = 38.65°C t_2 = 25°C F_A = 50 kg/sec

C_{pW} = 1,530 J/kg°C ϕ = 3.47 MW $\Delta\theta_1 = 39$°C $\Delta\theta_2 = 13.65$°C

$\Delta\theta_{ML} = 24.15$ °C and A_0 = 319.31 m²

Step 3. Types of plates that can be used

3.1- Calculating the numbers of transfer units:

$$NTU_h = \frac{T_1 - T_2}{\Delta\theta_{ML}} \text{ and } NTU_c = \frac{t_2 - t_1}{\Delta\theta_{ML}}$$

NUMERICAL APPLICATION.–

T_1 = 84°C t_1 = 45°C T_2 = 38.65°C t_2 = 25°C

NTU_h = 1.878 NTUc = 0.828

3.2- Calculating the average number of transfer units:

$$NTU_{Average} = \frac{NTU_h + NTU_c}{2}$$

NUMERICAL APPLICATION.–

$NTU_{Average}$ = 0.1353

3.3- Calculating the number of passes for each type of plate:

$$N_p = \frac{NTU_{Average}}{NTU_{Max}}$$

We have a Liquid/Liquid exchange. We take LL-type plates (see Table A1.24).

Plate type	LL1	LL2	LL3	LL4	LL5	LL6	LL7	LL8	LL9
NTU_{max}	6	6	6	5	5	5	4.9	4.5	4.2
N_p	0.23	0.23	0.23	0.27	0.27	0.27	0.28	0.30	0.32
Plate type	**LL10**	**LL11**	**LL12**	**LL13**	**LL14**	**LL6**	**LL15**	**LL16**	**LL17**
NTUmax	4	4	4	4	4	4	3.9	3.9	3.85
Np	0.34	0.34	0.34	0.34	0.34	0.35	0.35	0.35	0.34
Plate type	**LL18**	**LL19**	**LL20**	**LL21**	**LL22**	**LL23**	**LL24**	**LL25**	**LL26**
NTUmax	3.8	3.8	3.8	3.8	3.8	2.9	0.9	0.85	0.85
Np	0.36	0.36	0.36	0.36	0.36	0.47	*1.50*	*1.59*	*1.59*
Plate type	**LL27**	**LL28**	**LL29**	**LL30**	**LL31**	**L32**	**LL33**	**LL34**	**LL35**
NTUmax	0.85	0.8	0.8	0.8	0.8	0.8	0.8	0.8	0.8
Np	1.59	1.69	1.69	1.69	1.69	1.69	1.69	1.69	1.69

Table 6.13. *Calculation of the number of passes for each type of plate*

3.4- We only keep the plate types that give a number of passes below 1; i.e. plates LL1 to LL23.

Step 4. Hydraulic checks

4.1- Calculating the longitudinal and transverse gaps, flow length and hydraulic diameter.

Table 6.14 presents the gaps and the hydraulic diameters of the retained plates. These plates are all constructed from stainless-steel sheets of thickness $\Delta x = 0.4$ mm.

Plate type	a (mm)	b (mm)	e (mm)	δ (mm)	e_L (mm)	e_T (mm)	L (mm)	D_H (mm)
LL1	4,095	1,550	4.00	500	3,595	1,050	3,745	7.3
LL2	3,951	1,550	3.95	500	3,451	1,050	3,607	7.2
LL3	3,951	1,550	3.95	500	3,451	1,050	3,607	7.2
LL4	3,103	920	2.52	250	2,853	670	2,931	4.8
LL5	3,103	940	3.95	250	2,853	690	2,935	7.5
LL6	3,103	940	3.95	250	2,853	690	2,935	7.5
LL7	2,183	780	2.90	200	1,983	580	2,066	5.5
LL8	2,736	1,174	2.00	300	2,436	874	2,588	3.8
LL9	2,711	942	2.90	250	2,461	692	2,556	5.5

LL10	2,595	920	2.00	200	2,395	720	2,501	3.9
LL11	2,596	920	2.00	200	2,396	720	2,502	3.9
LL12	2,661	913	2.00	200	2,461	713	2,562	3.9
LL13	2,036	650	4.00	200	1,836	450	1,890	7.4
LL14	4,146	755	4.00	200	3,946	555	3,985	7.7
LL15	1,871	650	3.80	150	1,721	500	1,792	7.3
LL16	2,141	780	1.90	150	1,991	630	2,088	3.7
LL17	1,833	610	2.42	150	1,683	460	1,745	4.6
LL18	3,210	1,154	2.50	350	2,860	804	2,971	4.6
LL19	3,210	1,154	2.50	350	2,860	804	2,971	4.6
LL20	3,218	1,174	2.50	350	2,868	824	2,984	4.6
LL21	3,218	1,174	2.50	350	2,868	824	2,984	4.6
LL22	1,815	610	2.50	150	1,665	460	1,727	4.8
LL23	1,405	800	4.00	150	1,255	650	1,413	8.0

Table 6.14. *Gaps, flow lengths and hydraulic diameters*

4.2- Calculating the number of plates and the Reynolds number for the retained plate types.

For each of the retained plates, Table 6.15 gives the results of the calculations of the transfer area and the Reynolds number.

Plate type	A_p **(m²)**	**N**	*N'*	**Re Water circuit**	**Re Acid circuit**
LL1	11.14	29	31	2,411	434
LL2	10.70	30	32	2,343	421
LL3	10.70	30	32	2,343	421
LL4	5.25	61	63	1,971	354
LL5	5.36	60	62	1,963	353
LL6	5.36	60	62	1,963	353
LL7	3.09	104	106	1,401	252
LL8	5.72	56	58	1,735	312
LL9	4.64	69	71	1,727	311
LL10	4.41	73	75	1,683	303
LL11	4.41	73	75	1,683	303

LL12	4.49	72	74	1,714	308
LL13	2.39	134	136	1,295	233
LL14	5.96	54	56	2,656	478
LL15	2.24	143	145	1,229	221
LL16	3.11	103	105	1,424	256
LL17	2.05	156	158	1,196	215
LL18	6.60	49	51	1,963	353
LL19	6.60	49	51	1,963	353
LL20	6.73	48	50	1,971	354
LL21	6.73	48	50	1,971	354
LL22	2.03	158	160	1,182	213
LL23	2.01	160	162	966	174

Table 6.15. *Calculation of the transfer area and the Reynolds number for each circuit*

4.3- We keep the plates that give the largest Reynolds numbers.

These are the plates: LL1, LL2 and LL14.

For these plates, we calculate the following parameters:

$$C_{f_i} = \frac{2.5}{4Re_{E_i}^{0.3}};\ G_i = \frac{W_i}{A_{L_i}};\ \Delta P_1 = 2N_{p_1}\, C_{f_1}\frac{G_1^2}{\rho_1}\frac{L}{D_H};\ \Delta P_2 = 2N_{p_2}\, C_{f_2}\frac{G_2^2}{\rho_2}\frac{L}{D_H}$$

$$\Delta P_{ci} = 0.3\Delta P_i$$

$$\Delta P_{T_i} = \Delta P_i + \Delta P_{ci}$$

Tables 6.16 and 6.17 give the results of the pressure drop calculations.

Plate type	C_{f1}	C_{f2}	A_{L_i} (m²)	G_1	G_2
LL1	6.04E-02	1.01E-01	6.20E-03	2.17E+02	2.60E+02
LL2	6.09E-02	1.02E-01	6.12E-03	2.13E+02	2.55E+02
LL14	5.87E-02	9.82E-02	3.02E-03	2.47E+02	2.96E+02

Table 6.16. *Parameters for calculating pressure drops*

4.4- We calculate the total pressure drop for each circuit:

Plate type	ΔP_1	ΔP_2	ΔP_{c1}	ΔP_{c2}	ΔP_{T1}	ΔP_{T2}
LL1	0.0289	0.0388	0.0087	0.0117	0.0376	0.0505
LL2	0.0274	0.0368	0.0082	0.0110	0.0357	0.0479
LL14	0.0368	0.0493	0.0110	0.0148	0.0478	0.0641

Table 6.17. *Calculation of pressure drops*

Allowable pressure drops for the Water circuit: $\Delta P_1^{Adm} = 30\ \text{kPa/NTU}$.

Allowable pressure drops for the Acid circuit: $\Delta P_2^{Adm} = 30\ \text{kPa/NTU}$.

NUMERICAL APPLICATION.–

$NTU_1 = 0.828 \qquad \Delta P_1^{Adm} = 0.25\ \text{Atm}$

$NTU_2 = 1.878 \qquad \Delta P_1^{Adm} = 0.56\ \text{Atm}$

$\Delta P_{T_1} < \Delta P_1^{Adm}$ and $\Delta P_{T_2} < \Delta P_2^{Adm}$; all the plates kept meet pressure-drop conditions.

Step 5. Thermal check

Would the retained plates ensure transfer of the flux, ϕ?

5.1 and 5.2- Calculating the Nusselt numbers, Nu_i, the heat transfer coefficients, h_i, and the heat transfer coefficients of the clean heat exchanger, U^*, for each type of plate.

Without taking account of temperature correction, the Nusselt numbers are given by:

$$Nu_i = 0.374\, Re_{E_i}^{0.668}\, Pr_i^{\frac{1}{3}};\ Pr_i = \frac{Cp_i\,\mu_i}{\lambda_i};\ h_i = \frac{\lambda_i Nu_i}{DH_i};\ \frac{1}{U^*} = \frac{1}{h_1} + \frac{\Delta x}{\lambda_p} + \frac{1}{h_2}$$

NUMERICAL APPLICATION.–

$\lambda_P = 0.36\ \text{W/m°C} \qquad \Delta x = 0.4\ \text{mm}$

$Pr_1 = 5.03 \qquad Pr_2 = 22.50$

Table 6.18 gives the results of the calculations of the Nusselt numbers, Nu_i, the heat transfer coefficients, h_i, and the heat transfer coefficients of the clean heat exchanger, U^*, for each type of exchanger.

Plate type	Nu_1	Nu_2	h_1 W/m²°C	h_2 W/m²°C	U^* W/m²°C
LL1	116.44	60.98	9,867	2,834	643
LL2	114.25	59.83	9,821	2,821	642
LL14	124.21	65.05	10,016	2,877	646

Table 6.18. *Calculating U* for each of the retained plates*

5.3- to 5.5- Calculating the thermal capacity ratios, Z, correction factors, Y, and "fouled" heat transfer coefficients, U_F^*:

$$Z = \frac{T_1 - T_2}{t_1 - t_2}; \; X = \frac{t_1 - t_2}{T_1 - t_2}$$

- for $Z \neq 1$: $Y = \frac{1}{\text{NTU}\,(1-Z)} Ln\left[\frac{1-XZ}{1-X}\right]$;

- for $Z = 1$: $Y = \frac{X}{(1-X)\text{NTU}}$.

NUMERICAL APPLICATION.–

$T_1 = 84°C$ $\quad t_1 = 45°C$ $\quad T_2 = 38.65°C$ $\quad t_2 = 25°C$

$NTU_1 = 0.828$ $\quad NTU_2 = 1.878$

$\text{NTU}_{\text{Average}} = 1.353$

$X = 0.339$ $\quad Z = 2.268$

$$Z \neq 1: Y = \frac{1}{NTU\,(1-Z)} Ln\left[\frac{1-XZ}{1-X}\right] \qquad Y = 0.612$$

$$U_F^* = \frac{U_F A_0}{YA^*}$$

where: $A^* = NA_p$.

Table 6.19 gives the results of the calculations of A^* and U_F^*, for the retained heat exchangers.

Plate type	A^* m²	U_F^* W/m²°C
LL1	323.19	726.39
LL2	320.94	731.47
LL14	321.76	729.62

Table 6.19. *Calculating A^* and U_F^* for each of the retained plates*

5.6- Calculating the sum $R_A + R_E$ and tests on:

$R_W + R_A$ and $\frac{1}{U_F^*} - \frac{1}{U^*}$

$R_W + R_A = 6\,10^{-5}$

Table 6.20 gives: $\frac{1}{U_F^*} - \frac{1}{U^*}$, for the retained heat exchangers.

For all the retained plates, $\frac{1}{U_F^*} - \frac{1}{U^*}$ is *of the same order of magnitude* as the sum of the fouling resistances. The exchangers to be constructed from plates LL1, LL2 and LL14 will therefore assure the thermal schedule, even in the fouled state. Table 6.21 summarizes the characteristics of these exchangers.

Plate type	U^* W/m²°C	A^* m²	U_F^* W/m²°C	$\frac{1}{U_F^*} - \frac{1}{U^*}$	$\frac{\frac{1}{U_F^*} - \frac{1}{U^*}}{R_A + R_E}$
LL1	643.24	323.19	726.39	1.78E-04	2.97
LL2	642.37	320.94	731.47	1.90E-04	3.16
LL14	646.05	321.76	729.62	1.77E-04	2.95

Table 6.20. *Calculating A^* and U_F^* for each of the retained plates*

Exchanger	Plate type	A_p (m²)	N	N'
1	LL1	11.14	29	31
2	LL2	10.70	30	32
3	LL14	5.96	54	56

Table 6.21. *Characteristics of the retained plate heat exchangers*

The final choice of the exchanger to be used will be made based on cost considerations.

6.9. Calculation algorithm for spiral heat exchangers

As with plate heat exchangers, spiral heat exchanger design depends closely on manufacturers who willingly offer to apply the design that will enable the desired thermal schedule to be carried out.

The algorithm presented in this section is not linked to a given manufacturer. It enables the transfer area, the length, L, of the helix and its width, b, and the number of revolutions of the spiral to be determined.

It is based on a certain number of *rules of thumb* that stem from the recommendations of different manufacturers. It uses a set of "standard" spirals that make it possible to perform the various calculations without having to refer to a given manufacturer. The characteristics of these spirals are presented in Table A1.28.

RULES OF THUMB.–

R_1 – The gap, e, and the width of the spiral plates, b, are the most important parameters in the design of spiral heat exchangers. Indeed, they influence turbulence to a large extent, since they define the cross-section area perpendicular to the flow.

R_2 – In order to obtain good heat transfer efficiency, we will seek the spiral types that generate significant turbulence in the flow. However, particular importance will be given to the pressure drops in the spiral, keeping in mind that the allowable pressure drops are of the order of 20 kPa/NTU.

R_3 – As with other types of heat exchanger, the calculation principle is based on a compromise between heat transfer efficiency and pressure drops.

The following sizing algorithm is composed of four steps.

Step 1. Define the circulation mode and calculate the flux and the NTUs

Based on the problem data (fluid flow rate, hot fluid inlet temperature, cold fluid inlet and outlet temperature, allowable pressure drops, allowable fouling, etc.) and the physical properties:

1.1- Define the *circulation mode*: counter-flow is favored;

1.2- *Calculate the flux, ϕ, to be exchanged* and the missing temperature;

1.3- Calculate the numbers of transfer units:

$$NTU_h = \frac{T_1 - T_2}{\Delta\theta_{ML}};\ NTU_c = \frac{t_1 - t_2}{\Delta\theta_{ML}} \text{ and } NTU_{Average} = \frac{NTU_h + NTU_c}{2}$$

Step 2. Knowing ϕ

2.1- Choose a value of the "fouled" heat transfer coefficient (U_F) according to the experimental orders of magnitude presented in Table A1.25.

2.2- Calculate the approximate exchange area, A_0, from the expression: $A_0 = \frac{\phi}{U_F\,\Delta\theta_{ML}}$.

$$\Delta\theta_{ML} = \frac{\Delta\theta_1 - \Delta\theta_2}{Ln\left(\frac{\Delta\theta_1}{\Delta\theta_2}\right)};\ \Delta\theta_1 = T_1 - t_1\ ;\ \Delta\theta_2 = T_2 - t_2$$

Step 3. Hydraulic checks

For standard spirals (see Table A1.28), and for each circuit, calculate:

– the approximate length of the spiral: $L_0 = \frac{A_0}{b}$;

– the free flow area : $A_{0i} = e_i b$ (i=1; 2);

– the hydraulic diameter: $D_{H_i} = 2\ e_i$;

– the Reynolds number: $Re_i = \frac{D_{H_i} W_i}{\mu_i A_{0_i}}$;

– the expected pressure drops (see section A3.9)[45]: $\Delta P_i = \frac{2\,C_f\,L\,W_i^2}{\rho\,D_H\,A_0^2}$.

Step 4. Thermal checks

4.1- Calculate the Prandtl number for each circuit.

4.2- For standard spirals satisfying the constraint on allowable pressure drops, calculate:

- the Nusselt number of each circuit[46]: $Nu_i = 0.04 Re_i^{0.74} Pr^{0.4}$;

- the convection heat transfer coefficient on each side: $h_i = \frac{\lambda_i Nu_i}{D_{H_i}}$;

45 C_f is given as a function of the Reynold's number in Table A3.3.

46 See section A2.8.

- the overall coefficient, U: $\frac{1}{U^*} = \frac{1}{h_1} + \frac{\Delta x}{\lambda_p} + \frac{1}{h_2}$.

4.3- Calculate:

- the number of revolutions of the spiral[47]:

$$n = Ent\left[\frac{\sqrt{(r_s+e+2\Delta x)^2+(e+\Delta x)\frac{A_0}{\pi b}}-(r_s+e+2\Delta x)}{e+\Delta x}+1\right]$$

- the correction factor, Y[48]: $Y = \frac{\mathrm{n}}{\mathrm{NTU}}\tanh\frac{\mathrm{NTU}}{\mathrm{n}}$;

- the exact length and the exact transfer area for each of the retained heat exchangers:

$$L^* = \pi[(e + \Delta x)n^2 + 2(r_s + e + 2\Delta x)n] \text{ and } A^* = L^* b$$

4.4- For each of the retained heat exchangers, calculate the "fouled" heat transfer coefficient, U_F^*: $\phi = U_F A_0\, \Delta\theta_{ML} = Y\, U_F^*\, A^* \Delta\theta_{ML} \Rightarrow U_F^* = \frac{U_F A_0}{Y A^*}$.

4.5- Calculate the sum of the allowable fouling resistances: $R_{f_1} + R_{f_2}$.

4.6- Tests on $\frac{1}{U_F^*} - \frac{1}{U_e}$:

- If $\frac{1}{U_F^*} - \frac{1}{U^*} \cong R_{f_1} + R_{f_2}$: *the device is correctly sized.* It is retained among the satisfactory exchangers.

- If $\frac{1}{U_F^*} - \frac{1}{U^*} < 0$ or $\frac{1}{U_F^*} - \frac{1}{U^*} \ll R_{fi} + R_{fe}$: the device is undersized. We eliminate it from the list.

- If $\frac{1}{U_F^*} - \frac{1}{U^*} \gg R_{fi} + R_{fe}$: the device is oversized. We eliminate it from the list.

6.10. Illustration: implementation of the spiral heat exchanger algorithm

A production unit generates a high flowrate of a liquid waste in the form of an aqueous solution at temperature T_1. Prior to its regeneration, this waste needs to be cooled to temperature T_2. The cooling energy is recovered by heating a liquid

47 n_0 and L_0 are related by the second-degree equation in n given in Table 3.2, section 3.6.4.

48 See section 5.7.4.6.

reagent from the ambient temperature, t_2, to temperature t_1, required at the reactor inlet.

Both the aqueous solution to be regenerated and the liquid reagent to be heated are relatively charged with suspended matter. We thus decide to use a spiral heat exchanger to carry out the desired thermal schedule.

Your department manager asks you which heat exchanger to use.

Available data:

Aqueous solution:

F_g = 12.2 tonnes/hr T_1 = 92°C ρ_A = 1,250 kg.m^{-3} C_{pA} = 5,125 J/kg°C

λ_A = 0.87 W/m°C μ_A = 0.96 Cp Fouling resistance R_A = 5 10^{-5} m^2°C/W

Liquid reagent:

F_R = 4.33 tonnes/hr t_1 = 87°C t_2 = 21°C ρ_R = 850 kg.m^{-3}

λ_R = 0.65 W/m°C μ_R =1.65 Cp C_{pR} = 3,160 J/kg°C

Fouling resistance R_R= 5 10^{-5} m^2°C/W

Spiral plates:

Thermal conductivity : λ_p = 25.97 W/m°C.

Thickness, Δx, depending on the spiral type: see Table A1.28.

Allowable pressure drops for circuit i: $\Delta P_i^{Adm} = 20 \frac{\text{kPa}}{\text{NTU}}$.

Solution

Step 1. Define the circulation mode and calculate the flux and the NTUs

1.1- *Circulation mode*: counter-flow.

1.2- *Flux, ϕ, to be exchanged* and missing temperature:

$$\phi = F_R C_{p_R}(t_1 - t_2) \text{ and } T_2 = T_1 - \frac{F_R C_{pR}}{F_A C_{pA}}(t_1 - t_2)$$

NUMERICAL APPLICATION.–

F_R =4.33 kg/sec C_{pR} = 3,160 J/kg°C t_1 = 87°C t_2 = 21°C

F_A = 3.39 kg/sec C_{pA}= 5,125 J/kg°C T_1 = 92°C

ϕ = 903.76 kW T_2 =39.96°C

1.3- Calculating the numbers of transfer units:

$$NTU_h = \frac{T_1 - T_2}{\Delta\theta_{ML}};\ NTU_c = \frac{t_1 - t_2}{\Delta\theta_{ML}} \text{ and } NTU_{Average} = \frac{NTU_c + NTU_f}{2}$$

NUMERICAL APPLICATION.–

T_1 = 92°C t_1 = 87°C T_2 = 39.96°C t_2 = 21°C

NTU_c = 4.968 NTU_f= 6.301 $NTU_{Average}$ = 5.634

Step 2. Knowing ϕ

2.1- Choosing U_F: Table A1.21 (Low-visc liq/Low-visq liq): U_F between 700 and 1,800 kcal/hm²°C; thus U_F between 800 and 2,100 W/m²°C.

We take: U_F = 1,450 W/m²°C.

2.2- Calculating A_0:

$$A_0 = \frac{\phi}{U_S\,\Delta\theta_{ML}}$$

with $\Delta\theta_{ML} = \frac{\Delta\theta_1 - \Delta\theta_2}{Ln\left(\frac{\Delta\theta_1}{\Delta\theta_2}\right)};\ \Delta\theta_1 = T_1 - t_1;\ \Delta\theta_2 = T_2 - t_2.$

NUMERICAL APPLICATION.–

T_1 = 92°C t_1 = 87°C T_2 = 39.96°C t_2 = 21°C

ϕ = 903.76 kW

$\Delta\theta_1 = 5$°C $\Delta\theta_2 = 18.96$°C $\Delta\theta_{ML} = 10.47$°C A_0= 60 m²

Step 3. Hydraulic checks

Table 6.22 gives the geometric characteristics of standard spiral heat exchangers, as well as the approximate length of each spiral ($L_0 = \frac{A_0}{b}$), the free flow area (A_L= eb) and the hydraulic diameter (D_H = 2 e).

Spiral type	D_s (mm)	d_s (mm)	b (mm)	e (mm)	Δx (mm)	L_0 (m)	A_L (m²)	D_H (m)
SP1	813	203	102	4.76	3.18	583	4.86E-04	9.52E-03
SP2	813	203	152	4.76	3.18	391	7.24E-04	9.52E-03
SP3	813	203	305	4.76	3.18	195	1.45E-03	9.52E-03
SP4	1,473	305	305	4.76	3.18	195	1.45E-03	9.52E-03
SP5	813	203	457	6.35	3.18	130	2.90E-03	1.27E-02
SP6	1,473	305	457	6.35	4.76	130	2.90E-03	1.27E-02
SP7	813	203	610	6.35	4.76	98	3.87E-03	1.27E-02
SP8	1,473	305	610	6.35	4.76	98	3.87E-03	1.27E-02
SP9	1,473	305	762	6.35	4.76	78	4.84E-03	1.27E-02
SP10	1,473	305	914	6.35	4.76	65	5.80E-03	1.27E-02
SP11	1,473	305	1,219	6.35	4.76	49	7.74E-03	1.27E-02
SP12	1,473	305	1,524	7.94	6.35	39	1.21E-02	1.59E-02
SP13	1,473	305	1,778	9.53	7.94	33	1.69E-02	1.91E-02
SP14	1,473	305	1,778	10.25	7.94	33	1.82E-02	2.05E-02
SP15	1,473	305	1,778	11.53	7.94	33	2.05E-02	2.31E-02
SP16	1,473	305	1,219	12.03	7.94	49	1.47E-02	2.41E-02
SP17	1,473	305	1,524	12.03	7.94	39	1.83E-02	2.41E-02

Table 6.22. *Calculations of length, free area and hydraulic diameter*

Moreover, Table 6.23 presents the Reynolds numbers ($Re_i = \frac{D_H \; W_i}{\mu_i A_L}$) and the expected pressure drops ($\Delta P_i = \frac{2\,C_f\,L\,W_i^2}{\rho\,D_H\,A_L^2}$).

For each of the circuits, the allowable pressure drops are given by:

$$\Delta P_i^{Adm} = 20\,\frac{\text{kPa}}{\text{NTU}}$$

Spiral type	Re_R	Re_A	C_{fR}	C_{fA}	ΔP_R Atm
SP1	5.15E+04	6.92E+04	5.16E-03	4.83E-03	584.57
SP2	3.46E+04	4.64E+04	5.65E-03	5.28E-03	193.56
SP3	1.72E+04	2.31E+04	6.70E-03	6.22E-03	28.41
SP4	1.72E+04	2.31E+04	6.70E-03	6.22E-03	28.41
SP5	1.15E+04	1.54E+04	7.45E-03	6.89E-03	3.96
SP6	1.15E+04	1.54E+04	7.45E-03	6.89E-03	3.96
SP7	8.61E+03	1.16E+04	8.07E-03	7.44E-03	1.80
SP8	8.61E+03	1.16E+04	8.07E-03	7.44E-03	1.80
SP9	6.89E+03	9.27E+03	8.59E-03	7.90E-03	0.98
SP10	5.75E+03	7.72E+03	9.06E-03	8.32E-03	0.60
SP11	4.31E+03	5.79E+03	9.89E-03	9.04E-03	0.28
SP12	3.45E+03	4.63E+03	1.06E-02	9.67E-03	0.08
SP13	2.95E+03	3.97E+03	1.12E-02	1.01E-02	0.03
SP14	2.95E+03	3.97E+03	1.12E-02	1.01E-02	0.02
SP15	2.95E+03	3.97E+03	1.12E-02	1.01E-02	0.02
SP16	4.31E+03	5.79E+03	9.89E-03	9.04E-03	0.04
SP17	3.45E+03	4.63E+03	1.06E-02	9.67E-03	0.02

Table 6.23. *Calculation of the pressure drops for each circuit*

NUMERICAL APPLICATION.–

$NTU_A = 6.986$ $NTU_R = 7.931$

Hence: $\Delta P_A^{Adm} = 0.98$ Atm and $\Delta P_R^{Adm} = 1.24$ Atm.

Table 6.23 shows that only spirals of the SP9 to SP17 type give pressure drops less than the allowable pressure drops. We therefore subject them to thermal checks.

Step 4. Thermal checks

4.1- The Prandtl number for each circuit is calculated: $Pr_i = \frac{Cp_i \, \mu_i}{\lambda_i}$.

NUMERICAL APPLICATION.–

$C_{pA} = 5{,}125$ J/kg°C $\mu_A = 0.96$ Cp $\lambda_A = 0.87$ W/m°C

$C_{pR} = 3{,}160$ J/kg°C $\mu_R = 1.65$ Cp $\lambda_R = 0.65$ W/m°C

Hence: $\text{Pr}_A = 5.66$ and $\text{Pr}_\text{R} = 8.02$.

4.2- Calculating the overall coefficient.

Table 6.24 gives the Nusselt numbers ($\text{Nu}_i = 0.04\text{Re}_i^{0.74}\text{Pr}^{0.4}$), as well as the local heat transfer coefficients ($\text{h}_i = \frac{\lambda_i \text{Nu}_i}{\text{D}_{\text{H}_i}}$) and the overall heat transfer coefficient ($\frac{1}{U^*} = \frac{1}{h_1} + \frac{\Delta x}{\lambda_p} + \frac{1}{h_2}$), for each of the retained spiral types, and for each circuit.

Spiral type	Re_R	Re_A	Nu_R	Nu_A	h_R	h_A	U^*
SP9	6,893	9,265	63.71	68.95	3,260.61	4,723.14	1,425.03
SP10	5,747	7,724	55.68	60.26	2,850.00	4,128.36	1,287.95
SP11	4,309	5,792	45.00	48.70	2,303.04	3,336.07	1,090.17
SP12	3,447	4,633	38.14	41.28	1,561.31	2,261.63	753.46
SP13	2,954	3,971	34.03	36.83	1,160.58	1,681.16	567.45
SP14	2,954	3,971	34.03	36.83	1,079.06	1,563.07	534.10
SP15	2,954	3,971	34.03	36.83	959.27	1,389.55	483.58
SP16	4,309	5,792	45.00	48.70	1,215.66	1,760.93	589.53
SP17	3,447	4,633	38.14	41.28	1,030.49	1,492.72	513.84

Table 6.24. *Calculation of the overall coefficient*

4.3- to 4.6- Calculating the length, exact transfer area and the sum of the fouling resistances, then tests on $\frac{1}{U_F^*} - \frac{1}{U_e}$.

For each of the retained spirals, Table 6.25 gives the results of the calculations of the number of revolutions, $n = Ent\left(\frac{\sqrt{(r_s+e+2\Delta x)^2+(e+\Delta x)\frac{A_0}{\pi b}}-(r_s+e+2\Delta x)}{e+\Delta x} + 1\right)$, the correction factor ($\text{Y} = \frac{\text{n}}{\text{NTU}}\tanh\frac{\text{NTU}}{\text{n}}$), the exact length, $L^* = \pi[(e + \Delta x)n^2 + 2(r_s + e + 2\Delta x)n]$ and the exact surface area $A^* = L^* b$. This makes it possible to determine the "fouled" coefficient, $U_F^* = \frac{U_F A_0}{YA^*}$.

Spiral type	n	Y	L^* (m)	A^* (m²)	U_F^*	$\frac{1}{U_F^*}-\frac{1}{U^*}$
SP9	35	0.99	79.78	60.79	1,431.43	-3.13E-06
SP10	31	0.99	66.34	60.63	1,438.63	-8.13E-05
SP11	26	0.98	51.10	62.29	1,406.71	-2.06E-04
SP12	20	0.97	39.71	60.53	1,463.01	-6.44E-04
SP13	17	0.96	34.86	61.99	1,442.42	-1.07E-03
SP14	17	0.96	35.60	63.29	1,412.82	-1.16E-03
SP15	16	0.96	33.75	60.00	1,496.96	-1.40E-03
SP16	21	0.98	51.47	62.74	1,407.93	-9.86E-04
SP17	18	0.97	40.73	62.07	1,435.04	-1.25E-03

Table 6.25. *Spiral length and exact transfer area*

$$\mathrm{R_{d_1}} + \mathrm{R_{d_2}} = 6\ 10^{-5}\ \mathrm{m^2{}^\circ C/W}$$

For all of the spirals, the difference $\frac{1}{U_F^*}-\frac{1}{U^*}$ is negative, which means that $U^* < U_F^*$. They are therefore not suitable. We thus change the value of U_F.

We take: U_F = 750 W/m²°C and return to step 2.2.

2.2- Calculating A_0:

$$A_0 = \frac{\phi}{U_S\, \Delta\theta_{ML}}$$

with $\Delta\theta_{ML} = \frac{\Delta\theta_1 - \Delta\theta_2}{Ln\left(\frac{\Delta\theta_1}{\Delta\theta_2}\right)}$; $\Delta\theta_1 = T_1 - t_1$; $\Delta\theta_2 = T_2 - t_2$.

NUMERICAL APPLICATION.–

$T_1 = 92°C$ $t_1 = 87°C$ $T_2 = 39.96°C$ $t_2 = 21°C$

$\phi = 903.76$ kW

$\Delta\theta_1 = 5°C$ $\Delta\theta_2 = 18.96°C$ $\Delta\theta_{ML} = 10.47°C$ $A_0 = 115$ m²

Step 3. Hydraulic checks

$L_0 = \frac{A_0}{b}$, A_L= eb, $Re_i = \frac{D_H\ W_i}{\mu_i A_L}$ and $\Delta P_i = \frac{2\, C_f\, L\, W_i^2}{\rho\, D_H\, A_L^2}$

Table 6.26 presents, for each circuit, L_0, A_L, the Reynolds numbers and the expected pressure drops.

Spiral type	L_0 (m)	A_L (m²)	Re_R	Re_A	C_{fR}	C_{fA}	$\Delta P_{R\ Atm}$	$\Delta P_{A\ Atm}$
SP1	583	4.86E-04	5.15E+04	6.92E+04	5.16E-03	4.83E-03	1,130.16	440.36
SP2	391	7.24E-04	3.46E+04	4.64E+04	5.65E-03	5.28E-03	374.22	145.37
SP3	195	1.45E-03	1.72E+04	2.31E+04	6.70E-03	6.22E-03	54.92	21.21
SP4	195	1.45E-03	1.72E+04	2.31E+04	6.70E-03	6.22E-03	54.92	21.21
SP5	130	2.90E-03	1.15E+04	1.54E+04	7.45E-03	6.89E-03	7.65	2.94
SP6	130	2.90E-03	1.15E+04	1.54E+04	7.45E-03	6.89E-03	7.65	2.94
SP7	98	3.87E-03	8.61E+03	1.16E+04	8.07E-03	7.44E-03	3.48	1.33
SP8	98	3.87E-03	8.61E+03	1.16E+04	8.07E-03	7.44E-03	3.48	1.33
SP9	78	4.84E-03	6.89E+03	9.27E+03	8.59E-03	7.90E-03	1.90	0.73
SP10	65	5.80E-03	5.75E+03	7.72E+03	9.06E-03	8.32E-03	1.16	0.44
SP11	49	7.74E-03	4.31E+03	5.79E+03	9.89E-03	9.04E-03	0.53	0.20
SP12	39	1.21E-02	3.45E+03	4.63E+03	1.06E-02	9.67E-03	0.15	0.06
SP13	33	1.69E-02	2.95E+03	3.97E+03	1.12E-02	1.01E-02	0.06	0.02
SP14	33	1.82E-02	2.95E+03	3.97E+03	1.12E-02	1.01E-02	0.05	0.02
SP15	33	2.05E-02	2.95E+03	3.97E+03	1.12E-02	1.01E-02	0.03	0.01
SP16	49	1.47E-02	4.31E+03	5.79E+03	9.89E-03	9.04E-03	0.08	0.03
SP17	39	1.83E-02	3.45E+03	4.63E+03	1.06E-02	9.67E-03	0.04	0.02

Table 6.26. *Reiteration of hydraulic calculations*

Table 6.26 shows that only spirals of type SP10 to SP 17 give pressure drops under the allowable pressure drops. We therefore subject them to thermal checks.

Step 4: Thermal checks

The overall heat transfer coefficient stays the same.

For each of the retained spirals, Table 6.27 gives the results of the calculations of the number of revolutions, the correction factor, Y, the exact length, L^*, the exact surface area A^* and the "fouled" coefficient, $U_F^* = \frac{U_F A_0}{Y A^*}$.

Spiral type	n	Y	L^* (m)	A^* (m²)	U_F^*	$\frac{1}{U_F^*} - \frac{1}{U^*}$
SP10	47	1.00	126.82	115.92	747.89	5.61E-04
SP11	40	0.99	98.16	119.66	725.81	4.60E-04
SP12	31	0.99	76.87	117.14	744.61	1.58E-05
SP13	26	0.98	66.17	117.64	744.85	-4.20E-04
SP14	26	0.98	67.81	120.57	726.76	-4.96E-04
SP15	25	0.98	66.49	118.22	742.14	-7.20E-04
SP16	31	0.99	95.43	116.33	749.82	-3.63E-04
SP17	27	0.99	76.34	116.34	752.31	-6.17E-04

Table 6.27. *Spiral length and exact transfer area*

The difference $\frac{1}{U_F^*} - \frac{1}{U^*}$ is positive for spirals of type SP10 to SP12. However, only the SP12 spiral gives a difference $\frac{1}{U_F^*} - \frac{1}{U^*}$ in the order of magnitude of $R_A + R_R$: ($R_A + R_R = 6\ 10^{-5}\ \mathrm{m}^{2\circ}\frac{\mathrm{C}}{\mathrm{W}}$). It is therefore a SP12-type spiral exchanger that is suitable.

6.11. Calculation algorithm for printed-circuit heat exchangers

The modeling of compact heat exchangers and the evaluation of correlations giving the Nusselt numbers and friction factors help guide the sizing methods for this category of exchangers (Ma et al. 2021; Sarmiento et al. 2021). The degrees of freedom are:

– the diameter of the channels, d;

– the flow velocity in each circuit, v_i;

– the ratio between the channel diameter and the thickness of each module, t_p.

RULES OF THUMB.–

R_1 – The diameter, d, of the channels is the most important parameter in the design of printed-circuit heat exchangers. Generally, d is between 1 and 2 mm. For exchangers intended for nuclear applications, d = 1.2 mm is often retained.

R_2 – The flow velocity shall be set so as to have the maximum turbulence while limiting it to 5 m/sec.

R_3 – The pitch, *a*, is generally between 1.1 d and 1.22 d (Oh and Kim 2008; Ariu 2014).

R_4 – The thickness, t_p, of a module shall be of the order of 0.8 d.

R_5 – The gap between the channels, t_f, shall satisfy the constraint: $t_f = p - d$.

R_6 – In counter-flow PCHEs, we shall have $n_1 = n_2$; $N_1 = N_2$ and $L_3 = 2 N_1 t_p$.

R_7 – In a cross-flow PCHE, we shall have:

$$n_1 = \frac{L_2+t_f}{d+t_f};\ n_2 = \frac{L_1+t_f}{d+t_f};\ \mathrm{N}_2 = N_1\left(\frac{L_2+t_f}{L_1+t_f}\right)\left(\frac{L_1}{L_2}\right) \text{ and } L_3 = N_1\left[1+\left(\frac{L_2+t_f}{L_1+t_f}\right)\left(\frac{L_1}{L_2}\right)\right]t_p$$

R_8 – As with other types of heat exchanger, the calculation principle is based on a compromise between heat transfer efficiency and pressure drops.

The sizing algorithm for this category of compact heat exchangers is composed of five steps.

Step 1. Define the circulation mode and calculate the flux and the NTUs

Based on the problem available data (fluid flowrates, hot fluid inlet temperature, cold fluid inlet and outlet temperature, diameters, thicknesses of the modules and gap between the channels that are technically realizable, allowable pressure drops, allowable foulings, etc.) and the physical properties of the fluids considered:

1.1- Define the circulation mode: cross-flow or counter-flow. Cross-flow mode is favored.

1.2- Calculate the flux, ϕ, to be exchanged and the missing temperature.

1.3- Calculate the numbers of transfer units:

$$NTU_h = \frac{T_1-T_2}{\Delta\theta_{ML}};\ NTU_c = \frac{t_1-t_2}{\Delta\theta_{ML}} \text{ and } NTU_{Average} = \frac{NTU_h+NTU_c}{2}$$

Step 2. Knowing ϕ

2.1- Choose a value of the "fouled" heat transfer coefficient (U_F) according to the experimental orders of magnitude presented in Table A1.21.

2.2- Calculate the approximate exchange area, A_0, from the expression:

$$A_0 = \frac{\phi}{U_F\,\Delta\theta_{ML}}$$

$$\Delta\theta_{ML} = \frac{\Delta\theta_1 - \Delta\theta_2}{Ln\left(\frac{\Delta\theta_1}{\Delta\theta_2}\right)};\ \Delta\theta_1 = T_1 - t_1;\ \Delta\theta_2 = T_2 - t_2$$

Step 3. Hydraulic calculations

3.1- We set the channel diameter, d.

We deduce therefrom the pitch: $a = 1.1\ d$ to $1.22\ d$.

3.2- We calculate t_f and t_p: $t_f = p - d$ and t_p such that it satisfies the mechanical stiffness constraint[49].

3.3- With L_1, L_2 being set[50], calculate[51]:

- the number of channels in each circuit: $n_1 = Ent\left(\frac{L_2+t_f}{d+t_f} + 1\right);\ n_2 = Ent(\frac{L_1+t_f}{d+t_f} + 1)$;

- the number of modules in circuit 1: $N_1 = \frac{A_0}{n_1 d\left(\frac{\pi}{2}+1\right)L_1}$;

- the number of modules in circuit 2: $N_2 = N_1\left(\frac{L_2+t_f}{L_1+t_f}\right)\left(\frac{L_1}{L_2}\right)$;

- dimension L_3 of the exchanger: $L_3 = N_1\left[1 + \left(\frac{L_2+t_f}{L_1+t_f}\right)\left(\frac{L_1}{L_2}\right)\right]t_p$.

3.4- Calculate the flow velocity in each channel:

$$v_i = \frac{W_i}{N_i n_i \rho_i}\frac{8}{\pi d_i^2};\ i = 1; 2$$

3.5- Calculate the Reynolds numbers.

The hydraulic diameter is[52] $D_H = d$; the Reynolds number in each of the circuits is given by[53]: $Re_i = \frac{d_i v_i \rho_i}{\mu_i}$; $i = 1\text{-}2$.

49 See section 3.7.6: $t_f \geq \frac{p}{\left(\frac{\sigma_{max}}{\Delta P}+1\right)}$ and $t_p \geq \begin{cases} \frac{d}{2}\sqrt{\frac{\sigma_{max}+P_p}{\sigma_{max}+2P_s-P_p}} & \text{when } P_p \geq P_s \\ \frac{d}{2}\sqrt{\frac{\sigma_{max}-P_p}{\sigma_{max}-2P_s-P_p}} & \text{when } P_p < P_s \end{cases}$.

50 For a counter-flow $L_1 = L_2$.

51 See sections 3.7.2 and 3.7.3.

52 See section 3.7.4.

53 See section 3.7.5.

3.6- We calculate the pressure drops: $\Delta P_i = 4 f_i \frac{L_i}{d} \rho_i v_i^2$ (see section A3.10).

$$Re < 2{,}300: f_i = \frac{15.78}{Re_i}$$

$$Re \geq 8{,}200: f_i = 0.478\, Re_i^{-0.26}$$

3.7- Tests on ΔP_i: If $\Delta P_i \leq \Delta P_i^{Adm}$, proceed to step 4; otherwise, repeat step 3.1- with a larger diameter.

Step 4. Thermal calculations

At this level, we have a heat exchanger of surface area A^* (which may well be different from A_0). We now need to check that this exchanger will ensure the transfer of the flux, ϕ.

4.1- We calculate the areas: $A_1^* = N_1 n_1 d\left(\frac{\pi}{2} + 1\right) L_1$; $A_2^* = N_2 n_2 d\left(\frac{\pi}{2} + 1\right) L_2$ and $A_m^* = \frac{A_1^* + A_2^*}{2}$.

4.2- We calculate Nu_i: see Appendix 2, section A2.6.5.

We deduce therefrom: h_i: $h_i = \frac{\lambda_i Nu_i}{d_{H_i}}$; i = 1-2.

4.3- We calculate the correction coefficient[54]:

$$y_p = 0.4499\eta^2 + 0.3403\eta + 1 \text{ with } \eta = \frac{d}{t_p}.$$

4.4- We calculate the heat transfer coefficient for the clean heat exchanger, U^*:

$$\frac{1}{U^*} = \frac{1}{h_1} + \frac{t_p}{\lambda_p y_p} + \frac{1}{h_2}$$

λ_t being the heat conductivity of the material constituting the heat exchanger.

4.5- Calculate the correction factor, Y:

- for a counter-flow, we take $Y = 1$;
- for a cross-flow, we calculate Y using the expressions[55] given in Table 5.7:

54 See section A2.9.

55 See section 5.7.4.3.

$$(Z < 1): Y = \frac{1}{\text{NTU}\,(1-Z)} Ln\left[\frac{1-XZ}{1-X}\right]$$

with: $X = \frac{t_{out} - t_{in}}{T_{in} - t_{in}}$ and $Z = \frac{T_{in} - T_{out}}{t_{out} - t_{in}}$.

4.6- We re-evaluate the "fouled" heat transfer coefficient: U_F^*:

$$\phi = U_F A_0\, \Delta\theta_{ML} = Y\, U_F^*\, A_m^*\, \Delta\theta_{ML} \Rightarrow U_F^* = \frac{U_F\, A_0}{Y A_m^*}$$

Step 5. Checking the margin allowed for fouling

If $\frac{1}{U_F^*} - \frac{1}{U^*} \cong R_{fi} + R_{fe}$: the device is correctly sized.

If $\frac{1}{U_F^*} - \frac{1}{U^*} < 0$ or $\frac{1}{U_F^*} - \frac{1}{U^*} \ll R_{fi} + R_{fe}$: the device is undersized. We then repeat the calculation at step 2.1, choosing a more suitable value for U_F.

If $\frac{1}{U_S^*} - \frac{1}{U^*} \gg R_{fi} + R_{fe}$: the device is oversized. We then repeat the calculation at step 2.1, choosing a more suitable value for U_F.

6.12. Illustration: implementation of the PCHE calculation algorithm

The most recent technologies for concentrated solar power plant primary circuits use metal-ceramic printed-circuit heat exchangers, which can operate at very high temperatures.

Table 6.28 defines the thermal schedule to be assured by the printed-circuit heat exchanger to be designed to transfer energy between two molten salts: FLiBe[56] in the primary circuit and FliNaK[57] in the secondary circuit.

The maximum shear stress of the alloy used to make the exchanger is σ_{max}.

Fluids		Temperatures (°C)		Pressures (Pa)	
Primary	**Secondary**	**Primary**	**Secondary**	**Primary**	**Secondary**
FLiBe	FLiNaK	760/650	600/690	3.00E+05	1.00E+05

Table 6.28. *Temperatures and pressures of primary and secondary circuits*

56 Salt consisting of lithium and beryllium fluorides. See Volume 3, section A13.

57 Salt consisting of lithium, sodium and potassium fluorides. See Volume 3, section A13.

Available data:

Fluid	W (T/hr)	ρ (kg/m³)	μ (Pa.s)	C_p (J/kg°C)	λ (W/m°C)
FLiBe[58]	?	1,954	$6.30\ 10^{-3}$	2,383	1.10
FLiNaK[59]	100	1,999	$3.07\ 10^{-3}$	1,966	0.890

Table 6.29. *Data on molten salts used*

Construction alloy of the exchanger: $\lambda_p = 25.97$ W/m°C; $\sigma_{max} = 2.18\ 10^8$ Pa.

Allowable pressure drops: $\Delta P_{1\ Adm}$ and $\Delta P_{2\ Adm} = 20$ kPa/NTU.

The installation constraints impose the following dimensions: $L_1 = L_2 = 85$ cm.

We suggest starting with a channel diameter, $d = 5$ mm.

Solution

Step 1. Circulation mode, flux and NTU

1.1- Circulation mode: cross-flow.

1.2- Flux, ϕ, to be exchanged and missing datum.

We calculate the flux from the secondary circuit data: $\phi = W_2 C_{p_2}(t_1 - t_2)$.

NUMERICAL APPLICATION.–

$W_2 = 27.78$ kg/sec $\quad t_1 = 690$°C $\quad t_2 = 600$°C $\quad Cp_2 = 1{,}966$ J/kg°C

$\phi = 4{,}915$ kW and $W_1 = 18.75$ kg/sec

1.3- Calculating the numbers of transfer units:

$$NTU_h = \frac{T_1 - T_2}{\Delta\theta_{ML}};\ NTU_c = \frac{t_1 - t_2}{\Delta\theta_{ML}} \text{ and } NTU_{Average} = \frac{NTU_h + NTU_c}{2}$$

with $\Delta\theta_{ML} = \dfrac{\Delta\theta_1 - \Delta\theta_2}{\mathrm{Ln}\left(\frac{\Delta\theta_1}{\Delta\theta_2}\right)}$; $\Delta\theta_1 = T_1 - t_1$; $\Delta\theta_2 = T_2 - t_2$.

58 See Table A.23, Volume 3.

59 See Table A.20, Volume 3.

NUMERICAL APPLICATION.–

$T_1 = 760°C$ $T_2 = 650°C$ $t_1 = 690°C$ $t_2 = 600°C$

$\Delta\theta_1 = 70°C$ $\Delta\theta_2 = 50°C$ $\Delta\theta_{ML} = 59.44°C$

$NTU_h = 1.85$ $NTU_c = 1.51$ $NTU_{Average} = 1.68$

Step 2. Knowing ϕ

2.1- Choosing U_F: For a heat exchange between two low-viscosity liquids, Table A1.21 gives: U_F between 813 and 2,090 W/m²°C.

We take U_F = 1,000 W/m²°C.

2.2- Approximate exchange area, A_0:

$$A_0 = \frac{\phi}{U_F\,\Delta\theta_{ML}}$$

NUMERICAL APPLICATION.–

$\phi = 4{,}915$ kW $U_F = 1{,}000$ W/m²°C $\Delta\theta_{ML} = 59.44°C$

$A_0 = 82.69$ m²

Step 3. Hydraulic calculations

3.1- We set the channel diameter, $d = 5$ mm.

We deduce therefrom the pitch: $a = 1.1\ d = 5.5$ mm.

3.2- Calculating t_f and t_p:

$t_f = p - d$; $t_f = 0.5$ mm

Checking the mechanical stiffness constraints:

$$t_f \geq \frac{p}{\left(\frac{\sigma_{max}}{\Delta P}+1\right)} \text{ and } t_p \geq \begin{cases} \frac{d}{2}\sqrt{\frac{\sigma_{max}+P_1}{\sigma_{max}+2P_2-P_1}} & \text{when } P_1 \geq P_2 \\ \frac{d}{2}\sqrt{\frac{\sigma_{max}-P_1}{\sigma_{max}-2P_2-P_1}} & \text{when } P_1 < P_2 \end{cases}$$

NUMERICAL APPLICATION.–

$P_1 = 3\ 10^5$ Pa $\qquad P_2 = 10^5$ Pa $\qquad \sigma_{max} = 2.18\ 10^8$ Pa

Hence: $t_f \geq 0.00504$ mm; the constraint on t_f is therefore satisfied.

$P_1 > P_2$, therefore, we shall have: $t_p \geq \frac{d}{2}\sqrt{\frac{\sigma_{\max}+P_p}{\sigma_{\max}+2P_S-P_p}}$.

$t_p \geq 2.5$ mm. We take $t_p = 4$ mm.

3.3- L_1 and L_2 set:

$$n_1 = \frac{L_2+t_f}{d+t_f};\ n_2 = \frac{L_1+t_f}{d+t_f};\ N_1 = \frac{A_0}{n_1 d\left(\frac{\pi}{2}+1\right)L_1};$$

$$N_2 = N_1\left(\frac{L_2+t_f}{L_1+t_f}\right)\left(\frac{L_1}{L_2}\right);\ L_3 = N_1\left[1+\left(\frac{L_2+t_f}{L_1+t_f}\right)\left(\frac{L_1}{L_2}\right)\right]t_p$$

NUMERICAL APPLICATION.–

$L_1 = L_2 = 0.85$ m $\qquad t_f = 0.5$ mm $\qquad d = 5$ mm $\qquad A_0 = 82.69$ m²

$t_p = 4$ mm $\qquad n_1 = n_2 = 155 \qquad N_1 = N_2 = 49 \qquad L_3 = 39.1$ cm

3.4- Calculating the flow velocity in each channel:

$$v_i = \frac{W_i}{N_i n_i \rho_i}\frac{8}{\pi d_i^2};\ i = 1\text{-}2$$

NUMERICAL APPLICATION.–

$W_1 = 18.75$ kg/sec $\qquad \rho_1 = 1{,}954$ kg/m³ $\qquad n_1 = n_2 = 155 \qquad N_1 = N_2 = 49$

$W_2 = 27.78$ kg/sec $\qquad \rho_2 = 1{,}999$ kg/m³ $\qquad d = 5$ mm

$v_1 = 0.13$ m/sec $\qquad v_2 = 0.19$ m/sec

3.5- Calculating the Reynolds numbers and pressure drops:

$D_H = d$

$$Re_i = \frac{D_{H_i} v_i \rho_i}{\mu_i};\ i = 1\text{-}2$$

NUMERICAL APPLICATION.–

$v_1 = 0.13$ m/sec $\rho_1 = 1{,}954$ kg/m^3 $\mu_1 = 6.3\ 10^{-3}$ kgm^{-1}s^{-1}

$v_2 = 0.19$ m/sec $\rho_2 = 1{,}999$ kg/m^3 $\mu_2 = 3.7\ 10^{-3}$ kgm^{-1}s^{-1} $d = 5$ mm

$Re_1 = 200$ $Re_2 = 609$

$$\Delta P_i = 4f_i \frac{L_i}{d} \rho_i v_i^2$$

$$Re < 2{,}300:\ f_i = \frac{15.78}{Re_i}$$

NUMERICAL APPLICATION.–

$Re_1 = 200$ $v_1 = 0.13$ m/sec $\rho_1 = 1{,}954$ kg/m^3 $d = 5$ mm

$Re_2 = 609$ $v_2 = 0.19$ m/sec $\rho_2 = 1{,}999$ kg/m^3 $L_1 = L_2 = 0.85$ m

Hence: $f_1 = 0.0788$; $f_2 = 0.0259$; $\Delta P_1 = 1{,}746$ Pa and $\Delta P_2 = 1{,}232$ Pa.

3.6- Tests on ΔP_i:

$\Delta P_{1\,Adm}$ and $\Delta P_{2\,Adm} = 20$ kPa/NTU

NUMERICAL APPLICATION.–

$NTU_1 = 1.85$ $NTU_2 = 1.51$ $\Delta P_1 = 1{,}746$ Pa $\Delta P_2 = 1{,}232$ Pa

$\Delta P_{1\,Adm} = 37$ kPa therefore $\Delta P_1 < \Delta P_{1\,Adm}$.

$\Delta P_{2\,Adm} = 30$ kPa therefore $\Delta P_2 < \Delta P_{2\,Adm}$.

Step 4. Thermal calculations

4.1- Calculating the exact exchange areas:

$$A_1^* = N_1 n_1 d \left(\frac{\pi}{2} + 1\right) L_1;\ A_2^* = N_2 n_2 d \left(\frac{\pi}{2} + 1\right) L_2;\ A_m^* = \frac{A_1^* + A_2^*}{2}$$

NUMERICAL APPLICATION.–

$N_1 = N_2 = 49$ $n_1 = n_2 = 155$ $L_1 = L_2 = 0.85$ m $d = 5$ mm

$A_1^* = 82.69$ m² $A_2^* = 82.69$ m² $A_m^* = 82.69$ m²

4.2- Calculating the Nu_i values and h_i values:

$(h_i = \frac{\lambda_i Nu_i}{d})$; $i = 1\text{-}2$

For molten salts and a laminar flow ($100 < Re < 2{,}000$), Table A2.17 gives:

$$Nu_i = 4.82 + +0.0185(Re_i Pr_i)^{0.827}$$

with $Pr_i = \frac{Cp_i \mu_i}{\lambda_i}$.

NUMERICAL APPLICATION.–

$Cp_1 = 2{,}383$ J/kg°C $\mu_1 = 6.3\ 10^{-3}$ kgm^{-1}s^{-1} $\lambda_1 = 1.10$ W/m°C $Re_1 = 200$

$Cp_2 = 1{,}966$ J/kg°C $\mu_2 = 3.7\ 10^{-3}$ kgm^{-1}s^{-1} $\lambda_2 = 0.89$ W/m°C $Re_2 = 609$

$Pr_1 = 13.65$ $Nu_1 = 17.68$ $h_1 = 3{,}890$ W/m^2°C

$Pr_2 = 13.65$ $Nu_2 = 17.68$ $h_2 = 3{,}890$ W/m^2°C

4.3- Calculating the correction coefficient:

$$y_p = 0.4499\left(\frac{d}{t_p}\right)^2 + 0.3403\left(\frac{d}{t_p}\right) + 1$$

NUMERICAL APPLICATION.–

$d = 5$ mm $t_p = 4$ mm $\eta = \frac{d}{t_p}$

$\eta = 1.25$ $y_p = 2.13$

4.4- Calculating the heat transfer coefficient for the clean heat exchanger, U^*:

$$\frac{1}{U^*} = \frac{1}{h_1} + \frac{t_p}{\lambda_p y_p} + \frac{1}{h_2}$$

NUMERICAL APPLICATION.–

$h_1 = 3{,}890$ W/m^2°C $h_2 = 3{,}890$ W/m^2°C $t_p = 4$ mm

$\lambda_p = 25.97$ W/m°C $y_p = 2.13$

Hence: $U^* = 1{,}740$ W/m²°C.

4.5- Calculating the correction coefficient, Y:

$$Z = \frac{T_1 - T_2}{t_1 - t_2} \text{ and } X = \frac{t_1 - t_2}{T_1 - t_2}$$

NUMERICAL APPLICATION.–

$T_1 = 760°C$ $T_2 = 650°C$ $t_1 = 690°C$ $t_2 = 600°C$

Hence: $Z = 1.22$ and $X = 0.56$.

$$Z \neq 1: Y = \frac{1}{NTU\,(1-Z)} \text{Ln}\left[\frac{1-XZ}{1-X}\right] \quad Y = 0.90$$

4.6- Re-evaluating the "fouled" heat transfer coefficient: U_S^* :

$$U_F^* = \frac{U_F\,A_0}{Y A_m^*}$$

NUMERICAL APPLICATION.–

$U_F = 1{,}000$ W/m²°C $A_0 = 82.69$ m² $Y = 0.90$ $A_m^* = 82.69$ m²

Hence: $U_F^* = 1.111$ W/m²°C.

Step 5. Checking the margin allowed for fouling

$R_{fi} + R_{fe} = 6\ 10^{-5}$ m²°C/W and $\frac{1}{U_F^*} - \frac{1}{U^*} = 3.25\ 10^{-4}$ m²°C/W

$\frac{\frac{1}{U_S^*} - \frac{1}{U^*}}{R_{di} + R_{do}} = 5.42$; the device is oversized. We then return to step 2.1, choosing a more suitable value for U_F.

We take $U_F = 2{,}000$ W/m²°C.

We then find the heat exchanger as having the characteristics presented in Table 6.30, $\frac{1}{U_F^*} - \frac{1}{U^*}$ of the same order of magnitude as $R_{fi} + R_{fe}$.

A_0 (m²)	L_1 (cm)	L_2 (cm)	L_3 (cm)	d (mm)	p (mm)	t_f (mm)	t_p (mm)	n_1	n_2	N_1	N_2
41.34	85	85	19.5	5	5.5	0.5	4	155	155	24	24
v_1 (m/s)	**v_2 (m/s)**	**Re_1**	**Re_2**	**Δp_1 (Pa)**	**Δp_2 (Pa)**	**Nu_1**	**Nu_2**	**h_1 (W/m²°C)**	**h_2 (W/m²°C)**	**U^* (W/m²°C)**	**U_F^* (W/m²°C)**
0.26	0.37	401	1,218	3,492	2,464	27.64	36.92	6,081	6,571	2571	2,222

Table 6.30. *Characteristics of the retained PCHE*

Appendices

Appendix 1

Database

A1.1. Introduction

This appendix presents the physical data useful for the sizing of heat exchangers, including viscosities, densities, heat conductivities, heat transfer coefficients, fouling resistances, molten salt properties, etc.

Other data that have been recorded in Volumes 1 to 5 of the Energy Engineering series may prove necessary for certain calculations or estimates. In this case, the reader should refer to the volume concerned.

When the data sought concern uncommon materials or components, the reader is directed back to the source documents cited in the bibliographic references. But where the data sought are not indicated in the bibliographic sources, the values of the corresponding physical properties may be estimated from the methods and correlations presented in Volume 5 of this series.

This appendix also includes unit conversion tables, which are extremely useful when dealing with different unit systems, as well as a table of the fundamental constants.

A1.2. Densities

A1.2.1. *Densities for certain metals and alloys*

Metals and alloys	ρ (kg m^{-3})
Carbon steel	7,833
Stainless steel 15% Cr, 10% Ni	7,864
Stainless steel 18% Cr, 8% Ni	7,816

Stainless steel 25% Cr, 20% Ni	7,864
Aluminum	2,707
Silver	10,525
Bronze 75% Cu, 25% Sn	8,800
Bronze 92% Cu, 8% Al	7,900
Carbon graphite	2,250
Chrome	2,118
Constantan 60% Cu, 40% Ni	8,922
Copper	8,954
Cupronickel 70% Cu, 30% Ni	8,900
Duralumin	2,787
Tin	7,304
Iron	7,870
Cast iron	7,849
Brass 70% Cu, 30% Zn	8,522
Magnesium	1,740
Gold	19,300
Platinum	21,400
Lead	11,373
Liquid sodium	930
Titanium	4,500
Tungsten	19,350
Zinc	7,144

Table A1.1. *Densities for metals and alloys (in kg/m^3)*

A1.2.2. *Densities for certain construction materials*

Construction materials	ρ (kg m^{-3})
Asbestos[1]	575
Slate	2,400
Asphalt	2,115
Basalt	2,850
Cavernous concrete	1,900

1 Warning: inhaling the dust of this product can have serious consequences (carcinogenic effects).

Solid concrete	2,300
Bitumen (cardboard)	1,050
Low-density hardwood	525
Medium-density hardwood	675
Very-low-density hardwood	375
Low-density softwood	375
Medium-density softwood	500
Very-low-density softwood	375
Clay brick	1,800
Hard limestone	2,450
Soft limestone	1,650
Rubber (natural)	1,150
Rubber (vulcanized)	1,100
Tile	2,400
Cardboard	650
Okoume plywood	400
Pine plywood	500
Leather	998
Ice	920
Granite	2,600
Gravel (loose)	1,800
Sandstone	2,500
Lava	2,350
Marble	2,700
Plaster	1,440
Plexiglas	1,190
Porcelain	2,400
Polyethylene	929
PVC	1,459
Sand	1,515
Shale	2,400
Teflon	2,170
Wet earth	1,900
Dry earth	1,500
Glass	2,300
Pyrex glass	2,220

Table A1.2. *Densities for construction materials (in kg/m^3)*

A1.2.3. *Densities of certain thermal insulation materials*

Thermal insulation	Density	ρ (kg m^{-3})
Balsa	High	140
Cotton	High	80
Kapok	Medium	50
Rock wool	Low	20
	Medium	55
	High	135
Glass wool	Low	8
	Medium	10
	Medium	15
	High	40
Expanded cork	High	120
Carpet	High	200
Polyurethane (foam)	Low	32
	Medium	50
	High	85
PVC (rigid foam)	Medium	30
	High	40
Expanded polystyrene	Low	12
	Medium	14
	High	18
Styrofoam	Medium	30

Table A1.3. *Densities according to manufacture (in kg/m^3)*

A1.3. Heat capacities

A1.3.1. *Sensible heats of certain metals and alloys*

Materials	C_p (J kg^{-1} °C^{-1})
Carbon steel	465
Stainless steel 15% Cr, 10% Ni	460
Stainless steel 18% Cr, 8% Ni	460
Stainless steel 25% Cr, 20% Ni	460

Aluminum	896
Silver	234
Bronze 75% Cu, 25% Sn	377
Bronze 92% Cu, 8% Al	377
Carbon graphite	707
Chrome	7,160
Constantan 60% Cu, 40% Ni	410
Copper	383
Cupronickel 70% Cu, 30% Ni	377
Duralumin	883
Tin	226
Iron	452
Cast iron	460
Brass 70% Cu, 30% Zn	385
Magnesium	1,004
Gold	128
Platinum	140
Lead	130
Liquid sodium	1,381
Titanium	523
Tungsten	134
Zinc	384

Table A1.4. *Sensible heats for metals and alloys (in J kg^{-1}°C)*

A1.3.2. *Sensible heats of certain construction materials*

See (Benallou 2018, section A.6.2).

A1.3.3. *Sensible heats of certain thermal insulation materials*

See (Benallou 2018, section A.6.3).

A1.4. Heat conductivities

A1.4.1. *Heat conductivities of certain metals and alloys*

Metals and alloys	λ (W m^{-1}°C^{-1})
Carbon steel	54
Stainless steel 15% Cr, 10% Ni	20
Stainless steel 18% Cr, 8% Ni	16.3
Stainless steel 25% Cr, 20% Ni	13
Aluminum	204
Silver	407
Bronze 75% Cu, 25% Sn	188
Bronze 92% Cu, 8% Al	71
Carbon graphite	147
Chrome	449
Constantan 60% Cu, 40% Ni	22.7
Copper	386
Cupronickel 70% Cu, 30% Ni	29.3
Duralumin	164
Tin	64
Iron	73
Cast iron	59
Brass 70% Cu, 30% Zn	111
Magnesium	151
Gold	312
Platinum	69
Lead	35
Liquid sodium	84.5
Titanium	20.9
Tungsten	163
Zinc	112

Table A1.5. *Heat conductivities (metals and alloys)*

A1.4.2. *Heat conductivities of construction materials*

See (Benallou 2018, section A.7.2).

A1.4.3. *Conductivities of certain thermal insulation materials*

See (Benallou 2018, section A.7.3).

A1.5. Heat-lagging data

See (Benallou 2018, section A.8).

A1.6. Thermal conductivities of liquids

Liquids	**λ (W m^{-1}°C^{-1})**
Acetone	0.177
Acetic acid	0.171
Phosphoric acid	0.193
30% sulfuric acid	0.519
60% sulfuric acid	0.433
90% sulfuric acid	0.364
Ammonia (26% in water)	0.452
Aniline	0.173
Benzene	0.159
Bromobenzene	0.128
Butanol	1.679
CCl4	0.168
Chlorobenzene	0.144
Chloroform	0.138
12.5% sodium chloride	0.589
25% sodium chloride	0.571
CS_2	0.161
Dichloroethane	0.142
Dichloromethane	0.166
Sulfur dioxide	0.192

Water	0.628
Gasoline	0.135
Ethanol	0.182
Heptane	0.140
Hexane	0.138
Olive oil	0.168
Kerosene	0.149
Mercury	8.361
Methanol	0.215
Nitrobenzene	0.164
Nitromethane	0.022
Octane	0.144
Pentane	0.135
Perchloroethylene	0.159
Propanol	0.171
Toluene	0.149
Trichloroethylene	0.138
Petroleum jelly	0.183
Xylene	0.156

Table A1.6. *Heat conductivities of liquids*

A1.7. Physical properties of water

θ	ρ	C_p	λ	$10^5\mu$	$10^7\alpha$	*Pr*
°C	$kg.m^{-3}$	$J.kg^{-1}.°C^{-1}$	$W.m^{-1}.°C^{-1}$	$Pa.s^{-1}$	$m^2.s^{-1}$	dimensionless
0	1,002	4,218	0.552	17.90	1.31	13.06
20	1,001	4,182	0.597	10.10	1.43	7.02
40	995	4,178	0.628	6.55	1.51	4.34
60	985	4,184	0.651	4.71	1.55	3.02
80	974	4,196	0.668	3.55	1.64	2.22
100	960	4,216	0.680	2.82	1.68	1.74
120	945	4,250	0.685	2.33	1.71	1.45
140	928	4,283	0.684	1.99	1.72	1.24
160	910	4,342	0.680	1.73	1.73	1.10

180	889	4,417	0.675	1.54	1.72	1.00
200	867	4,505	0.665	1.39	1.71	0.94
220	842	4,610	0.652	1.26	1.68	0.89
240	816	4,756	0.635	1.17	1.64	0.88
260	786	4,949	0.611	1.08	1.58	0.87
280	753	5,208	0.580	1.02	1.48	0.91
300	714	5,728	0.540	0.96	1.32	1.02

Table A1.7. *Density (ρ), sensible heat (C_p), heat conductivity (λ), thermal diffusivity (α) and Prandtl number (Pr) as a function of temperature*

T (°C)	σ (N/m)
0	0.0757
20	0.0727
40	0.0696
60	0.0662
80	0.0627
100	0.0589
120	0.055
140	0.0509
160	0.0466
180	0.0422
200	0.0377
220	0.0331
240	0.0284
260	0.0237
280	0.019
300	0.0144
320	0.0099
340	0.0056
360	0.0019
374	0

Table A1.8. *Surface tensions at the water/water-vapor interface*

A1.8. Physical properties of air

θ	ρ	C_p	λ	10^5 μ	10^5 α	*Pr*
°C	kg.m^{-3}	J.kg^{-1}.°C^{-1}	W.m^{-1}.°C^{-1}	PA1.s^{-1}	m^2.s^{-1}	dimensionless
0	1.292	1,006	0.0242	1.72	1.86	0.715
20	1.204	1,006	0.0257	1.81	2.1	0.709
40	1.127	1,007	0.0272	1.90	2.40	0.703
60	1.059	1,008	0.0287	1.99	2.69	0.699
80	0.999	1,010	0.0302	2.09	3.00	0.699
100	0.946	1,012	0.0318	2.18	3.32	0.694
120	0.898	1,014	0.0333	2.27	3.66	0.691
140	0.854	1,016	0.0345	2.34	3.98	0.689
160	0.815	1,019	0.0359	2.42	4.32	0.687
180	0.779	1,022	0.0372	2.50	4.67	0.687
200	0.746	1,025	0.0386	2.57	5.05	0.682
220	0.700	1,028	0.0399	2.64	5.43	0.680
240	0.688	1,032	0.0412	2.72	5.80	0.681
260	0.662	1,036	0.0425	2.79	6.20	0.680
280	0.638	1,040	0.0437	2.86	6.59	0.681
300	0.616	1,045	0.0450	2.93	6.99	0.680
350	0.566	1,055	0.0485	3.12	8.10	0.679
400	0.524	1,068	0.0516	3.28	9.18	0.679
450	0.488	1,080	0.0543	3.44	10.30	0.684
500	0.456	1,092	0.0570	3.86	11.41	0.739
600	0.404	1,114	0.0621	3.98	13.80	0.714
700	0.363	1,335	0.0667	4.12	16.20	0.825
800	0.329	1,356	0.0705	4.37	18.55	0.841
900	0.321	1,372	0.0741	4.59	21.05	0.850
1,000	0.277	1,385	0.0770	4.80	23.48	0.863

Table A1.9. *Density (ρ), sensible heat (C_p), heat conductivity (λ), thermal diffusivity (α) and Prandtl number (Pr) as a function of temperature*

A1.9. Physical properties of nanofluids

See (Benallou 2019, section A1.12).

A1.10. Physical properties of molten salts

See (Benallou 2019, section A1.13).

A1.11. Physical properties of liquid metals

See (Benallou 2019, section A1.14).

A1.12. Dimensions and thicknesses of tubes for exchangers

Commercially, tube dimensions are standardized. Thus, the diameters cannot be chosen arbitrarily, at the risk of not finding the tube sought when ordering. The diameters of the tubes of an exchanger will therefore be defined from the tables proposed in the standards.

Several standards exist: the American standards (ASME B36-10 for carbon-steel tubes and ASME B36-19 for stainless-steel tubes), the imperial system standards (BWG, *Birmingham Wire Gauge*), the European standards (EN10-220) and the international standard ISO4200.

For heat-exchanger design, tube dimensions are based on the North American nomenclature. Table A1.10 presents the tube characteristics set by this standard. A tube is thus designated by a series of parameters (see Table A1.10), including *the diameter category*. This parameter can be defined either by the *ND (nominal diameter)*, or by the NPS *(Nominal Pipe Size)*. The ND is defined (standard ISO 6708) as a *dimensionless number* that is unequivocally linked to the outer diameter of the tube. Thus, a tube of ND 25 has an outer diameter of 33.4 mm. The NPS also defines the diameter category, but does not give the outer diameter. Thus, a tube of ND 25 has an NPS of 1 inch, but an outer diameter of 33.4 mm, which is different from 24.5 mm, i.e., 1 inch. In summary, by defining the diameter category (by the ND or the NPS) we define the *outer diameter* of the tube, but neither the NPS nor the ND give it directly.

It should also be noted that for the same diameter category – therefore for the same outer diameter – there are several wall thicknesses commercially available. For category ND10, for example, we find tubes with thicknesses of 1.65 to 3.20 mm. The choice of which thickness to use depends on the working pressure. The higher the pressure, the thicker the tube wall will be. This thickness/pressure relationship is given by a parameter called *schedule (SCH)*, which is a dimensionless number (between 5 and 160) defined as follows:

$$SCH = 1{,}000\frac{P}{\tau}$$

where:

– P is the working pressure;

– τ is the allowable stress.

The number (between 5 and 160) that defines the SCH may be followed by the letter "S", when the tube is made of stainless steel, and the letters "XS" for "Extra Strong" or "XXS" for even larger wall thicknesses.

Thus, although SCH does not directly designate a wall thickness, it is linked to the latter since it *defines the ability of the tube considered to withstand pressure.*

Table A1.10 presents the characteristics of the tubes to be used for the sizing of heat exchangers. Note that, for a given diameter category (set NPS or ND), knowledge of the *SCH* unequivocally defines the *inner diameter* of the tube.

Category of dimensionless diameter (ND)	Size: NPS (in inches)	Outer diameter (mm)	Schedule	Thickness (mm)	Inner diameter (mm)	Weight (kg/m)
6	1/8	10.3	10	1.24	7.82	0.28
			30	1.45	7.40	0.32
			40	1.73	6.84	0.37
			80	2.41	5.48	0.47
8	1/8	13.7	10	1.65	10.40	0.49
			30	1.85	10.00	0.54
			40	2.24	9.22	0.63
			80	3.02	7.66	0.80
10	3/8	17.1	10	1.65	13.80	0.63
			30	1.85	13.40	0.70
			40	2.31	12.48	0.84
			80	3.20	10.70	1.10
15	1/2	21.3	5	1.65	18.00	0.80
			10	2.11	17.08	1.00
			30	2.41	16.48	1.12
			40	2.77	15.76	1.27
			80	3.73	13.84	1.62
			160	4.78	11.74	1.95

20	3/4	26.7	5	1.65	23.40	1.03
			10	2.11	22.48	1.28
			30	2.41	21.88	1.44
			40	2.87	20.96	1.69
			80	3.91	18.88	2.20
			160	5.56	15.58	2.90
25	1	33.4	5	1.65	30.10	1.29
			10	2.77	27.86	2.09
			30	2.90	27.60	2.18
			40	3.38	26.64	2.50
			80	4.55	24.30	3.24
			160	6.35	20.70	4.29
32	1 ¼	42.2	5	1.65	38.90	1.65
			10	2.77	36.66	2.69
			30	2.97	36.26	2.87
			40	3.56	35.08	3.39
			80	4.85	32.50	4.47
			160	6.35	29.50	5.61
40	1 ½	48.3	5	1.65	45.00	1.90
			10	2.77	42.76	3.11
			30	3.18	41.94	3.53
			40	3.68	40.94	4.05
			80	5.08	38.14	5.41
			160	7.14	34.02	7.25
50	2	60.3	5	1.65	57.00	2.39
			10	2.11	56.08	3.03
			10	2.77	54.76	3.93
			30	3.18	53.94	4.48
			30	3.58	53.14	5.01
			40	3.91	52.48	5.44
			40	4.37	51.56	6.03
			40	4.78	50.74	6.54
			80	5.54	49.22	7.48
			80	6.35	47.60	8.45
			80	7.14	46.02	9.36
			160	8.74	42.82	11.11
			160	11.07	38.16	13.44

65	2 ½	73	5	2.11	68.78	3.69
			10	2.77	67.46	4.80
			10	3.05	66.90	5.26
			30	3.18	66.64	5.48
			30	3.58	65.84	6.13
			30	3.96	65.08	6.74
			30	4.37	64.26	7.40
			40	4.78	63.44	8.04
			40	5.16	62.68	8.63
			80	5.49	62.02	9.14
			80	6.35	60.30	10.44
			80	7.01	58.98	11.41
			160	9.53	53.94	14.92
			160	14.02	44.96	20.39
80	3	88.9	5	2.11	84.68	4.52
			10	2.77	83.36	5.88
			10	3.05	82.80	6.46
			30	3.18	82.54	6.72
			30	3.58	81.74	7.53
			30	3.96	80.98	8.29
			40	4.37	80.16	9.11
			40	4.78	79.34	9.92
			80	5.49	77.92	11.29
			80	6.35	76.20	12.93
			80	7.14	74.62	14.40
			80	7.62	73.66	15.27
			160	11.13	66.64	21.35
			160	15.24	58.42	27.68
90	3 ½	101.6	5	2.11	97.38	5.18
			10	2.77	96.06	6.75
			10	3.05	95.50	7.41
			10	3.18	95.24	7.72
			10	3.58	94.44	8.65
			30	3.96	93.68	9.53
			30	4.37	92.86	10.48
			30	4.78	92.04	11.41
			40	5.74	90.12	13.57
			40	6.35	88.90	14.92
			80	7.14	87.32	16.63
			80	8.08	85.44	18.63

			5	2.11	110.08	5.84
			10	2.77	108.76	7.62
			30	3.05	108.20	8.37
			30	3.18	107.94	8.71
			30	3.58	107.14	9.77
			30	3.96	106.38	10.78
			30	4.37	105.56	11.85
			40	4.78	104.74	12.91
100	4	114.3	40	5.16	103.98	13.89
			40	5.56	103.18	14.91
			40	6.02	102.26	16.07
			80	6.35	101.60	16.90
			80	7.14	100.02	18.87
			80	7.92	98.46	20.78
			80	8.56	97.18	22.32
			120	11.13	92.04	28.32
			160	13.49	87.32	33.54

Tables A1.10. *Dimensions of tubes for heat exchangers*

For heat-exchanger manufacturing, tubes with diameters corresponding to NPS ≥ ¾" are generally preferred. Tubes of NPS ≤ ½" are difficult to handle in the manufacturing processes, moreover their cleaning is not straightforward, particularly in the case where the fluids used are likely to be charged with impurities. They are nevertheless used in refrigeration heat exchangers given that refrigerants flow in closed loop.

Tubes of NPS ≥ 1 ¼" are used in boilers, evaporators, reboilers and direct-combustion furnaces.

A1.13. Exchanger shell diameters

The shell diameter is determined based on heat-transfer and pressure-drop considerations (see Chapter 6). On the one hand, heat-transfer requirements determine the transfer area necessary to accomplish a thermal program, and therefore the number of tubes of which the heat exchanger should be composed. On the other hand, flow considerations determine the diameter of the shell capable of accommodating the determined number of tubes, without creating unacceptable pressure drops.

For a known number of tubes, the shell diameter can be determined as a function of the number of tube-side passes and the specifications of the parameters that characterize the chosen array, such as the pitch, array type, etc. Thus, Tables A1.11 to A1.20 gives the total number of tubes insertable into a shell of diameter, D_{sh}, as a function of the number of passes.

But, the shell diameter – or indeed the number of tubes insertable into a shell of a given diameter – can also be calculated directly based on procedures developed for numerical calculations (see section 4.6.1.8). This type of procedure is extremely useful in computerized shell-diameter calculations.

Type of array: Square			**Pitch: 1 inch**		
Tubes: NPS 3/4 inches			**Inner diameter of the shell: D_{sh}**		
D_{sh} (in inches)	**1 pass**	**2 passes**	**4 passes**	**6 passes**	**8 passes**
8	32	26	20	20	
10	52	52	40	36	
12	81	76	68	68	60
13 1/4	97	90	82	76	70
15 1/4	137	124	116	108	108
17 1/4	177	166	158	150	142
19 1/4	224	220	204	192	188
21 1/4	277	270	246	240	234
23 1/4	341	324	306	302	292
25	413	394	370	356	346
27	461	460	432	420	408
29	553	526	480	462	456
31	657	640	600	580	560
33	749	712	638	676	648
35	845	824	780	766	748
37	934	914	886	866	838
39	1,049	1,024	982	968	948

Table A1.11. *Total number of tubes insertable into a shell of diameter D_{sh} as a function of the number of tube-side passes for a square array measuring 1 inch*

Array: Square			Pitch: 1 1/4 inches		
Tubes: NPS 1 inch			Inner diameter of the shell: D_{sh}		
D_{sh} (in inches)	1 pass	2 passes	4 passes	6 passes	8 passes
8	21	16	14		
10	32	32	26	24	
12	48	45	40	38	36
13 1/4	61	56	52	46	44
15 1/4	81	76	63	63	64
17 1/4	112	112	96	90	82
19 1/4	138	132	120	122	116
21 1/4	177	166	158	152	148
23 1/4	213	208	192	184	184
25	260	252	238	226	222
27	300	268	278	26	260
29	341	326	300	294	286
31	406	398	360	360	358
33	465	460	432	420	414
35	522	510	400	484	472
37	596	574	562	544	532
39	665	644	624	612	600

Table A1.12. *Total number of 1-inch tubes insertable into a shell of inner diameter D_{sh} as a function of the number of tube-side passes for a square array measuring 1 1/4 inches*

Array: Square			Pitch: 1 9/16 inches		
Tubes: NPS 1 1/4 inches			Inner diameter of the shell: D_{sh}		
D_{sh} (in inches)	1 pass	2 passes	4 passes	6 passes	8 passes
10	16	12	10		
12	30	24	22	16	16
13 1/4	32	30	30	22	22
15 1/4	44	40	37	35	31
17 1/4	56	53	51	48	44
19 1/4	78	73	71	64	56
21 1/4	96	90	86	82	78

23 1/4	127	112	106	102	96
25	140	135	127	123	115
27	166	160	151	146	140
29	193	159	170	174	166
31	226	220	209	202	193
33	258	252	244	238	226
35	293	287	275	268	253
37	334	322	311	304	293
39	370	362	340	342	336

Table A1.13. *Total number of 1-1/4-inch tubes of insertable into a shell of inner diameter D_{sh} as a function of the number of tube-side passes for a square array measuring 1 1/26 inches*

Array: Square			**Pitch: 1 7/8 inches**		
Tubes: NPS 1 1/2 inches			**Inner diameter of the shell: D_{sh}**		
D_{sh} (in inches)	**1 pass**	**2 passes**	**4 passes**	**6 passes**	**8 passes**
12	16	16	12	12	
13 1/4	22	22	16	16	
15 1/4	29	29	25	24	
17 1/4	39	39	34	32	22
19 1/4	50	48	45	43	29
21 1/4	62	60	57	54	39
23 1/4	78	74	70	66	50
25	94	90	36	64	78
27	112	108	102	98	94
29	131	127	120	116	112
31	151	146	141	132	131
33	176	170	164	160	151
35	202	196	138	182	176
37	224	220	217	210	202
39	252	246	237	230	224

Table A1.14. *Total number of 1-1/2-inch tubes insertable into a shell of inner diameter D_{sh} as a function of the number of tube-side passes for a square array measuring 1 7/8 inches*

Array: Triangular			Pitch: 15/16 inches		
Tubes: NPS 3/4 inches			Inner diameter of the shell: D_{sh}		
D_{sh} (in inches)	1 pass	2 passes	4 passes	6 passes	8 passes
8	36	32	26	24	18
10	62	56	47	42	36
12	109	98	96	82	78
13 1/4	127	114	96	90	86
15 1/4	170	160	140	136	128
17 1/4	239	224	194	189	173
19 1/ 4	301	282	252	244	234
21 1/4	361	342	314	306	290
23 1/4	442	420	386	378	364
25	532	506	468	446	434
27	637	602	550	536	524
29	721	692	640	620	594
31	847	822	766	722	720
33	974	938	878	852	826
35	1,102	1,068	1,004	988	958
37	1,240	1,200	1,144	1,104	1,072
39	1,377	1,330	1,258	1,248	1,212

Table A1.15. *Total number of 3/4-inch tubes insertable into a shell of diameter D_{sh} as a function of the number of tube-side passes for a square array measuring 15/16 inches*

Array: Triangular			Pitch: 1 inch		
Tubes: NPS ¾ inches			Inner diameter of the shell: D_{sh}		
D_{sh} (in inches)	1 pass	2 passes	4 passes	6 passes	8 passes
8	37	30	24	24	
10	61	52	40	36	
12	92	62	76	74	70
13 1/4	109	106	86	86	74
15 1/4	151	138	122	118	110
17 1/4	203	196	173	172	166
19 1/4	626	250	226	216	210
21 1/4	316	302	270	272	260

23 1/4	384	376	252	342	328
25	470	452	422	394	362
27	559	534	488	474	464
29	630	604	556	538	508
31	745	728	678	666	640
33	856	830	774	760	732
35	970	938	882	864	848
37	1,074	1,044	1,012	986	870
39	1,206	1,176	1,128	1,100	1,078

Table A1.16. *Total number of tubes insertable into a shell of diameter D_{sh} as a function of the number of tube-side passes for a triangular array measuring 1 inch*

Array: Triangular			Pitch: 1 1/4 inches		
Tubes: NPS 1 inch			Inner diameter of the shell: D_{sh}		
D_{sh} (in inches)	1 pass	2 passes	4 passes	6 passes	8 passes
8	21	16	16	14	
10	32	32	26	24	
12	55	52	48	46	44
13 1/4	68	66	58	54	50
15 1/4	91	86	80	74	72
17 1/4	131	118	106	104	94
19 1/4	163	152	140	136	128
21 1/4	199	188	170	164	160
23 1/4	241	232	212	212	202
25	294	282	256	252	242
27	349	334	302	296	286
29	397	376	338	334	316
31	472	454	430	424	400
33	538	522	486	470	454
35	603	592	562	548	532
37	674	664	632	614	598
39	766	736	700	688	672

Table A1.17. *Total number of tubes insertable into a shell of diameter D_{sh} as a function of the number of tube-side passes for a triangular array measuring 1 1/4 inches*

Array: Triangular			Pitch: 1 9/16 inches		
Tubes: NPS 1 1/4 inches			Inner diameter of the shell: D_{sh}		
D_{sh} (in inches)	1 pass	2 passes	4 passes	6 passes	8 passes
10	20	18	14		
12	32	30	26	22	20
13 1/4	38	36	32	28	26
15 1/4	54	51	45	42	38
17 1/4	69	66	62	58	54
19 1/4	95	91	86	79	69
21 1/4	117	112	105	101	95
23 1/4	140	136	130	123	117
25	170	164	155	150	140
27	202	196	185	179	170
29	235	228	217	212	202
31	275	270	255	245	235
33	315	305	297	288	275
35	357	348	335	327	315
37	407	390	380	374	357
39	449	436	425	419	407

Table A1.18. *Total number of tubes 1 1/4 inches in diameter, insertable into a shell of diameter D_{sh} as a function of the number of tube-side passes for a square array measuring 1 9/16 inches*

Array: Triangular			Pitch: 1 7/8 inches		
Tubes: NPS 1 1/2 inches			Inner diameter of the shell: D_{sh}		
D_{sh} (in inches)	1 pass	2 passes	4 passes	6 passes	8 passes
12	18	14	14	12	12
13 1/4	27	22	18	16	14
15 1/4	36	34	32	30	27
17 1/4	48	44	42	38	36
19 1/4	61	58	55	51	48
21 1/4	76	72	70	66	61
23 1/4	95	91	86	80	76
25	115	110	105	98	95
27	136	131	125	118	115

29	160	154	147	141	136
31	194	177	172	165	160
33	215	106	200	190	184
35	246	238	230	220	215
37	275	268	260	252	246
39	307	299	290	284	275

Table A1.19. *Total number of tubes insertable into a shell of diameter D_{sh} as a function of the number of tube-side passes for a square array measuring 1 7/8 inches*

A1.14. Fouling coefficients and resistances

Fluid	h_f(Wm^{-2}°C^{-1})		Resistance (m^2°CW^{-1})	
	from	to	from	to
Dust-free air	4,000	7,000	2.50E-04	1.43E-04
Undusted air	3,000	5,000	3.33E-04	2.00E-04
Vapor condensate	1,500	5,000	6.67E-04	2.00E-04
Seawater	1,800	4,500	5.56E-04	2.22E-04
Cooling water (towers)	3,000	6,000	3.33E-04	1.67E-04
Filtered river water	2,500	9,000	4.00E-04	1.11E-04
Unfiltered river water	1,500	4,500	6.67E-04	2.22E-04
City water (soft)	3,000	5,000	3.33E-04	2.00E-04
City water (hard)	1,000	2,000	1.00E-03	5.00E-04
Distilled water	9,500	11,000	1.05E-04	9.09E-05
Brackish water	1,900	2,850	5.26E-04	3.51E-04
Gasoline, kerosene	4,900	5,000	2.04E-04	2.00E-04
Fuel oil No. 2	2,800	3,000	3.57E-04	3.33E-04
Fuel oil No. 6	10,000	11,000	1.00E-04	9.09E-05
Heat-transfer fluids	4,800	5,200	2.08E-04	1.92E-04
Combustion gas	4,500	5,000	2.22E-04	2.00E-04
Industrial gases	2,000	5,000	5.00E-04	2.00E-04
Lubricating oils	5,000	5,500	2.00E-04	1.82E-04
Transformer oil	5,500	6,000	1.82E-04	1.67E-04
Light hydrocarbons	4,800	5,000	2.08E-04	2.00E-04
Heavy hydrocarbons	1,850	2,100	5.41E-04	4,76E-04
Organic liquids	1,800	5,500	5.56E-04	1.82E-04
Liquid refrigerants	5,000	6,000	2.00E-04	1.67E-04

Organic condensation matter	4,500	5,600	2.22E-04	1.79E-04
Organic matter under boiling conditions	1,900	2,600	5.26E-04	3.85E-04
Fuel oil	980	1,100	1.02E-03	9.09E-04
Cracking residues	400	500	2.50E-03	2.00E-03
Caustic solutions	2,800	3,000	3.57E-04	3.33E-04
Brine	3,000	5,000	3.33E-04	2.00E-04
Aqueous saline solutions	3,500	5,200	2.86E-04	1.92E-04
Methanol, ethanol or ethylene glycol solutions	2,800	3,000	3.57E-04	3.33E-04
Water vapor without traces of oils	4,000	10,000	2.50E-04	1.00E-04
Water vapor with traces of oils	2,500	5,000	4.00E-04	2.00E-04
Organic vapors	1,200	4,300	8.33E-04	2.33E-04

Table A1.20. *Exchanger fouling coefficients and resistances. For plate heat exchangers, take 20 to 25% of the fouling resistances shown in this table (source: Marriott (1979; Cooper et al. (1983)*

A1.15. Overall heat transfer coefficient for "fouled" heat exchangers

Exchangers with no phase change					
Finless devices		**U_F (kcal/hr m^2°C)**		**U_F (W /m^2°C)**	
Hot fluid	**Cold fluid**	**from**	**to**	**from**	**to**
Gas	Gas	10	50	12	58
Gas	Viscous liquid	20	50	23	58
Gas	Low-viscosity liquid	20	80	23	93
Viscous liquid	Gas	20	50	23	58
Low-viscosity liquid	Gas	20	80	23	93
Viscous liquid	Viscous liquid	100	200	116	232
Viscous liquid	Low-viscosity liquid	100	300	116	348
Low-viscosity liquid	Viscous liquid	200	550	232	639
Low-viscosity liquid	Low-viscosity liquid	700	1,800	813	2,090
Vapor	Viscous liquid	200	400	232	464
Vapor	Low-viscosity liquid	1,000	2,000	1.161	2.322
Vapor	Boiling liquid	700	1,500	813	1,742
Devices with fins		**U_F (kcal/hr m^2°C)**		**U_F (W /m^2°C)**	
Hot fluid	**Cold fluid**	**from**	**to**	**from**	**to**
Gas	Gas	30	85	35	100
Gas	Viscous liquid	35	90	40	105

Gas	Low-viscosity liquid	45	150	50	175
Viscous liquid	Gas	35	100	40	120
Low-viscosity liquid	Gas	45	150	50	175

Table A1.21. *Orders of magnitude of U_F for exchangers with no phase change*

Boilers and evaporators	**U_F (kcal/hr m^2°C)**		**U_F (W /m^2°C)**	
Boilers with long, vertical tubes	**from**	**to**	**from**	**to**
Natural-circulation vertical evaporators	860	2,320	1,000	2,700
Forced-circulation vertical evaporators	1,720	6,450	2,000	7,500
Boilers with short, vertical tubes and stirred-film evaporators	640	2,150	750	2,500
Low to medium viscosities: < 1 Poise	1,550	2,320	1,800	2,700
High viscosities: > 1 Poise	1,200	1,370	1,400	1,600
Falling-film evaporators (viscosities < 0.1 Poise)	430	2,150	500	2,500
Rising-film evaporators	1,720	4,300	2,000	5,000

Table A1.22. *Orders of magnitude of U_F for boilers and evaporators*

Helix-in-tank heat exchangers					
Coil type	**Operation**	**U_F (kcal/hr m^2°C)**		**U_F (W /m^2°C)**	
		from	**to**	**from**	**to**
Immersed	In heating mode	510	1,290	600	1,500
	In cooling mode	210	680	250	800
External	In heating mode	510	940	600	1,100
	In cooling mode	170	600	200	700
Immersed	In heating mode	340	770	400	900
	In cooling mode	170	510	200	600
External	In heating mode	210	550	250	650
	In cooling mode	120	380	150	450

Table A1.23. *Orders of magnitude of U_F for helix-in-tank heat exchangers (source: Fletcher (1987) and personal data)*

A1.16. Characteristics of heat exchanger plates

Plate type	A_p (m^2)	a (mm)	b (mm)	e (mm)	δ (mm)	Δx (mm)	NTU_{max}	NTU_{min}	Similar plates		
									Universal Hydraulik	APV	Alfa Laval
LL1	5.57	4,095	1,550	4.00	500	0.4	6	1			AQ20-FM
LL2	5.35	3,951	1,550	3.95	500	0.4	6	1.6		AR4-2	AQ20-FG
LL3	5.35	3,951	1,550	3.95	500	0.4	6	1.6		AR4-2	AQ20-FD
LL4	2.62	3,103	920	2.52	250	0.4	5	1.1		AR12-4	AQ10-FG
LL5	2.68	3,103	940	3.95	250	0.4	5	1.1		AR12-4	AQ10-FD
LL6	2.68	3,103	940	3.95	250	0.4	5	1.1		AR12-4	AQ10-FS
LL7	1.55	2,183	780	2.90	200	0.4	4.9	1.5	TL 850	AR4-3	AQ8-P
LL8	2.86	2,736	1,174	2.00	300	0.4	4.5	1.5			AQ14
LL9	2.32	2,711	942	2.90	250	0.4	4.2	1.6			AQ10T-P
LL10	2.20	2,595	920	2.00	200	0.4	4	1.5			AQ10-FMS
LL11	2.20	2,596	920	2.00	200	0.4	4	1.6			AQ10-FGS
LL12	2.25	2,661	913	2.00	200	0.4	4	1.5			AQ10T-B
LL13	1.19	2,036	650	4.00	200	0.4	4	1.1			AQ6-M
LL14	2.98	4,146	755	4.00	200	0.4	4	1.5			AQ8-M
LL15	1.12	1,871	650	3.80	150	0.4	3.9	1.2	TL 650	AR4-2	AQ6T-M
LL16	1.55	2,141	780	1.90	150	0.4	3.9	1.2	TL 850		AQ8-B
LL17	1.03	1,833	610	2.42	150	0.4	3.85	1	TL 500; TL 650	AR4-2	AQ6T-B
LL18	3.30	3,210	1,154	2.50	350	0.4	3.8	1.2		AR12-5	AQ14L-TL35-FM
LL19	3.30	3,210	1,154	2.50	350	0.4	3.8	1.2		AR12-5	AQ14L-TL35-FG

LL20	3.37	3,218	1,174	2.50	350	0.4	3.8	1.2		AR16-1	AQ14L-TL35-FD
LL21	3.37	3,218	1,174	2.50	350	0.4	3.8	1.2		AR16-1	AQ14-TL35-LFS
LL22	1.02	1,815	610	2.50	150	0.4	3.8	1	TL 500	AR4-1	AQ6-B
LL23	1.00	1,405	800	4.00	150	0.4	2.9	0.9		AR4-1	AQ8S
LL24	1.53	2,752	610	4.00	250	0.4	0.9	0.1		AR4-3	AQ6L
LL25	0.45	1,054	470	3.95	100	0.4	0.85	0.1	TL 250	AR2-1	AQ4T-M
LL26	0.46	1,084	470	2.60	100	0.4	0.85	0.1	TL 250	AR2-1	AQ4-B
LL27	0.46	1,084	470	4.00	100	0.4	0.85	0.1	TL 250	AR2-1	AQ4-M
LL28	0.34	890	400	4.00	50	0.4	0.8	0.1			AQ3
LL29	0.36	957	420	2.52	100	0.4	0.8	0.1			AQ4T-B
LL30	0.28	920	320	2.00	50	0.4	0.8	0.1			AQ2-M6
LL31	0.29	940	330	3.00	50	0.4	0.8	0.1			AQ2-M6-M
LL32	0.39	1,264	320	1.80	50	0.4	0.8	0.1	TL 250	AR2-1	AQ2L
LL33	0.26	704	400	4.00	65	0.4	0.8	0.1	PWT 90		AQ2S-TS6-FG
LL34	0.26	870	320	1.80	50	0.4	0.8	0.1	PWT 100		AQ2T-B
LL35	0.28	890	330	3.00	50	0.4	0.8	0.1	PWT 100		AQ2T-P

Table A1.24. *Characteristics of heat exchanger plates for liquid/liquid heating or cooling*

Plate type	A_p (m²)	a (mm)	b (mm)	e (mm)	δ (mm)	Δx (mm)	NTU_{max}	NTU_{min}	Similar plates		
									Universal Hydraulik	APV	Alfa Laval
CV1	0.74	1,405	610	2.5	200	0.4	5.5	1.2		AR4-1	A15-B
CV2	2.68	3,210	920	4.0	300	0.4	5	2			AX35
CV3	1.61	2217	780	1.90	150	0.4	4.2	1		AR4-3	AX30-B
CV4	2.25	2661	913	2.0	200	0.4	4	1.5			A45
CV5	1.06	1,833	650	2.4	200	0.4	4	1.5	TL 500; TL 650	AR4-2	A20-B
CV6	0.90	1,405	780	2.5	250	0.4	2.9	0.6		AR3-1	AM20-B
CV7	0.34	890	400	2.3	50	0.4	2.9	0.6	TL 90; PWT 100		T8-B
CV8	0.34	890	400	3.9	50	0.4	2.9	0.9	TL 90; PWT 100		T8-M
CV9	0.26	870	320	1.8	50	0.4	2.8	1	TL 90; PWT 100		AM10-B
CV10	2.00	2,752	800	2.0	250	0.4	1.8	0.6			A35
CV11	0.78	1,815	470	2.5	150	0.4	1.2	0.4	TL 400; TL 500	AR4-2	AM20
CV12	0.42	1,054	470	2.52	150	0.4	1	0.3	TL 150	AR2-1	AM10
CV13	0.47	1,054	470	2.52	50	0.4		0.3	TL 1560	AR2-1	AK20
CV14	0.45	1,054	470	2.52	100	0.4		0.1	TL 150	AR2-1	T10-B
CV15	0.45	1,054	470	3.95	100	0.4		0.1	TL 150	AR2-1	T10-M
CV16	0.05	311	173	3.95	50	0.4		0.1	PWT 40		AQ1

Table A1.25. *Characteristics of vapor-heated exchanger plates*

Plate type	A_p (m²)	a (mm)	b (mm)	e (mm)		δ (mm)			Δx (mm)	NTU_{max}	NTU_{min}	Similar Alfa Laval plates
				Vap	Coolant	Vap	Condensates	Coolant				
CD1	1.12	2,217	617	11	3	400	100	150	0.5	7	1	AlfaCond400 Semi-welded AC400-W
CD2	2.01	2,860	888	11	3	600	150	250	0.5	7	1	AlfaCond60 Semi-welded AC600-W
CD3	1.64	3,191	686	11	3	800	200	300	0.5	7	1	AlfaCond600 Semi-welded AC800-W

Table A1.26. *Characteristics of plates for plate condensers*

Plate type	A_p (m²)	a (mm)	b (mm)	e (mm)	δ (mm)				Δx (mm)	NTU_{max}	NTU_{min}	Similar plates	
					Product Inlet	Product Outlet	Vapor Inlet	Condensate Outlet				Universal Hydraulik	Alfa Laval
BEV1	1.98	2,610	806	6	150	350	350	100	0.6	7	1		AlfaVap 350 Semi-welded EC350-W
BEV2	1.41	1,903	806	6	150	500	500	100	0.6	7	1		AlfaVap 500 Semi-welded EC500-W
BEV3	2.79	3,210	913	6.5	150	650	450	150	0.6	7	1		AlfaVap 650 Semi-welded EC650-W
BEV4	2.90	2,710	1,156	6	200	700	400	200	0.6	7	1		AlfaVap 700 Semi-welded EC700-W

Table A1.27. *Characteristics of plates for plate evaporators and boilers*

A1.17. Characteristics of standard spiral heat exchangers

Spiral type	Large diameter D_s (mm)	Small diameter d_s (mm)	Width b (mm)	Gap e (mm)	Wall thickness Δx (mm)
SP1	813	203	102	4.76	3.18
SP2	813	203	152	4.76	3.18
SP3	813	203	305	4.76	3.18
SP4	1,473	305	305	4.76	3.18
SP5	813	203	457	6.35	3.18
SP6	1,473	305	457	6.35	4.76
SP7	813	203	610	6.35	4.76
SP8	1,473	305	610	6.35	4.76
SP9	1,473	305	762	6.35	4.76
SP10	1,473	305	914	6.35	4.76
SP11	1,473	305	1,219	6.35	4.76
SP12	1,473	305	1,524	7.94	6.35
SP13	1,473	305	1,778	9.53	7.94
SP14	1,473	305	1,778	10.25	7.94
SP15	1,473	305	1,778	11.53	7.94
SP16	1,473	305	1,219	12.03	7.94
SP17	1,473	305	1,524	12.03	7.94

Table A1.28. *Dimensions of standard spiral heat exchangers (source: Picon et al. (2007))*

A1.18. Unit conversion tables

The different units are listed alphabetically.

Initial unit	Final unit	Multiply by
Acres	ha	4.074E-01
	m^2	4.074E+03
Acre-feet	m^3	1.233E+03
Ah	Cb	3.600E+03
Angstrum	m	1.000E-10

Atmosphere	Bar	1.013E+00
	cm H_2O	1.000E+02
	g/m^2	1.033E+03
	lb/in^2 (psi)	1.470E+01
	mmHg	7.600E+02
	N/m^2	1.013E+05
	Pa	1.013E+05
Bags (cement)	kg (cement)	4.264E+01
Bar	Atm	9.869E-01
	cm H_2O	9.869E+01
	lb/in^2 (psi)	1.450E+01
	mm Hg	7.500E+02
	N/m^2	1.000E+05
	Pa	1.000E+05
Barrel (cement)	kg (cement)	1.705E+02
Barrel (oil)	liters	1.590E+02
Btu	cal	2.520E+02
	Joules	1.055E+03
	kWh	2.930E-04
Btu/hr	W	2.931E+00
Btu/hr-ft-°F	W/m-°C	1.731E+00
Btu/hr-ft^2-°F	kcal/hr-m^2-°C	4.880E+00
	cal/s-cm^2-°C	1.355E-04
	W/cm^2-°C	5.680E-06
	W/m^2-°C	5.678E+00
Btu/lb	J/kg	2.326E+03
Btu/lb°F	cal/g°C	1.000E+00
Btu/sec	W	1.054E+03
Bushels (US dry)	m^3	3.524E-02
cal	Btu	3.968E-03
	kWh	1.163E-06
	J	4.187E+00
	liters-Atm	4.130E-02

Carat	kg	2.000E-04
	lb	4.409E-04
Centipoises (Cp)	g/sec-cm or Poise	1.000E-02
	kg/hr-m	3.600E+00
	kg/sec-m	1.000E-03
	lb/sec-ft	6.720E-04
	lb/hr-ft	2.420E+00
	N-s/m, i.e. Pa s or Poiseuil	1.000E-03
Ch or Hp	W	735.49875
cm	in	3.937E-01
	ft	3.281E-02
	yd	1.094E-02
cm H_2O	Atm	1.000E-02
	Bar	1.013E-02
	mm Hg	7.600E+00
	N/m^2	1.013E+03
	Pa	1.013E+03
cm^2	ft^2	1.076E-03
Curie	Disintegrations/min.	2.200E+12
Dynes	N	1.000E-05
Ergs	J	1.000E-07
eV	Btu	1.520E-22
Faraday (F)	Coulomb (Cb)	9.650E+04
ft	m	3.048E-01
	in	1.200E+01
ft/min	cm/sec	5.080E-01
ft^2	m^2	9.290E-02
ft^2/hr	m^2/sec	2.581E-05
ft^3	m^3	2.832E-02
Gallons	liters	3.785E+00
	m^3	3.785E+03
Hectares (ha)	Acres	2.471E+00
	m^2	1.00E+04
in	cm	2.540E+00
	m	2.540E-02
in^2	cm^2	6.452E+00
in^3	m^3	1.639E-05

Joule	Btu	9.478E-04
	cal	2.388E-01
	Wh	2.778E-04
	kWh	2.778E-07
kcal	kWh	1.163E-03
kcal/hm °C	W/m °C	1.161E+00
kcal/hm^2 °C	W/m^2 °C	1.161E+00
kg	lb	2.205E+00
lb	kg	4.536E-01
lb/ft^2	Atm	4.725E-04
lb/ft^2	kg/m^2	4.882E+00
lb/ft^3	kg/m^3	1.602E+01
lb/in^2 (psi)	Atm	6.805E-02
	kg/cm^2	7.031E-02
	N/m^2	6.895E+03
lb/sec-ft	N-sec/m^2	1.448E+00
lb$_f$	N	4.448E+00
lb$_f$/ft^2	N/m^2	4.788E+01
Lumens	W	1.496E-03
m	ft	3.281E+00
	in	3.937E+01
	yd	1.094E+00
	mile	6.214E-04
	rods	1.988E-01
mile	m	1.609E+03
	rods	3.200E+02
mile/hr	m/sec	4.470E-01
	km/hr	1.609E+00
mm Hg	Atmosphere	1.316E-03
	Bar	1.333E-03
	cm H_2O	1.316E-01
	N/m^2	1.333E+02
	Pa	1.333E+02
Pa	N/m^2	1.000E+00
	Atmosphere	9.869E-06
	Bar	1.000E-05
	cm H_2O	9.869E-04
	mm Hg	7.501E-03

Poise	Centipoises (Cp)	1.000E+02
	g/sec-cm	1.000E+00
	lb/sec-ft	6.720E-02
	lb/hr-ft	2.419E+02
	kg/hr-m	3.600E+02
	N-sec/m, i.e. Pa s	1.000E-01
Poiseuil	Poise	1.000E+01
	Centipoises	1.000E+03
	N-sec/m^2	1.000E+00
Quarts (US, liquid)	liters	9.464E-01
Rod (rd)	ft	1.650E+01
Slugs	kg	1.459E+01
Stokes	m^2/sec	1.000E-04
toe	Btu	3.981E+07
	Calories	1.003E+10
	kWh	1.167E+04
	Joules	4.200E+10
Torr	N/m^2	1.333E+02
W	Btu/hr	3.413E+00
	Btu/sec	1.411E-04
	Uranium nuclei fissions/sec	3.100E+10
W/m °C	cal/hr-m-°C	8.612E+02
	kcal/hr-m-°C	8.612E-01
	Btu/hr-ft-°F	5.777E-01
W/m^2 °C	cal/hr-m-°C	2.116E+06
	kcal/hr-m-°C	2.116E+03
	Btu/hr-ft-°F	1.761E+05

Table A1.29. *Unit conversions*

Appendix 2

Calculation of Convection Heat Transfer Coefficients in Exchangers

A2.1. Introduction

The calculation of the overall heat transfer coefficient, U, is an important step in the sizing of a heat exchanger. It should be recalled that the determination of this coefficient requires the calculation of the *convective coefficients,* h_i and h_e (see section 5.3.1).

This appendix presents the correlations available in the literature for the calculation of the convective coefficients, also called convective heat transfer coefficients, *film coefficients* or *local coefficients*. These correlations are suitable for the calculations on the tube side, h_i, and on the shell side, h_e.

Sections A2.2 to A2.11 proceed by type of exchanger:

– shell-and-tube heat exchangers;

– coaxial single-tube heat exchangers;

– tank-and-coil heat exchangers;

– fin-and-tube heat exchangers;

– plate-fin heat exchangers;

– plate heat exchangers;

– spiral heat exchangers;

– printed-circuit heat exchangers;

For a color version of all figures in this chapter, see www.iste.co.uk/benallou/energy6.zip.

– heat exchangers operating with two-phase fluids: condensers, boilers and evaporators;

– heat exchangers operating with metallic liquids or molten salts.

The different presentations pay particular attention to compact heat exchangers.

Furthermore, the correlations are presented in *Nusselt number* form. Note, however, that in the literature, correlations may be encountered in the form of the *Colburn factor, j*. Such presentations are equivalent since the *j* factor is related to the Nusselt number as follows: $j = St\ Pr^{\frac{2}{3}} = \frac{Nu}{Re\ Pr} Pr^{\frac{2}{3}} = \frac{Nu}{Re\ (Pr)^{\frac{1}{3}}}$.

A2.2. Calculation of coefficients for shell-and-tube heat exchangers

Multitube heat exchangers are very widely used in industry thanks to their fairly simple design and relatively low production costs.

Section A2.2 presents the correlations available for the calculation of convective heat transfer coefficients (film coefficients) for single-phase flow multitube heat exchangers. The calculation of these coefficients is presented both for the tube-side and the shell-side heat exchange. The different correlations in sections A2.2.1 to A2.3 are valid both for co-flow configurations and for counter-flow circulations.

A2.2.1. *Coefficients inside the tubes, h_i*

On the tube side, laminar and turbulent regimes are distinguished (see Volume 3, sections 2.2.2.1 and 2.4.2). The recommended correlations for each case are presented in Table A2.1.

Reynolds domain	Regime	Recommended correlation
$Re < 2{,}300$	Laminar	$h_i = \frac{48}{11} \frac{\lambda_i}{d_i}$
$Re \geq 2{,}300$	Turbulent	$Nu_i = 0.023\ Re^{0.8}\ Pr_i^{\frac{1}{3}}$

Table A2.1. *Nusselt number for flow inside tubes*

where $Nu = \frac{h_i D_i}{\lambda_i}$; $Re = \frac{d_i v_i \rho_i}{\mu_i}$; $Pr = \frac{Cp_i \mu_i}{\lambda_i}$.

A2.2.2. *Shell-side coefficient, h_e*

In order to determine the heat transfer coefficient between the outside of the tubes and the fluid circulating in the shell, we should take into consideration the fact that, in this space, the fluid does not always circulate parallel or perpendicular to the tubes. The flow is parallel to the tubes only when the exchanger is not equipped with transverse baffles and when far from the inlets. Otherwise (transverse-baffle devices), the flow is rather complex. As we have seen in section 2.5, it sometimes occurs parallel to the tubes (passage through a baffle), and sometimes intersects the tubes between two consecutive baffles. Consequently, a distinction will be made between shells without baffles and shells with baffles.

A2.2.2.1. *Shell space without baffles*

A2.2.2.1.1. Equivalent diameter

Even without baffles, the flow in the shell is not simple. The presence of tubes imposes the use of the equivalent diameter, D_E, defined by (see Volume 3, section 3.2):

$$D_E = \frac{4S}{P}$$

where S is the free cross-section area perpendicular to the flow and P is the wetted perimeter.

The expression of the equivalent diameter will therefore depend on the array type, and more specifically on the longitudinal (e_L) and transverse (e_T) gaps, which themselves are expressed as a function of the pitch, a (see section 2.4.6.4 for the definition of e_L and e_T, and their relationships with the pitch, a). Indeed, we can see (Figure A.21) that for a square array, the cross-section area, S, is equal to that of the square, minus the cross-section area of a tube. Likewise for a triangular array, the area is that of the triangle, minus the cross-section area of a tube. Whereas in both situations, the perimeter is that of the circle that defines the tube.

We will thus obtain:

– for an in-line or staggered square array: $D_E = \dfrac{4\left(a^2 - \frac{\pi d_o^2}{4}\right)}{\pi d_o} = \dfrac{4a^2 - \pi d_o^2}{\pi d_o}$;

– for an in line triangular array: $D_E = \dfrac{4\left[a^2 \sin\left(\frac{\pi}{3}\right) - \frac{\pi d_o^2}{4}\right]}{\pi d_o} = \dfrac{2a^2\sqrt{3} - \pi d_o^2}{\pi d_o}$.

More generally, for an arbitrary array: $D_E = \dfrac{4 e_T e_L - \pi d_o^2}{\pi d_e}$.

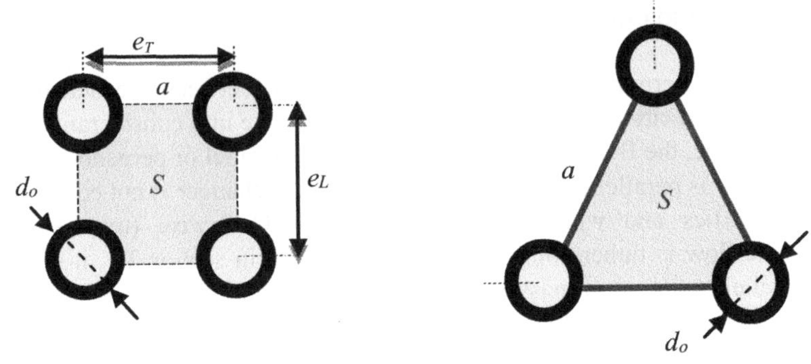

Figure A2.1. *S and P for square and triangular arrays of pitch a*

A2.2.2.1.2. Reynolds number

Two Reynolds numbers are used in the calculations:

– a Reynolds number to describe the flow in the baffle space, Re_E, based on the equivalent diameter;

– a Reynolds number used in the correlations enabling calculation of the coefficient of the film, Re_d, based on the outer diameter of the tubes, d_o:

$$Re_E = \frac{D_E G}{\mu};\ Re_d = \frac{d_o G}{\mu}$$

where:

– $G = \frac{4W}{\pi\left(D_{sh}^2 - N_t d_o^2\right)}$;

– W is the mass flow rate in the shell;

– N_t is the total number of tubes placed in the shell of diameter D_{sh};

– μ is the viscosity of the fluid circulating on the shell side.

A2.2.2.1.3. Calculation of the film coefficient

For the shell side, the following correlations are recommended (Zukauskas 1972; 1987; Yvanovich et al. 1988; 2005). They incorporate the influence of the array type in a factor F, such that the average Nusselt number in the shell is expressed as follows, as a function of the Reynolds and Prandtl numbers:

$$Nu_d = C\ F\ Re_d^n Pr^m$$

where C is a constant. It is given, together with exponents m and n, in Tables A2.2 and A2.4, according to the array type and the equivalent Reynolds number, Re_E.

The *array factor*, F, is given in Tables A2.3 and A2.5, as a function of the number of rows in the bundle:

$$Nu_d = \frac{h_e d_o}{\lambda};\ \mathrm{Re_d} = \frac{\mathrm{d_o G}}{\mu};\ G = \frac{4W}{\pi(D_c^2 - N_t d_o^2)}$$

where:

– d_o is the outer diameter of the tubes;

– W is the mass flow rate in the shell;

– N_t is the total number of tubes in the shell of diameter D_{sh};

– μ is the viscosity of the fluid circulating on the shell side;

– λ is the heat conductivity of the fluid circulating on the shell side.

Taking into account the correction for large temperature differences between the fluid and the tube walls (Rohsenow et al. 1985), the correlation becomes:

$$Nu_d = C\ F\ Re_d^n Pr^m \left(\frac{\mu}{\mu_p}\right)^{0.25}$$

μ_p being the viscosity of the shell fluid expressed at the temperature of the tube wall.

Reynolds number	Recommended correlation
10-100	$Nu = 0.9F\ Re_d^{0.4} Pr^{0.36} \left(\frac{\mu}{\mu_p}\right)^{0.25}$
100-1,000	$Nu = 0.52F\ Re_d^{0.5} Pr^{0.36} \left(\frac{\mu}{\mu_p}\right)^{0.25}$
1,000 to 2 10^5	$Nu = 0.27F\ Re_d^{0.63} Pr^{0.36} \left(\frac{e_T}{e_L}\right)^{0.2} \left(\frac{\mu}{\mu_p}\right)^{0.25}$
> 2 10^5	$Nu = 0.21F\ Re_d^{0.84} Pr^{0.4} \left(\frac{\mu}{\mu_p}\right)^{0.25}$

Table A2.2. *Correlations to be used for in-line tube bundles (finless tubes)*

NOTE.– The correlation presented in this section is reserved for *bare tubes*, i.e. without fins. The case where the tubes are fitted with surface extensions, such as those shown in Figures 1.4 and 3.2, is covered in sections A2.5.2.2 and A2.5.2.3.

Number of rows	1	2	3	4	5	7	10	13	≥16
F	0.70	0.80	0.86	0.90	0.93	0.96	0.98	0.99	1

Table A2.3. *F as a function of the number of rows for Table A2.2 (in-line bundles, finless tubes)*

Reynolds number	Recommended correlation
10-500	$Nu = 1.04F\ Re_d^{0.4} Pr^{0.36} \left(\frac{\mu}{\mu_p}\right)^{0.25}$
500-1,000	$Nu = 0.71F\ Re_d^{0.5} Pr^{0.36} \left(\frac{\mu}{\mu_p}\right)^{0.25}$
1,000 to 2 10^5 and $\frac{e_T}{e_L} < 2$	$Nu = 0.35F\ Re_d^{0.6} Pr^{0.36} \left(\frac{e_T}{e_L}\right)^{0.2} \left(\frac{\mu}{\mu_p}\right)^{0.25}$
1,000 - 2 10^5 and $\frac{e_T}{e_L} \geq 2$	$Nu = 0.40F\ Re_d^{0.6} Pr^{0.36} \left(\frac{\mu}{\mu_p}\right)^{0.25}$
> 2 10^5	$Nu = 0.022F\ Re_d^{0.84} Pr^{0.36} \left(\frac{\mu}{\mu_p}\right)^{0.25}$

Table A2.4. *Correlations to be used for staggered tube bundles (finless tubes)*

Number of rows	1	2	3	4	5	7	10	13	≥16
F	0.64	0.76	0.84	0.89	0.93	0.96	0.98	0.99	1

Table A2.5. *F as a function of the number of rows for Table A2.4 (staggered bundles, finless tubes)*

A2.2.2.2. *Shell space equipped with baffles*

A2.2.2.2.1. Donohue method

When the shell comprises transverse baffles, we propose using the Donohue (1949) correlation to estimate the heat transfer coefficient in the shell. Indeed, this correlation is simple to use. In addition, it is valid for any baffle type (segmental or disc-and-crown). The Donohue correlation, for the calculation of h_e, is given through the Nusselt number:

$$Nu_e = 0.22\, Re_d^{0.6} Pr^{0.33} \left(\frac{\mu}{\mu_p}\right)^{0.14}$$

where $Nu_e = \frac{h_e d_o}{\lambda}$, $Re_d = \frac{d_o G}{\mu}$ (G being the mass flowrate per unit area), $Pr = \frac{C_p \mu}{\lambda}$ and the correction $\left(\frac{\mu}{\mu_p}\right)^{0.14}$ is to be used only in the presence of large temperature differences between the wall and the fluid circulating in the shell.

A2.2.2.2.2. Bell-Delaware method

The *Bell-Delaware method* (see section 2.7.2) is more accurate, but is still more complex to use for the calculation of h_e. In this method, the film coefficient on the shell side, h_e, is calculated from an *ideal coefficient*, corresponding to a *perfect cross-flow*. Deviation from ideality is taken into consideration using *correction factors* to handle the particularities of the actual flow. This approach leads to the following formulation for the shell-side film coefficient:

$$h_e = J_C J_L J_B J_R J_S h_{ideal}$$

where:

– J_C, J_L, J_B, J_R, J_S are the *correction factors*;

– h_{ideal} is the *ideal* transfer coefficient on the shell side, corresponding to a perfect cross-flow.

Computation of the "ideal" transfer coefficient

The *ideal* heat transfer coefficient, on the shell side, is calculated from the Colburn number, j, for an ideal tube bundle, defined by: $j_{ideal} = \frac{Nu_{ideal}}{Re_d\,(Pr)^{\frac{1}{3}}}$.

Thus, j_{ideal} is given by:

$$j_{ideal} = K_1 \left(\frac{1.33}{\frac{a}{d_o}}\right)^K Re_d^{K_2}.$$

where a is the pitch and d_o is the outer diameter of the tubes.

$K = \frac{K_3}{1+0.14Re_d^{K_4}}$; the K_i values (i = 1 to 4) being constants that depend on the angle of arrangement of the tubes. They are given by Table A2.6. The coefficient h_{ideal} is then expressed by:

$$h_{ideal} = j\,\frac{\lambda}{d_o} Re_d\,(Pr)^{\frac{1}{3}}\left(\frac{\mu}{\mu_p}\right)^{0.14} = j\,C_p W\,(Pr)^{-\frac{2}{3}}\left(\frac{\mu}{\mu_p}\right)^{0.14}$$

where:

– λ is the heat conductivity;

– C_p is the specific heat;

– W is the mass flow rate in the shell;

– μ_p is the viscosity of the shell fluid expressed at the temperature of the tube wall.

Array angle	*Re*	K_1	K_2	K_3	K_4
30° arrays	< 10	1.400	-0.667	1.450	0.519
	10-100	1.360	-0.657		
	100-1,000	0.593	-0.477		
	$10^3 – 10^4$	0.321	-0.388		
	$10^4 – 10^5$	0.321	-0.388		
45° arrays	< 10	1.550	-0.667	1.930	0.500
	10-100	0.498	-0.656		
	100-1,000	0.730	-0.500		
	$10^3 – 10^4$	0.370	-0.396		
	$10^4 – 10^5$	0.370	-0.396		
90° arrays	< 10	0.970	-0.667	1.187	0.370
	10-100	0.900	-0.631		
	100-1,000	0.408	-0.460		
	$10^3 – 10^4$	0.107	-0.266		
	$10^4 – 10^5$	0.370	-0.396		

Table A2.6. *Constants for the Bell-Delaware method*

Computation of the configuration correction factor

The *configuration factor*, J_C, takes into account the heat transfer through the baffle window (see Figure 2.21 and section 2.6.2). It is given as a function of the number of tubes in the window of the baffle as follows.

$J_C = 1$ when no tube passes through the window.

Otherwise, it is given by: $J_C = 0.55 + 0.72(1 - 2F_W)$

with $F_W = \frac{\theta_{lim} - sin(\theta_{lim})}{2\pi}$; $\theta_{lim} = 2\cos^{-1}\left[\frac{D_S}{D_{lim} - d_o}(1 - 2\epsilon)\right]$.

$$D_{lim} = \sqrt{\frac{C_1 N_t a^2}{0.78}} + d_o \text{ (see section 2.4.6.5)}$$

ε is the cut ratio: for example, a 30% cut corresponds to a cut ratio, $\varepsilon = 0.3$.

Computation of correction factors for leaks

The *leakage factor*, J_L, takes into account the influence on heat transfer of leaks through the *tube passages* and through the *clearances between the baffle and the shell* (see section 2.6.4 where Figure 2.22 shows the clearances between the shell and the tubes). It is given by:

$$J_L = 0.44(1 - r_s) + [1 - 0.44(1 - r_s)]\text{Exp}(-2.2r_L)$$

with $r_s = \frac{a_{sb}}{a_{sb} + a_{tb}}$ and $r_L = \frac{a_{sb} + a_{tb}}{a_{Min}}$.

a_{Min} is the minimum clearance area at the center of the bundle (see section 2.7.2.1).

a_{sb}, a_{tb} are the leakage clearance areas (see sections 2.7.2.2 and 2.7.2.3).

Computation of correction factors for bypasses

The *bypass factor*, J_B, takes into account the flow between the last tubes, located at D_{lim}, and the shell. It is defined as a function of the ratio r_{ss}:

– for $r_{ss} < 0.5$: $J_B = Exp\left\{-C_j \frac{a_b}{S_m}\left[1 - (2r_{SS})^{\frac{1}{3}}\right]\right\}$;

– for $r_{ss} \geq 0.5$: $J_B = 1$.

with:

$C_j = 1.35$ for $Re_S < 100$ and $C_j = 1.25$ for $Re_S \geq 100$.

a_b is the bypass leakage clearance cross-section area (see section 2.7.2.4): $a_b = e_c\left(D_{sh} - D_{lim} - \frac{d_o}{2}\right)$.

$r_{ss} = \frac{N_{SS}}{N_{tcc}}$, $Re_s = \frac{d_o G_S}{\mu}$ and $G_s = \frac{W}{a_{Min}}$; W being the mass flow rate on the shell side.

a_{Min} is the minimum clearance area at the center of the bundle (see section 2.7.2.1): $a_{Min} = e_c\left[D_{sh} - D_{lim} + \frac{D_{lim}-d_o}{e_E}(a - d_o)\right]$.

a is the pitch and e_E is the effective gap, given in Table 2.4, as a function of the arrangement of the array.

N_{tcc} is the effective number of cross-flow rows between baffles (see section 2.7.2.5): $N_{tcc} = \frac{D_{sh}}{p_p}(1 - 2\varepsilon)$; p_p being given by Table 2.4.

N_{ss} is the number of pairs of sealing strips possibly attached to each baffle[1].

Computation correction factors for flow regime variation

The *flow regime factor*, J_R, takes into account the tendency to evolve toward a laminar regime, far from the inlet. It is given in Table A2.7 as a function of the Reynolds number.

Reynolds domain	Expression of J_R
$Re_S > 100$	$J_R = 1$
$20 < Re_S \leq 100$	$J_R = \left(\frac{10}{N_C}\right)^{0.18} + \frac{(20 - Re_S)(J_{RL} - 1)}{80}$
$Re_S \leq 20$	$J_R = \left(\frac{10}{N_C}\right)^{0.18}$

Table A2.7. *Expressions of the regime factor as a function of the Reynolds number*

where:

– $N_C = (N_{tcw} + N_{tcc})(1 + N_B)$;

– $N_{tcw} = \frac{0.8}{P_p}\left[\varepsilon D_{sh} - \frac{D_{sh}-(D_{lim}-d_o)}{2}\right]$;

1 These strips are placed between the baffles and the shell with a view to limiting leaks through the baffle-shell clearances, by redirecting the flow toward the tubes. Except for cases where the clearances are too large, make $N_{ss} = 1$.

– N_{tcc} is the effective number of cross-flow rows between baffles (see section 2.7.2.5);

– N_B is the number of baffles, calculated by: $N_B = 1 + int\left[\frac{L-d_o+d_i-e_{cin}-e_{cout}}{e_c}\right]$;

– L is the length of the tube;

– d_o and d_i are the outer and inner diameters of the tube;

– e_{cin} and e_{cout} are the spacings of the first baffles: at the inlet and outlet of the exchanger.

Computation of correction factors for nonuniform baffle spacings

The *correction factor for nonuniform baffle spacings*, J_S, is given by:

$$J_S = \frac{(N_B-1)+\left(\frac{e_{cin}}{e_c}\right)^{1-n_1}+\left(\frac{e_{cin}}{e_c}\right)^{1-n_1}}{(N_B-1)+\frac{e_{cin}}{e_c}+\frac{e_{cout}}{e_c}}$$

where:

– $n_1 = \frac{1}{3}$ for $Re_S < 100$ and $n_1 = 0.6$ for $Re_S \geq 100$;

– N_B is the number of baffles, calculated by: $N_B = 1 + int\left[\frac{L-d_o+d_i-e_{cin}-e_{cout}}{e_c}\right]$;

– e_{cin} and e_{cout} are the gaps between the first baffles: at the inlet and outlet of the exchanger.

Final expression for the calculation of h_e

Taking into account all of these corrections, the shell-side heat transfer coefficient is then calculated from the following expression:

$$h_e = j_{ideal}\, C_p W\, (Pr)^{-\frac{2}{3}}\left(\frac{\mu}{\mu_p}\right)^{0.14} J_C J_L\, J_B J_R J_S$$

A2.3. Coefficients for coaxial single-tube heat exchangers

For this type of device, a distinction shall be made between three heat transfer coefficients (see Figure A2.2):

– h_i, for heat exchange between the inner fluid and the internal pipe;

– h_1, for heat exchange between the annular fluid and the inner tube;

– h_2, for heat exchange between the annular fluid and the outer tube.

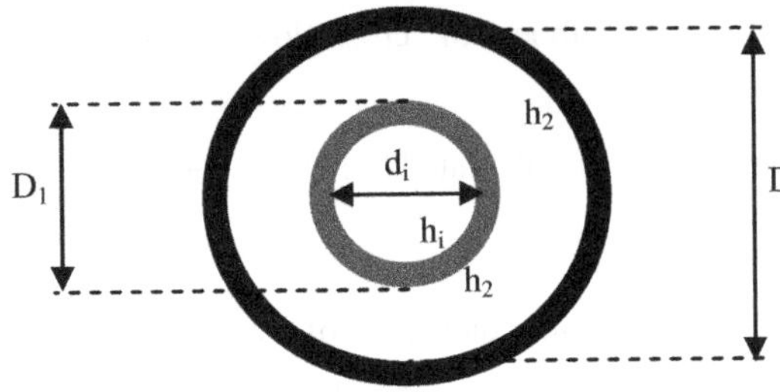

Figure A2.2. *Coaxial single-tube heat exchanger*

Table A2.8 presents the correlations to be used for each case, as a function of the flow regime and the hydraulic Reynolds number defined by (see Volume 2, Chapter 3, example 3.3):

$$Re_H = \frac{D_H v \rho}{\mu}$$

where $D_H = D_2 - D_1$.

The heat transfer coefficients are calculated from the Nusselt numbers by:

$$h_k = \frac{\lambda D_k}{D_2^2 - D_1^2} Nu_k \text{ with k = 1; 2}$$

$$Re_{H_1} = \frac{D_{H_1} v \rho}{\mu}; D_{H_1} = \frac{D_2^2 - D_1^2}{D_1}$$

$$Re_{H_2} = \frac{D_{H_2} v \rho}{\mu} \text{ and } D_{H_2} = \frac{D_2^2 - D_1^2}{D_2}$$

Flow regime	Exchange with the inner tube	Exchange with the outer tube
Laminar	$Nu_1 = 4.4438\left(\frac{D_1}{D_2}\right)^{-0.43}$	$Nu_2 = 0.8363\left(\frac{D_1}{D_2}\right) + 4.0202$
Turbulent	$Nu_1 = 0.02 Re_{H_1}{}^{0.8} Pr^{\frac{1}{3}}\left(\frac{D_2}{D_1}\right)^{0.53}$	$Nu_2 = 0.023 Re_{H_2}{}^{0.8} Pr^{\frac{1}{3}}\left(\frac{\mu}{\mu_p}\right)^{0.53}$

All physical properties are evaluated at the film temperature: $T_f = \frac{T_{pm}+T_m}{2}$.
T_{pm} and T_m are the average temperatures of the wall and the fluid, respectively.

Table A2.8. *Nusselt numbers for coaxial heat exchangers (source: Kays and Perkins (1972))*

A2.4. Heat transfer coefficients for coil-in-tank exchangers

We need to calculate two heat transfer coefficients:

– between the internal fluid and the inner wall of the coil, h_i;

– between the outer wall of the coil and the fluid of the tank, h_e.

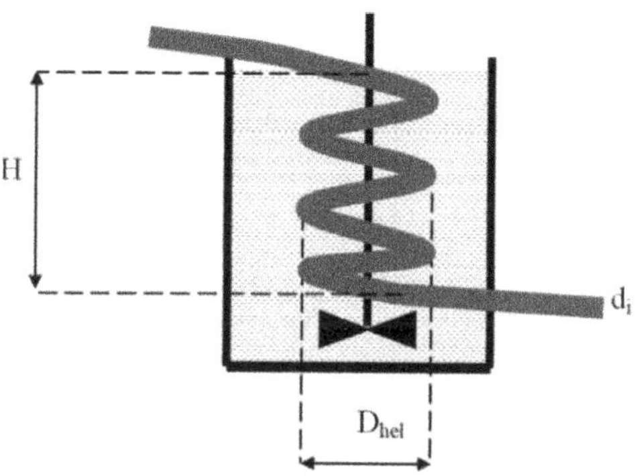

Figure A2.3. *Helix-in-tank heat exchanger*

A2.4.1. *Convection heat transfer coefficient inside the coil*

The calculation of the heat transfer coefficients between the inner wall of a coil and the fluid circulating through it, h_i, depends of course on the flow regime; that is to say, on the Reynolds number: $Re = \frac{d_i v \rho}{\mu}$, where d_i is the inner diameter of the tube constituting the helix.

Range of *Re*	Correlation to be used
$Re < 2{,}300$	$Nu_i = 1.86\left(\frac{d_i}{D_{hel}} Re\, Pr\right)^{\frac{1}{3}}\left(\frac{\mu}{\mu_p}\right)^{0.14}$
$2{,}300 \leq Re \leq 10^4$	$Nu_i = 0.116\left(Re^{\frac{2}{3}} - 125\right)Pr^{\frac{1}{3}}\left[1 + \left(\frac{d_i}{D_{hel}}\right)^{\frac{2}{3}}\right]\left(\frac{\mu}{\mu_p}\right)^{0.14}$
$Re > 10^4$	$Nu_i = 0.023Re^{0.8}Pr^{\frac{1}{3}}\left(1 + 3.5\frac{d_i}{D_{hel}}\right)\left(\frac{\mu}{\mu_p}\right)^{0.14}$

D_{hel} is the diameter of the helix (see Figure A2.3); $Nu_i = \frac{h_i d_i}{\lambda}$; $Re = \frac{d_i v \rho}{\mu}$; $Pr = \frac{C_p \mu}{\lambda}$.

The physical properties are taken at the average temperature of the fluid circulating in the coil.

Table A2.9. *h_i coefficients for the inside of the tube constituting the helix*

A2.4.2. Convection heat transfer coefficient on the tank side

Heat exchange between the outer wall of the coil and the fluid in which it is immersed depends, to a large extent (Debab 2011), on the type of turbine used to stir the tank (see Volume 3, Chapter 4, section 4.9.2). Thus, to estimate the heat transfer coefficients, h_e, we use the following correlations, where D_i is the inner diameter of the tank and d_o is the outer diameter of the tube.

Turbine type	Turbine diagram	Correlation to be used
Blade disc **$400 \leq Re_m \leq 2.10^5$**	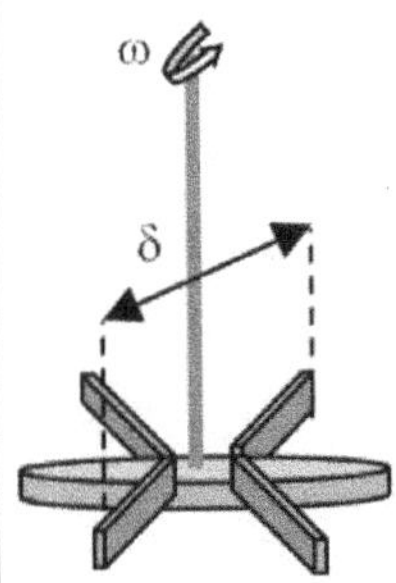	$Nu_e = 0.17 Re_m^{0.67} Pr^{0.37} \left(\frac{\delta}{D_i}\right)^{0.1} \left(\frac{d_o}{D_i}\right)^{0.5}$ (Oldshue and Gretton 1954)
Straight-bladed turbine **$300 \leq Re_m \leq 4.10^5$**	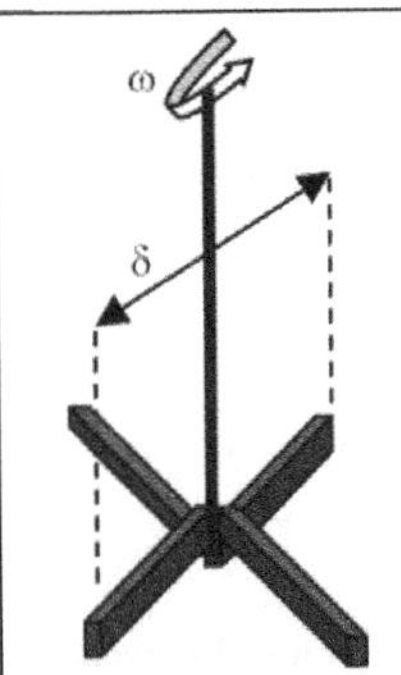	$Nu_e = 0.87 Re_m^{0.62} Pr^{\frac{1}{3}} \left(\frac{\delta}{D_i}\right)^{0.1} \left(\frac{d_o}{D_i}\right)^{0.5}$ (Chilton Drew and Jebens 1944)

D_i: inner diameter of the tank; δ: diameter of the turbine; ω: turbine rotational velocity (in revolutions/sec); μ: viscosity of the tank fluid; ρ: density of the tank fluid, d_o: outer diameter of the tube. $Nu_e = \frac{h_e d_o}{\lambda}$, $Re_m = \frac{\delta^2 \omega \rho}{\mu}$; $Pr = \frac{C_p \mu}{\lambda}$.

Table A2.10. *h_e coefficients for helix-tank heat exchange*

A2.5. Coefficients for cross-flow tubular heat exchangers

Cross-flow heat exchangers are often used as air preheaters or economizers placed after boilers or furnaces, etc.

There are several types of cross-flow exchangers. The most efficient can be found among the compact heat exchangers that are covered in section A2.6, but the category of *bare-tube and cross-flow heat exchangers* can prove useful in certain situations where costs need to be limited. In this case, the sleeves are generally parallelepipedal and the tube bundle is perpendicular to the flow.

The heat transfer depends very much on the type of flow and turbulence, which are functions of the fluid velocity and the size and arrangement of the tubes.

A2.5.1. *Coefficients inside tubes, h_i*

The calculation of the heat transfer coefficients inside the tubes is conducted using the same expressions presented in section A2.2.1.

A2.5.2. *Heat transfer coefficients in the shell space*

Heat transfer in the sleeve space depends on the nature of the tube bundle considered: bare tubes, individually-finned tubes or collectively-finned tubes.

A2.5.2.1. *Bare-tube cross-flows*

Table A2.11 gives the correlations to be used for the calculation of the coefficients of the film, in laminar and turbulent regimes, for in-line and staggered bundles, of cross-flow heat exchangers with bare tubes (Bergelin et al. 1952). Finned-tube heat exchangers are examined in sections A2.5.2.2 and A2.5.2.3.

In the following correlations, Re_{max} is defined as a function of G_{max} by (see section 2.6.7):

$$Re_{max} = \frac{G_{max} D_{eq}}{\mu}$$

$$G_{Max} = \frac{4\,W\,e_T}{\pi D_{sh}^2 (e_T - d_o)} \text{ for an in-line bundle}$$

$$G_{Max} = \frac{4\,W\,e_T}{\pi D_{sh}^2 \sqrt{\frac{e_T^2}{4} + e_T^2 - d_o}} \text{ for a staggered bundle}$$

α and α' are functions of the number of rows, N, in the bundle. They are given in Table A2.12.

Bundle type	Flow regimes	Rows in the bundle	Expression to be used
In-line	Laminar and intermediate $Re_{max} < 1{,}500$	10	$h_e(10) = 0.72\frac{\lambda_e}{d_o}Re_{max}^{0.43}Pr^{\frac{1}{3}}\left(\frac{\mu_e}{\mu_{e,p}}\right)^{0.14}$
		$N \neq 10$	$h_e(N) = \left(\frac{10}{N}\right)^{0.18} h_e(10)$
	Turbulent $Re_{max} \geq 5{,}000$	10	$h_e(10) = 0.26\frac{\lambda_e}{d_o}Re_{max}^{0.6}Pr^{\frac{1}{3}}$
		$N \neq 10$	$\mathrm{h_e(N) = \alpha(N)\, h_e(10)}$
Staggered	Laminar and intermediate $Re_{max} < 1{,}500$	10	$h_e(10) = 1.3\frac{\lambda_e}{d_o}Re_{max}^{0.43}Pr^{\frac{1}{3}}\left(\frac{\mu_e}{\mu_{e,p}}\right)^{0.14}$
		$N \neq 10$	$h_e(N) = \left(\frac{10}{N}\right)^{0.18} h_e(10)$
	Turbulent $Re_{max} \geq 5{,}000$	10	$h_e(10) = 0.33\frac{\lambda_e}{d_e}Re_{max}^{0.6}Pr^{\frac{1}{3}}$
		$N \neq 10$	$h_e(N) = \alpha'(N)\, h_e(10)$

Table A2.11. *h_e coefficients for cross-flow heat exchangers with bare tubes*

N	1	2	3	4	5	6	7	8	9	10	$N > 10$
$\alpha(N)$	0.64	0.80	0.87	0.90	0.92	0.94	0.96	0.98	0.99	1	1
$\alpha'(N)$	–	0.70	0.82	0.87	0.92	0.94	0.96	0.98	0.99	1	1

Table A2.12. *The functions α and α'*

A2.5.2.2. *Individually-finned tube cross-flows*

As we saw in Chapters 1 and 3, finned-tube heat exchangers can consist of *individually-finned tubes*, as shown in Figures 1.4 and 3.3. But the tubes can also *collectively* pass through the same series of plates, which act as collective fins. This case is covered in section A2.5.2.3.

The different types of fins used in *individually-finned* configurations are presented in Figure A2.4.

For the calculation of the fin-side heat transfer coefficient, the following correlation is recommended (Webb 1994):

$$Nu = 0.134Re^{0.681} (Pr)^{\frac{1}{3}} \left(\frac{s}{\delta_a}\right)^{0.2} \left(\frac{s}{e_a}\right)^{0.11}$$

where:

- s is the distance, at the base, between two fins; $s = a_a - e_a$;
- a_a is the pitch of the fins;
- e_a is the thickness of the fin;
- δ_a is the height of the fin (in the radial direction).

The heat transfer coefficient thus obtained is associated with the *total surface area* (see section 5.3.3) and with a fin efficiency. The expressions of the efficiencies of the fins are given by Table 5.1 (section 5.3.3.1).

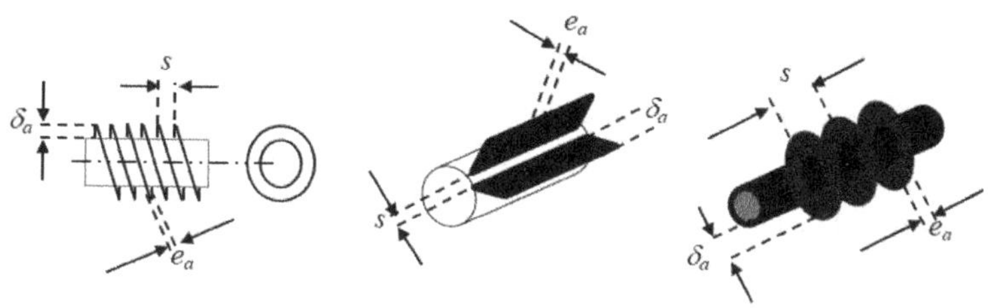

Figure A2.4. *Different types of individual fins*

A2.5.2.3. *Collectively-finned bundle cross-flows*

For finned bundles where the tubes *collectively* pass through a series of flat plates acting as fins (see Figure A2.5), the convection heat transfer coefficients depend on the number of rows of tubes perpendicular to the flow.

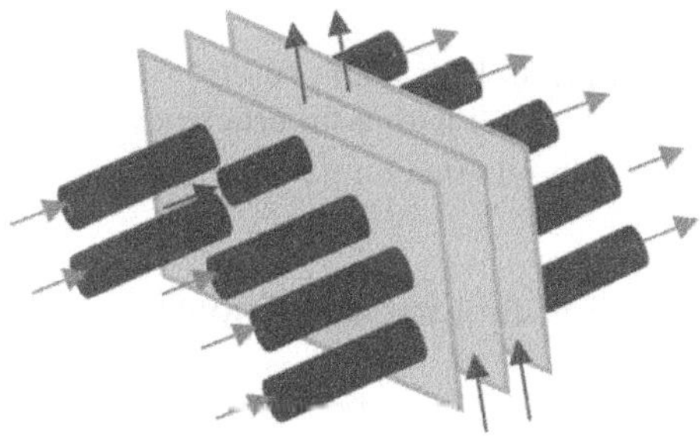

Figure A2.5. *Finned bundle*

The following correlations are then to be used according to the number of rows of tubes in the bundle (Webb 1994).

A2.5.2.3.1. For a bundle with four or more rows

The Nusselt number for a bundle with four or more rows is given by (Rohsenow 1998):

$$Nu_{4+} = 0.14 Re_d^{0.672} (Pr)^{\frac{1}{3}} \left(\frac{e_L}{e_T}\right)^{0.502} \left(\frac{s}{d_o}\right)^{0.0312}$$

where:

– $Nu_{4+} = \frac{h_e d_o}{\lambda}$; $Re_d = \frac{d_o v \rho}{\mu}$;

– d_o is the outer diameter of the tubes;

– λ is the heat conductivity of the shell fluid, r its density and μ its viscosity;

– e_L and e_T are the longitudinal and transverse gaps, respectively, given by Table 2.1;

– s is the distance between two fins, $s = a_a - e_a$; a_a being the pitch of the fins and e_a the thickness of a fin.

A2.5.2.3.2. For a bundle of one to three rows

The Nusselt number for a bundle of N rows, $N < 4$, is obtained from Nu_{4+} (Gray 1986):

$$\frac{Nu_N}{Nu_{4+}} = 0.991 \left[Re^{-0.092} \left(\frac{N}{4}\right)^{-0.031}\right]^{0.607(4-N)}$$

where N is the number of rows.

The heat transfer coefficients thus obtained are associated with the *equivalent surface area* (see section 5.3.3) and with a fin efficiency. The expressions of the fin efficiencies are given by Table 5.1 (section 5.3.3.1).

A2.6. Coefficients for plate-fin heat exchangers

These exchangers are described in section 3.4. They can be used counter-flow or cross-flow. However, in both flow configurations, it is necessary to determine the two local coefficients (of the film), one for each side. The description of the geometric parameters of this type of heat exchanger is given in Chapter 3.

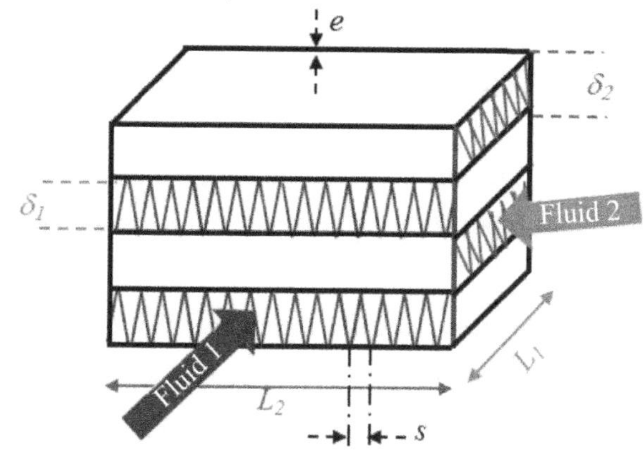

Figure A2.6. *Parameters for plate-fin heat exchangers*

The Nusselt number, for side i (i = 1; 2), is given by (Manglik et al. 1995):

$$Nu_i = 0.6522\, Re_{E_i}^{0.4597} \eta_i^{0.1} \,(Pr_i)^{\frac{1}{3}} \left(\frac{s}{\delta}\right)_i^{0.1541} - \left(\frac{e_a}{l_a}\right)_i^{0.1499} \left(\frac{e_a}{s}\right)_i^{-0.0678}$$

where:

– $Re_{E_i} = \frac{4W_i}{N_i\, \delta_i \beta_i \mu_i L_j}$; j = 2 if i = 1, and j = 1 if i = 2;

– $\eta_i = 1 + 5.269\; 10^{-5} Re_{E_i}^{1.340} \left(\frac{s}{\delta}\right)_i^{0.504} \left(\frac{e_a}{l_a}\right)_i^{0.456} \left(\frac{e_a}{s}\right)_i^{-1.055}$; ($i$ = 1; 2);

– $Pr_i = \frac{C_{p_i}\mu_i}{\lambda_i}$; $Nu_i = \frac{h_i D_{H_i}}{\lambda_i}$; $D_{H_i} = \frac{4\sigma_i}{\beta_i}$ (if σ_i is not given, see section 3.4.5)[2];

– N_i is the number of passes on side i;

– l_{ai}, e_{ai} and s_i are, respectively, the length[3], thickness and width at the base of the fin on side i (distance at the base, between two fins);

– δ is the height of a module on side i; this is the clearance height beneath the fin;

– ρ_i, μ_i, λ_i and C_{pi} are the density, viscosity, heat conductivity and sensible heat, respectively, of the fluid on side i;

– β_i, σ_i are the compactness and the porosity, respectively, on side i.

---

2 $D_{H_i} = \frac{4N_i L_i}{\beta_i V_i}\left(L_j \delta_i - 2N_{c_i} s_i e_{c_i}\right)$, with $V_i = N_i\, L_i L_j \delta_i$; j = 2 if i = 1, and j = 1 if i = 2.

3 Solely in the case where the latter does not continuously cover the entire length of flow. In the case of Figure 3.6: $l_{a1} = L_1$ and $l_{a2} = L_2$.

A2.7. Coefficients for plate heat exchangers

For plate heat exchangers, heat transfer coefficients will be calculated for each circuit (hot and cold). We will use h_1 for the hot side heat transfer coefficient and h_2 for the cold one.

Few correlations are proposed in the literature, so the designer often relies on the manufacturers' proposals.

For each of the circuits (i = 1; 2), we propose using the following equation (Marriott 1979):

$$Nu_i = 0.374\, Re_{E_i}^{0.668}\, Pr_i^{\frac{1}{3}} \left(\frac{\mu}{\mu_p}\right)_i^{0.15}$$

where $Nu_i = \frac{h_i D_{H_i}}{\lambda_i}$; $D_{H_i} = \frac{2\,e\sqrt{e_L^2+e_T^2}}{a-\delta}$; $Re_{E_i} = \frac{D_{H_i} v_i \rho_i}{\mu_i}$; $Pr_i = \frac{Cp_i\, \mu_i}{\lambda_i}$.

For Alfa Laval plate heat exchangers, the manufacturer offers the following dimensional correlation (Cahiers Alfa Laval 1987):

$$h_i = 234\, \lambda_i\, Pr^{\frac{1}{3}} \left(\frac{\Delta P\, \rho}{N_p\, \mu^2}\right)^{0.3275} \quad i = 1;\, 2$$

where N_p is the number of passes (different from the number of plates), defined by:

$$N_p = \frac{NTU_m}{NTU_{max}}$$

NTU_m is the arithmetic mean of the numbers of transfer units of the two circuits: $\mathrm{NTU_m} = \frac{NTU_1+NTU_2}{2}$.

NTU_{max} is the maximum number of transfer units for the type of plate considered: manufacturer's datum (see Appendix 1, Tables A1.24 to A1.26).

A2.8. Coefficients for spiral heat exchangers

For Reynolds numbers in the range 400 to 30,000, the convection heat transfer coefficient, h, is calculated from the following correlation (Picón-Núñeza et al. 2012):

$$Nu = 0.04 Re^{0.74} Pr^{0.4}$$

where:

– Pr is the Prandtl number of the fluid: $Pr = \frac{C_p \mu}{\lambda}$;

– Nu is the hydraulic Nusselt number: $Nu = \frac{h\, D_H}{\lambda}$;

– Re is the hydraulic Reynolds number: $Re = \frac{D_H\, v\, \rho}{\mu} = \frac{D_H\, W}{\mu\, A_0}$;

– A_0 is the cross-section area perpendicular to the flow (see section 3.6): $A_0 = e\, b$;

– W is the mass flow rate.

A2.9. Coefficients for printed-circuit heat exchangers

Printed-circuit heat exchangers are increasingly used for nuclear applications, in 4th-generation reactors, and solar applications, in concentrated thermal power plants.

In this type of application, the heat-transfer fluids are liquid metals, nanofluids or molten salts. Thus, as a function of the fluid considered, the suitable correlations will be used for the estimation of local heat transfer coefficients.

Naturally, these coefficients depend heavily on the flow regime. In addition, the small size of the channels in printed-circuit heat exchangers means that the frictional dissipation on the walls can be high in turbulent regime.

This particularity becomes significant when ordinary fluids are used in this type of exchanger. The estimation of local convective coefficients will then take into account the viscous dissipation effects in turbulent regime (see Volume 3, section 2.4.4.2).

The relationships to be used as a function of the heat-transfer fluid and the nature of the flow are presented in Table A2.13.

Type of fluids	Nature of the flow	Correlation
Ordinary (see Volume 3, section 2.4.4.2.2)	$Re < 2{,}300$ $Re \geq 3{,}000$	$Nu = 4.36$ $Nu = \frac{\left(\frac{f}{8}\right)(Re - 1{,}000)Pr}{1 + 12.7\left(\frac{f}{8}\right)^{\frac{1}{2}}\left(Pr^{\frac{2}{3}} - 1\right)}$ where $f = [1.82\, Log_{10}(Re) - 1.64]^{-2}$

Nanofluids **(see Volume 3, sections 6.2.2.1 and 6.2.2.2)**	$Re \leq 3{,}000$ $3{,}000 < Re < 16{,}000$	$Nu = 0.566(1+100\omega)^{0.57}\, Re^{0.20}\, Pr^{0.18}$ $Nu = 0.065(Re^{0.65} - 60.22)(1 + 0.0169\omega^{0.15})Pr^{0.542}$ ω being the volume fraction of the nanoparticles
Liquid metals **(see Volume 3, section 6.3.2)**	$Re \leq 100$ $Re > 100$	$Nu = 4.8 + 0.025\, Pe^{0.8}$ $Nu = 7 + 0.025\, Pe^{0.8}$
Molten salts **(see Volume 3, section 6.4)**	$Re \leq 100$ $100 < Re \leq 2{,}000$ $2{,}000 < Re \leq 10{,}000$ $Re > 10{,}000$	$Nu = 4.364$ $Nu = 4.82 + 0.0185(RePr)^{0.827}$ $Nu = 0.12(Re^{0.87} - 280) + 0.0185Pr^{0.4}\left[1+\left(\frac{d}{L}\right)^{\frac{2}{3}}\right]$ $Nu = 0.024\, Re^{0.807}\, Pr^{0.301}$

Table A2.13. *h coefficients for printed-circuit heat exchangers*

The overall coefficient, U^*, is calculated by considering the entire thickness of the modules, t_p, although the channels here constitute voids in the form of semicircles of diameter d. To take this particularity into account, the use of a correction factor, y_p, in the calculation of the overall coefficient is proposed:

$$\frac{1}{U^*} = \frac{1}{h_1} + \frac{t_p}{\lambda_p y_p} + \frac{1}{h_2}$$

where λ_p is the heat conductivity of the material constituting the heat exchanger.

The correction coefficient, y_p, depends on the ratio η defined by: $\eta = \frac{d}{t_p}$. It is given by (Ariu 2014):

$$y_p = 0.4499\eta^2 + 0.3403\eta + 1.$$

A2.10. Heat-transfer coefficients for two-phase heat exchangers

In this section, we present the correlations to be used to calculate heat transfer coefficients in exchangers where condensations or evaporations occur. In both situations (condensation and evaporation), we witness heat transfers with phase change. Indeed, condensation takes place when a vapor, in contact with a cold wall, changes to the liquid state. Likewise, in boilers or evaporators, a liquid that is in contact with a hot wall passes to the vapor state.

The study of heat transfers in two-phase environments is relatively complex, as large heat fluxes are generated by condensation or absorbed by vaporization.

A2.10.1. *Heat-transfer coefficients for condensers*

Although condensation can sometimes occur in droplets (in a situation where the condensate does not wet the surface), we are rather interested in situations where energy transfer occurs through a continuous film of condensate that forms on the wall. In a condenser, heat will therefore transfer through a liquid film on the transfer area.

Condensation occurs due to the existence of a heat potential difference: the average temperature of the liquid film covering the wall is lower than that of the vapor. Subsequently, the vapor pressure of the condensing substance is higher than the saturated vapor pressure of the liquid film. This induces a mass driving potential difference between the vapor phase and the liquid phase, inducing diffusion from the vapor phase to the liquid phase, followed by condensation in the liquid film, releasing the latent heat, which is then transferred to the wall. The condensation phenomenon therefore occurs through two resistances:

– a resistance to mass transfer between the vapor and the condensate film;

– a resistance to heat transfer through the condensate film.

For condensation of a pure saturation vapor, heat transfer can be described by an average heat transfer coefficient. The correlations to be used for its calculations are given, as a function of the flow regime, in Table A2.14, where Γ is defined in Table A2.15 for different condensations (Kirkbride 1933; Badger 1937; Moon-Hyun et al. 1990; Fujii 1991; Wang et al. 1993; Thome 2016).

RULES OF THUMB.–

R_1 – When the vapor velocity is high (Carpenter et al. 1951; Rohsenow et al. 1956), the friction between the vapor and the condensate film becomes significant. The average heat-transfer coefficient will then be estimated using the following correlation:

$$\frac{h}{C_p^L G_M^V} = 0.046 Re^{-\frac{1}{3}} \left(\frac{\rho_V}{\rho_L} f\right)^{\frac{1}{2}} Pr_L^{\frac{1}{2}} \text{ with } G_M^V = \sqrt{\frac{G_1^2 + G_1 G_2 + G_2^2}{3}}$$

where:

– G_1 and G_2 are the vapor flow rates at the ends of each tube;

– C_p^L, ρ_L and Pr_L are the specific heat, the density and the Prandtl number of the condensate, respectively;

– ρ_v is the density of the vapor;

– f is the Fanning friction factor evaluated at the average vapor velocity.

R_2 – When the condensate consists of two immiscible phases (e.g. mixtures of water vapor and hydrocarbons), the average heat-transfer coefficient, h, is given by (Fumimaru et al. 1988):

$$h = \text{v}_1\, h_1 + \text{v}_2\, h_2$$

where v_1 and v_2 have the volume fractions of each phase in the condensate, respectively, and h_1 and h_2 are the film condensation coefficients of each phase.

Exchanger	Flow regime	Correlation[4]
Tubular	Laminar ($Re \leq 2{,}100$)	$h = 1.51 Re^{-\frac{1}{3}} \left(\frac{\lambda^3 \rho^2 g}{\mu^2}\right)^{\frac{1}{3}}$
	Turbulent ($Re > 2{,}100$)	$h = 0.0077 Re^{0.4} \left(\frac{\lambda^3 \rho^2 g}{\mu^2}\right)^{\frac{1}{3}}$
Plate	Laminar and turbulent	$Nu = 0.00115 \left(\frac{Re}{Ja}\right)^{0.983} Pr_L^{\frac{1}{3}} \left(\frac{\rho_L}{\rho_G}\right)^{0.248}$ $Ja = \frac{C_{pL}(T_s - T_p)}{[\Lambda + 0.68 C_{pL}(T_s - T_p)]}$

Table A2.14. *Average heat-exchange coefficients for condensers*

4 *Ja* is the Jacob number. It takes into account the influence of the condensate film on heat transfer and possible under-cooling of the condensate (Wang et al. 1993; Jokar et al. 2006; Hayes et al. 2011). C_p is the specific heat; T_s is the saturation temperature; T_p is the wall temperature; m is the viscosity; g is the acceleration of gravity; L is the latent heat of condensation; ρ_L and ρ_G are the densities of the liquid and the vapor; d_o is the outer diameter of the tubes; d_i is the inner diameter of the tubes; N_t is the total number of tubes in the shell; W is the shell-side mass flow rate; L is the length of the tubes.

<table>
<tr><th rowspan="3">Place of condensation</th><th colspan="4">Type of condenser</th></tr>
<tr><th colspan="2">Vertical
$Re = \frac{4\Gamma}{\mu}$</th><th colspan="2">Horizontal
$Re = \frac{2\Gamma}{\mu}$</th></tr>
<tr><th>Single tube</th><th>Bundle of N tubes</th><th>Single tube</th><th>Bundle of N tubes</th></tr>
<tr><td>Outside the tubes</td><td>$\Gamma = \frac{W}{\pi d_o}$</td><td>$\Gamma = \frac{W}{N_t \pi d_o}$</td><td>$\Gamma = \frac{W}{L}$</td><td>$\Gamma = \frac{W}{N_t^{\frac{2}{3}} L}$</td></tr>
<tr><td>Inside the tubes</td><td>$\Gamma = \frac{W}{\pi d_i}$</td><td>$\Gamma = \frac{W}{N_t \pi d_i}$</td><td>$\Gamma = \frac{2W}{L}$</td><td>$\Gamma = \frac{2W}{NL}$</td></tr>
</table>

Table A2.15. *Re and Γ for the expressions of h given in Table A2.14*

A2.10.2. Heat transfer coefficients for boilers and evaporators

A2.10.2.1. *Boiling mechanism*

The mechanism of vapor formation is quite complex and can occur in different regimes, depending on the heat flux.

Without forced flow and for small fluxes, natural convection between the wall and the fluid leads to an increase in its temperature, which results in a low release of vapor (Sathyabhama 2019). For moderate fluxes, bubbles form, grow, then detach from the surface and are replaced by liquid (nucleate boiling).

However, when the heat flux is quite intense and becomes greater than a flux, ϕ_c, called the critical flux, the bubbles can join together and form a vapor film, which remains permanently on the wall (film boiling). The critical flux depends on the surface tension, σ, of the liquid concerned (Zuber and Tribus 1958):

$$\varphi_c = \rho_v^{0.5}(\sigma\rho_l g)^{0.25}\left[0.13 + 4\mu_l^{0.8}\left(\frac{\rho_l\sigma^3}{g}\right)^{-0.2}\right]\Lambda$$

where:

- ρ_v and ρ_l are the densities of the vapor and the liquid, respectively;
- σ is the surface tension of the liquid;
- g is the acceleration of gravity;
- Λ is the latent heat of vaporization of the liquid.

In terms of heat transfer, the motions generated by the bubbles favor nucleate boiling thanks to the stirring of the liquid, which leads to higher heat transfer coefficients.

In the case of film boiling, the latter represents an additional thermal resistance that needs to be incorporated. This resistance becomes all the more significant as vapor production increases; it then acts as a brake on heat transfer and could lead to an overheating of the wall. We will then seek to avoid the formation of a vapor film in an evaporator as much as possible. Note, however, that this risk is low since the fluid is in motion within the evaporators.

A2.10.2.2. Two-phase flow boiling in a heat exchanger

In a boiler, not only is vapor formed, but we also have a two-phase flow; liquid and the vapor produced flow through, and each contribute to the heat transfer.

Heat transfer phenomena in two-phase flow has been studied largely (Jens and Lottes 1951; Forester and Zuber 1955; Chen 1966; Brown 1967; Collier 1981; Klimenko 1988; Kannengieser 2009; Abadie et al. 2010), and it is currently established (Klimenko 1988) that two mechanisms contribute to this transfer: nucleate boiling and convection induced by film boiling. This is referred to as two-phase flow forced convection, also called convective boiling.

In convective boiling, heat transfer coefficients can reach multiples of the orders of magnitude achievable in the most effective nucleate boilings. They depend little on the flux value (Dengler and Addoms 1956; Guerrieri and Talty 1956; Bandel and Schlünder 1974; Mohr and Runge 1977; Aounallah et al. 1982). Rather, they are related to the mass vapor flowrates in the two-phase medium[5], to the ratio between the density of the vapor and that of the liquid ($\frac{\rho_v}{\rho_l}$), and to the ratio between the heat conductivities of the wall and of the liquid ($\frac{\lambda_p}{\lambda_l}$).

Heat transfer in boilers is in fact the result of two contributions (Chen 1966; Edelstein et al. 1984; Klimenko 1984; Kacac 2002): a convective contribution corresponding to the liquid phase flowing alone in forced convection and whose heat transfer coefficient is noted h_L; and, according to the authors, a component that takes into account nucleate boiling in two-phase convective flow, whose heat transfer coefficient is noted h_{EC}. The superposition of these two contributions depends on the authors, but the one that best represents the available experimental data gives the heat transfer coefficient on the shell side by (Klimenko 1988):

$$h_e = \sqrt[3]{h_L^3 + h_{EC}^3}$$

5 The vapor rate in the two-phase medium is defined by: $\tau = \frac{Vapor\ mass\ flow\ rate}{Total\ mass\ flow\ rate}$ $\tau = 0$ for a saturated liquid and $\tau = 1$ for a superheated vapor.

The single-phase convective component, h_L, is obtained from the Dittus-Boelter correlation (see Volume 3, section 2.4.1), independently of the two-phase flow regime:

$$h_L = 0.023 Re^{0.8} Pr^{0.4} \frac{\lambda_l}{d_o} \text{ Where } Re = \frac{4W}{\pi \mu_l d_o}; \; Pr = \frac{C_{p_l} \mu_l}{\lambda_l}$$

The contribution, h_{EC}, of two-phase convective nucleate boiling is calculated from the Nusselt number, Nu_{EC}, both for horizontal boilers and for vertical arrangements (Klimenko 1988):

– for $N_{EC} < 1.6\ 10^4$: $Nu_{EC} = 7.4\ 10^{-3} Pe_*^{0.6} K_p^{0.5} Pr_l^{-\frac{1}{3}} \left(\frac{\lambda_p}{\lambda_l}\right)^{0.15}$;

– for $N_{EC} > 1.6\ 10^4$: $Nu_{EC} = 0.087 Re_m^{0.6} Pr_l^{\frac{1}{6}} \left(\frac{\rho_v}{\rho_l}\right)^{0.2} \left(\frac{\lambda_p}{\lambda_l}\right)^{0.09}$.

where

– $Nu_{EC} = \frac{h_{EC} b}{\lambda_l}$ and N_{EC} is a number that characterizes convective boiling;

– $N_{EC} = \frac{\Lambda W}{\varphi}\left[1 + \tau\left(\frac{\rho_l}{\rho_v} - 1\right)\right]\left(\frac{\rho_v}{\rho_l}\right)^{\frac{1}{3}}$;

– b is the characteristic Laplace distance: $b = \sqrt{\frac{\sigma}{g(\rho_l - \rho_v)}}$; [b]=[L];

– σ is the surface tension $[\sigma] = [M]\,[T]^{-2}$;

– g is the acceleration of gravity;

– ρ_l is the density of the liquid expressed at its saturation temperature;

– ρ_v is the density of the vapor;

– λ_l is the heat conductivity of the fluid expressed at its saturation temperature;

– Λ is the latent heat;

– W is the total mass flow rate;

– Pe_* is the modified Peclet number: $Pe_* = \frac{b\varphi}{\Lambda \rho_v \lambda_l}$;

– φ is the density of the thermal flux;

– K_p is the dimensionless pressure given by: $K_p = \frac{p}{\sqrt{\sigma g(\rho_l - \rho_v)}}$;

– Pr_l is the Prandtl number of the liquid expressed at its saturation temperature: $Pr_l = \frac{C_{p_l}\mu_l}{\lambda_l}$;

– λ_p is the heat conductivity of the wall[6];

– Re_m is the Reynolds number of the two-phase mixture: $Re_m = \frac{bv_m\rho_l}{\mu_l}$;

– v_m is the flow velocity of the two-phase mixture: $v_m = \frac{w}{\rho_l}\left[1 + \tau\left(\frac{\rho_l}{\rho_v} - 1\right)\right]$.

A2.11. Coefficients for heat exchangers operating with nanofluids, liquid metals or molten salts

These are special exchangers that are increasingly used in concentrated solar power plants and in nuclear applications.

Nanofluids and liquid metals are mainly used in shell and multitube heat exchangers. However, more recent applications use compact heat exchangers.

For molten salts, two types of exchangers are generally used: shell-and-tube heat exchangers and printed-circuit heat exchangers (Ariu 2014). Experience has shown that limiting the volume of shell-and-tube heat exchangers can become a serious disadvantage. Printed-circuit heat exchangers are thus preferred, thanks to the very small diameters of the semicircular channels (typically around 1.5 mm), which makes it possible to have a very large exchange area per unit volume.

The analysis of convective transfers in nanofluids, liquid metals or molten salts is detailed in Volume 3 of this series (Benallou 2019), where the physical properties of these fluids can also be found.

A2.11.1. *Heat transfer coefficients inside the tubes*

A2.11.1.1. *For nanofluids*

The Nusselt number is given as a function of the flow regime and volume fraction, both for simple cylindrical pipes and for fluids circulating in an annulus (Benallou 2019), as shown in Table A2.16.

6 Note: this is not the heat conductivity of the liquid expressed at the wall temperature.

Flow	Nusselt number
Turbulent	$Nu_{nf} = 0.065(Re_{nf}^{0.65} - 60.22)(1 + 0.0169\omega^{0.15})Pr_{nf}^{0.542}$
Laminar	$Nu_{nf} = 0.566(1 + 100\omega)^{0.57} Re_{nf}^{0.20} Pr_{nf}^{0.18}$
Annular	$Nu_{nf} = 1.7 Re_H^{0.136} Pr^{0.8}(0.4 + 0.003\omega)$

Table A2.16. *Nusselt number for nanofluids*

where:

– $\omega = \frac{V_{np}}{V_{nf}}$; V_{np} being the volume of nanoparticles and V_{nf} the volume of the nanofluid;

– $Pr_{nf} = \frac{c_{p_{nf}} \mu_{nf}}{\lambda_{nf}}$;

– $Re_{nf} = \frac{d_i v \rho_{nf}}{\mu_{nf}}$; d_i being the pipe inner diameter and v the average flow velocity;

– the indices *nf* and *np* refer to the nanofluid and the nanoparticles, respectively;

– $Re_H = \frac{D_H v \rho_{nf}}{\mu_{nf}}$ is the Reynolds number for annular flow, calculated using the hydraulic diameter, D_H: $D_H = D_2 - D_1$;

– D_1 is the outer diameter of the inner tube and D_2 is the inner diameter of the outer tube.

A2.11.1.2. *For liquid metals*

The Nusselt number is given in Table A2.17 for a metallic liquid circulating in cylindrical tubes (Lyon 1952) or in an annulus.

Flow	Nusselt number
Cylindrical tube	$Nu = 7 + 0.025Pe^{0.8}$
Annulus	$Nu_{H1} = 0.70\left(\frac{D_2}{D_1}\right)^{0.53}\left[5 + 0.025Pe_{H_1}^{0.8}\right]$ $Nu_{H2} = 0.70\left(\frac{D_1}{D_2}\right)^{0.53}\left[7 + 0.025\ Pe_{H_2}^{0.8}\right]$
Between flat plates	$Nu = 5.8 + 0.020\ Pe^{0.8}$

Table A2.17. *Nusselt number for liquid metals*

where:

$$Nu_{H_1} = \frac{h_1 D_{H_1}}{\lambda} \quad Pe_{H_1} = Re_{H_1} Pr \quad Re_{H_1} = \frac{D_{H_1} \, v \, \rho}{\mu} \quad Pr = \frac{C_p \mu}{\lambda} \quad D_{H_1} = \frac{D_2^2 - D_1^2}{D_1}$$

$$Nu_{H_2} = \frac{h_2 D_{H_2}}{\lambda} \quad Pe_{H_2} = Re_{H_2} Pr \quad Re_{H_2} = \frac{D_{H_2} \, v \, \rho}{\mu} \quad D_{H_2} = \frac{D_2^2 - D_1^2}{D_2}$$

D_1 is the outer diameter of the inner tube and D_2 is the inner diameter of the outer tube.

A2.11.1.3. *For molten salts*

Table A2.18 gives the heat transfer coefficients for molten salts in different flow configurations (Benallou 2019).

Flow	Nusselt number
Turbulent	$Nu = 0.024Re^{0.807}Pr^{0.301}$
Laminar	$Nu = 4.82 + 0.0185(RePr)^{0.827}$
Transition	$Nu = 0.12(Re^{0.87} - 280) + 0.0185Pr^{0.4}\left[1 + \left(\frac{D}{L}\right)^{\frac{2}{3}}\right]\left(\frac{Pr}{Pr_w}\right)^{0.11}$

Table A2.18. *Nusselt number for molten salts*

where $Nu = \frac{hD}{\lambda}$; $\boldsymbol{Re} = \frac{Dv\rho}{\mu}$ and $\boldsymbol{Pr} = \frac{C_p\mu}{\lambda}$.

Pr_w is the Prandtl number at the wall temperature.

D is the tube inner diameter.

v is the flow velocity of the molten salt in the tube.

Appendix 3

Calculation of Pressure Drops in Exchangers

A3.1. Introduction

Hydraulic design is an important step in the sizing of heat exchangers. Indeed, the design of an exchanger is often carried out through an iterative procedure seeking to *optimize the heat flux transferred*, while *satisfying pressure drop constraints*.

Thus, knowledge of the pressure drops is essential in the process of sizing heat exchangers. It also makes it possible to determine the pumping cost, given that this is directly related to the pressure drops across the vessel.

Appendix 3 is reserved for the presentation of correlations available in the literature permitting calculation of pressure drops, for the different types of exchangers:

– shell-and-tube heat exchangers operating co- or counter-flow;

– cross-flow heat exchangers;

– coaxial single-tube heat exchangers;

– coil-in-tank heat exchangers;

– plate heat exchangers;

– spiral heat exchangers;

– printed-circuit heat exchangers;

– boilers and evaporators;

For a color version of all figures in this chapter, see www.iste.co.uk/benallou/energy6.zip

– condensers;

– heat exchangers operating with metallic liquids or molten salts.

A3.2. Pressure drops and operating costs

The operating cost of a heat exchanger is directly related to the electricity consumption of pumps used to circulate the fluids. This cost shall be examined in detail during the design phase, since, at times, the additional cost generated by higher pressure drops outweighs the advantage achievable by a more-efficient heat transfer. Such situations are encountered notably for gases and fluids with low densities or very high viscosities. In such situations, the designer shall *consider pressure drops and heat transfer with the same degree of importance.*

If P is the total power of the heat exchanger pumps, the annual operating cost will be given by:

$$C = C_E P\, D_F$$

where:

– C_E is the electricity cost;

– D_F is the annual operating time of the heat exchanger.

For each side of the exchanger, the pumping power, P_i, is proportional to the pressure drops ΔP_i:

$$P_i = -\frac{W_i\, \Delta P_i}{\eta_i \rho_i}$$

where:

– W_i is the mass flow rate on side i;

– ρ_i is the fluid density on side i;

– η_i is the efficiency of the pump (or fan) on the side i.

A3.3. Pressure drops in shell-and-tube heat exchangers

Pressure drop analysis is carried out for the two flows: inside the tubes and in the shell.

A3.3.1. *Pressure drops inside the tubes*

A3.3.1.1. *Single-pass tube-side devices*

$$\Delta P = \frac{\rho v^2}{2}\left(\frac{L}{d_i} f + 1.5\right)$$

where:

- ΔP is the pressure drop in Pascal;
- v: the velocity of the fluid within the tubes;
- ρ: the fluid density;
- L: the length of the tubes;
- d_i: the inner diameter of the tubes;
- f: the friction factor (see Volume 3, section 2.4.4.2.1)[1].

The first term in brackets relates to the friction pressure drop, while the second takes into account the pressure drop at the device inlet and outlet.

A3.3.1.2. *Multiple-pass tube-side devices*

$$\Delta P = \frac{\rho v^2}{2}\left(N_p \frac{L}{d_i} f + +4 N_p + 1.5\right)$$

where N_p is the number of tube-side passes.

The first term in brackets relates to the friction pressure drop, the second to the pressure drop at the change of pass and the third to the pressure drop at the device inlet and outlet.

A3.3.2. *Pressure drop in the shell*

In the shell, the presence of baffles induces a complex flow. As we have seen in section 2.5, the complexity of this flow can be comprehended by breaking it down symbolically into:

---

1 In laminar regime: $f = \frac{64}{Re}$; $Re = \frac{d_i v \rho_c}{\mu_c}$.

In turbulent regime: for smooth tubes (Moody 1944): $f = \frac{1}{[1.82 log_{10}(Re) - 1.64]^2}$. For tubes presenting roughness: (Haaland 1983): $\frac{1}{\sqrt{f}} \cong -1.8\, log_{10}\left\{\left[\frac{\left(\frac{\varepsilon}{D}\right)}{3.7}\right]^{1.11} + \frac{6.9}{Re}\right\}$.

– a transverse flow between the baffles;

– a longitudinal flow through a baffle.

The pressure drop analysis follows the same symbolic breakdown by defining transverse pressure drops (representing the decrease in momentum during flow between two baffles) and longitudinal pressure drops that correspond to the passage through the baffles.

Therefore, for a shell equipped with N_B baffles, the total pressure drop will be given by:

$$\Delta P = (N_B - 1)\Delta P_T + N_B \Delta P_L$$

where: ΔP_T and ΔP_L represent the transverse and longitudinal pressure drops, respectively.

A3.3.2.1. *Transverse pressure drop: flow between two baffles*

Between two baffles, we essentially have a flow in a cylindrical pipe that is not entirely free, but where there are a number of tubes. The flow in this space occurs essentially perpendicular to the tubes.

The pressure drop between two baffles is given by:

$$\Delta P_T = 1.5 \frac{D_{sh} - \delta_L}{\sqrt{e_T e_F}} \left[\frac{\mu}{G_T \sqrt{e_T e_L} - d_o}\right]^{0.2} \frac{G_T^2}{\rho}$$

where:

δ_F is the free height above the baffle (see section 2.6.2). δ_F may be approximately related to the cutting ratio of the baffle, ε, by (see Table 2.3, section 2.6.3): $\delta_F \cong \varepsilon D_{sh}$

where:

– G_T is the transverse flow rate: $G_T = \frac{W}{A_T}$;

– W is the mass flow rate of the fluid circulating in the shell;

– A_T is the transverse free area: $A_T = \frac{e_c D_{sh}\left(\sqrt{e_T e_L} - d_o\right)}{\sqrt{e_T e_L}}$; e_c being the spacing between two baffles.

A3.3.2.2. *Longitudinal pressure drop: flow through a baffle*

As it passes through a baffle, the fluid flows, as a whole, parallel to the tubes (see Figure 1.14). In this case, the pressure drop is said to be longitudinal. It is given by:

$$\Delta P_L = \frac{G_L^2}{\rho}$$

where:

– G_L is the longitudinal mass flow rate : $G_L = \frac{W}{A_L}$;

– W is the mass flow rate on the shell side;

– A_L is the longitudinal free area : $A_L = \frac{\epsilon\pi}{4}(D_{sh}^2 - nd_o^2)$;

– D_{sh} is the shell inner diameter;

– ε is the cutting ratio: $\epsilon = \frac{A_L}{S}$ (where S is the overall cross-section area of the shell);

– d_o is the outer diameter of the tubes.

A3.4. Pressure drops in coaxial single-tube heat exchangers

A3.4.1. *Pressure drops in the inner tube*

Calculation of the pressure drops in the inner tube is conducted based on the expression given in section A3.3.1.1.

A3.4.2. *Pressure drops in the annulus*

Calculation of pressure drops in the annulus is performed based on the expression given in section A3.3.1.1; using the hydraulic Reynolds number defined by (see Volume 3, Chapter 3, example 3.3): $Re_H = \frac{D_H v \rho}{\mu}$, where $D_H = D_2 - D_1$.

A3.5. Pressure drops in coil-in-tank heat exchangers

In this type of heat exchanger, the flow into the stirred tank usually occurs by overflow. The flow of the cooling liquid (generally water) occurs to maintain a constant temperature in the tank. The question of pressure drops is therefore not significant for the flow in the tank.

Conversely, the pressure drop in the helical tube is significant. It is given by:

$$\Delta P = \frac{\rho v^2}{2}\frac{L}{d_i}f$$

where: L is the length of the coil and f is the friction factor given by Table A3.1, as a function of the Dean number[2].

Flow	Friction factor
Laminar (Srinivasan et al. 1970 ; Kakaç and Liu 2002)	$De < 30: f = \frac{64}{Re}$
	$30 \leq De < 300: f = \frac{26.816}{Re} De^{0.275}$
	$De \geq 300: f = \frac{7.2}{Re} De^{0.5}$
Turbulent (Rohsenow 1998)	$f = 0.084 \mathrm{Re}^{-0.2} \left(\frac{d_o}{D_{hel}}\right)^{0.1}$

d_o: outer diameter of the tube

Table A3.1. *Friction factor for coil-in-tanks*

A3.6. Pressure drops for cross-flow tubular heat exchangers

A3.6.1. *Pressure drops inside the tubes*

The calculation of the pressure drops inside the tubes is conducted using the same expressions presented in section A3.3.1.

A3.6.2. *Pressure drop in the shell*

For the calculation of pressure drops in the shell, a distinction shall be made between bare tube heat exchangers and finned-tube heat exchangers.

A3.6.2.1. *Bare-tube cross-flows*

To calculate the pressure drops in the shell of an exchanger with bare, in-line or staggered tubes, operating in cross-flow, we recommend using the following correlation (Ariu 2014):

$$\Delta P_c = N_n K_f \left(\frac{1}{2} \rho v_{\max}^2\right)$$

2 Dean number: $De = \left(\frac{d_o}{D_{hel}}\right)^{0.5} Re$.

where:

– $v_{\max} = \frac{W}{\rho\, a_{Min}}$;

– W is the mass flow rate and ρ is the density of the fluid;

– a_{Min} is the minimum clearance area at the center of the bundle (see section 2.7.2.1);

– N_n is the number of layers perpendicular to the flow; calculated from Table 3.1;

– L_c is the length of tube that is cross-flow;

– a is the pitch of the tubes;

– K_f is given by Table A3.2 below, depending on Reynolds number and array type[3];

– D_H is the hydraulic diameter of the shell.

Array	K_f	
	$Re < 2{,}300$	$2{,}300 < Re < 2\ 10^6$
Square	$K_f = 0.272 + \frac{0.207\ 10^3}{Re} + \frac{0.102\ 10^3}{Re^2} - \frac{0.286\ 10^3}{Re^3}$	$K_f = 0.267 + \frac{0.249\ 10^4}{Re} - \frac{0.927\ 10^7}{Re^2} + \frac{10^{10}}{Re^3}$
Triangular	$K_f = 0.795 + \frac{0.247\ 10^3}{Re} + \frac{0.335\ 10^4}{Re^2} - \frac{0.155\ 10^4}{Re^3} + \frac{0.241\ 10^4}{Re^4}$	$K_f = 0.245 + \frac{0.339\ 10^4}{Re} - \frac{0.984\ 10^7}{Re^2} + \frac{0.133\ 10^{11}}{Re^3} - \frac{0.599\ 10^{13}}{Re^4}$

Table A3.2. *K_f coefficients for cross-flow heat exchangers*

A3.6.2.2. *Individually-finned tube cross-flows*

The pressure drop is given by:

$$\Delta P = \frac{\rho v_\infty^2}{2}\frac{L}{D_H} f$$

3 The pitch, a, is assumed to be at 1.25 d_o.

where:

– v_∞ is the approach velocity of the gas;

– D_H is the hydraulic diameter;

– L is the length of the flow perpendicular to the tubes;

– f is the friction factor given by (Robinson and Briggs 1966; Shah and Sekulic 2003):

$$f = \frac{9.465}{Re_{d_o}^{0.316}}\left(\frac{e_T}{d_o}\right)^{-0.927}\left(\frac{e_T}{\sqrt{e_T^2+e_L^2}}\right)^{0.515}$$

where:

– Re_{do} is the Reynolds number calculated with respect to the outer diameter of the tubes;

– e_L and e_T are the longitudinal and transverse gaps, respectively (see Table 2.1);

– d_o is the outer diameter of the tubes.

A3.6.2.3. *Collectively-finned bundle cross-flows*

The pressure drop in bundles of collectively finned tubes is given by:

$$\Delta P = \frac{\rho v_\infty^2}{2}\frac{L}{D_H}f$$

where: f is given by (Wang et al. 2000):

$$f = 0.0267 Re_{d_o}^{k_1}\left(\frac{e_T}{e_L}\right)^{k_2}\left(\frac{a_a}{d_o}\right)^{k_3}$$

with:

$$k_1 = -0.764 + +0.739\left(\frac{e_T}{e_L}\right) + 0.177\left(\frac{a_a}{d_o}\right) - \frac{0.00758}{N_n}$$

$$k_2 = -15.689 + \frac{64.021}{Ln(Re_{d_o})}$$

$$k_3 = 1.696 - \frac{15.695}{Ln(Re_{d_o})}$$

where:

– Re_{do} is the Reynolds number calculated with respect to the outer diameter of the tubes;

– d_o is the outer diameter of the tubes;

– a_a is the pitch of the fins;

– e_T and e_L are the transverse and longitudinal gaps, respectively (see Table 2.1);

– N_n is the number of layers of tubes perpendicular to the flow, calculated from Table 3.1.

A3.7. Pressure drops in plate-fin heat exchangers

These heat exchangers are described in section 3.4. Figure A3.1 recalls the characteristic dimensions of this type of exchanger.

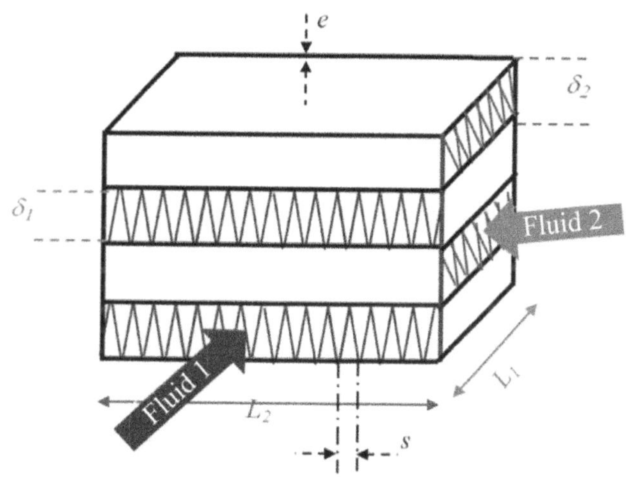

Figure A3.1. *Characteristic dimensions of a plate-fin heat exchanger*

The pressure drop on side i is given by:

$$\Delta P_i = 2\, f_i \frac{G_i^2}{\rho_i} \frac{L_i}{D_{Hi}}$$

where:

– G_i is the mass flow rate in a channel on side i: $G_i = \frac{W_i}{N_i N_{ci}}$;

– ρ_i is the fluid density on side i;

– L_i is the flow length on side i;

– D_{Hi} is the hydraulic diameter on side i;

– f_i is the friction factor on side i, given by (Manglik et al. 1995):

$$f = 9.6243 \frac{\xi_i^{0.1}}{Re_{E_i}^{0.7422}} \left(\frac{S}{\delta}\right)_i^{-0.1856} \left(\frac{e_a}{l_a}\right)_i^{0.3053} \left(\frac{e_a}{s}\right)_i^{-0.2659}$$

where:

– $\xi_i = 1 + +7.669\ 10^{-8} Re_{E_i}^{4.429} \left(\frac{s}{\delta}\right)_i^{0.920} \left(\frac{e_a}{l_a}\right)_i^{3.767} \left(\frac{e_a}{s}\right)_i^{0.236}$;

– $Re_{E_i} = \frac{4W_i}{N_i\ \delta_i \beta_i \mu_i L_j}$; $j = 2$ if $i = 1$, and $j = 1$ if $i = 2$;

– N_i is the number of passes on side i;

– l_{ai}, e_{ai} and s_i are, respectively, the length[4], thickness and width at the base of the fin on side i (distance at the base, between two fins);

– δ_i is the height of an exchange module on side i; this is the clearance height beneath the fin;

– β_i is the compactness on side i.

A3.8. Pressure drops in plate heat exchangers

The pressure drop in a plate heat exchanger is given by (Shah and Focke 1988; Picón-Núñez et al. 2014):

$$\Delta P = 2N_p\, C_f \frac{G^2}{\rho} \frac{L}{D_H}$$

where:

– C_f is the friction coefficient;

– N_p is the number of passes;

4 Solely in the case where the latter does not continuously cover the entire flow length. In the case of Figure 3.6, $l_{a1} = L_1$ and $l_{a2} = L_2$.

– G is the mass flux in each plate. This is the mass flow rate per unit of free flow area in each plate: $G = \frac{W}{N' A_L}$;

– N' is the total number of plates: $N' = N+2$; N being the number of plates excluding input and output plates;

– A_0 is the free flow area;

– L is the clearance length.

The friction coefficient is given by (Kuppan 2000): $C_f = \frac{2.5}{4Re^{0.3}}$.

A3.9. Pressure drops in spiral heat exchangers

For Reynolds numbers within the range 400 to 30,000, the pressure drop in each of the circuits is calculated from the following correlation (Picón-Núñez et al. 2012):

$$\Delta P = \frac{2\, C_f\, L\, W^2}{\rho\, D_H\, A_0^2}$$

where:

– C_f is the friction coefficient given by Table A3.1, as a function of the flow regime (Hesselgreaves 2001);

– W is the mass flow rate;

– ρ is the fluid density;

– L is the flow length;

– D_H is the hydraulic diameter;

– A_0 is the cross-section area perpendicular to flow.

Regime	Correlation to be used for the friction coefficient
Laminar $Re \leq 2{,}300$	$C_f = \frac{24}{Re}\left\{1 - 1.3553\frac{e}{b} + 1.9467\left(\frac{e}{b}\right)^2 - 1.7012\left(\frac{e}{b}\right)^3 + 0.9564\left(\frac{e}{b}\right)^4 - 0.2537\left(\frac{e}{b}\right)^5\right\}$
Intermediate $2\,300 < Re \leq 4{,}000$	$C_f = 0.0054 + \frac{2.3\ 10^8}{Re^{-\frac{3}{2}}}$
Turbulent $Re > 4{,}000$	$\frac{1}{\sqrt{C_f}} = 1.56Ln(Re) - 3$

Table A3.3. *Friction coefficients for spiral heat exchangers*

e is the gap between the plates and b is the spiral width (see section 3.6 and Figure 3.19). The Reynolds number for side i is given by $Re_i = \frac{D_{H_i} W_i}{\mu_i A_{0_i}}$, where D_{H_i} is the hydraulic diameter on side i : $D_{H_i} = 2e$.

A3.10. Pressure drops in printed-circuit heat exchangers

For printed-circuit heat exchangers, the correlations giving the pressure drops in cylindrical pipes cannot be used, since the fluids circulate in channels with semi-circular cross-section areas. Consequently, we refer to correlations specific to non-cylindrical pipes using the hydraulic diameter, D_H (see section 3.7.4).

For each channel, the pressure drop is obtained from the following correlation (Yoon et al. 2014):

$$\Delta P_i = 4 f_i \frac{L_i}{D_{H_i}} \rho_i v_i^2$$

where:

– f_i is the friction factor of channel i, given in Table A3.2 as a function of the Reynolds number;

– L_i is the length of channel i;

– v_i is the velocity of the fluid within channel i.

Reynolds domain	Correlation to be used
$Re < 2{,}300$	$f_i = \frac{15.78}{Re_i}$
$8{,}200 < Re < 58{,}000$	$f_i = 0.478\, Re_i^{-0.26}$

Table A3.4. *Friction factor for PCHE*

A3.11. Pressure drops in condensers

Generally, condensers are designed so that the vapor to be condensed circulates in the shell space. Condensation then takes place on the outer walls of the tubes located in the shell.

However, for special cases, condensations may be encountered inside the heat exchanger tubes. For this reason, both situations are presented in sections A3.11.1.1 and A3.11.1.2.

A3.11.1. *Pressure drops in tubular condensers*

A3.11.1.1. *Condensation in the shell space*

This is the most common case.

The pressure drop in this case is given by:

$$\Delta P = \frac{1}{2}\frac{G_{mv}^2 D_{sh}(N_c + 1)}{D_{eq_{sh}}\rho_v} f$$

where:

– D_c is the inner diameter of the shell;

– N_c is the number of transverse baffles;

– ρ_v is the vapor density;

– D_{eqsh} is the equivalent diameter of the shell;

– G_{mv} is the vapor average unit flow rate;

– f is the given friction factor, as a function of the Reynolds number, as follows:

- for $10 < Re = \frac{D_{eq_c} G_{mv}}{\mu_v} < 250 : f = \frac{0.36}{Re^{0.925}}$,

- for $250 < Re < 10^6 : f = \frac{1.19\ 10^{-2}}{Re^{0.187}}$.

A3.11.1.2. *Condensation inside the tubes*

$$\Delta P = \frac{1}{2}\frac{L G_{mv}^2 N_p}{d_i \rho_v} f$$

where:

– L is the length of the tubes;

– G_{mv} is the vapor average unit mass flow rate, defined by:

$$G_{mv} = \frac{\left(G_1^2 + G_1 G_2 + G_2^2\right)^{1/2}}{3}$$

where:

– G_1 and G_2 are the vapor unit mass flow rates at both ends of the tube;

– N_p is the number of tube-side passes;

– d_i is the inner diameter of the tubes;

– ρ_v is the vapor density;

– f is the friction factor given, as a function of the Reynolds number, by:

- for $50 < Re = \frac{D_{eqc} G_{mv}}{\mu_v} < 1{,}000$: $f = \frac{0.51}{Re}$,

- for $1{,}000 < Re < 10^5$: $f = \frac{2.79\ 10^{-3}}{Re^{0.248}}$.

A3.11.2. *Pressure drops in plate heat condensers*

For plate heat condensers (Tao and Infante-Ferreira 2019; Zhang et al. 2019b), the equation giving the friction pressure drop in the plates is similar to that of plate heat exchangers with single-phase flow, except that the two-phase flow friction factor expression needs to be used.

In a two-phase medium, f_{DP} is the friction factor, given as follows, depending on the type of plates considered:

- for ordinary plates (Shon et al. 2018): $f_{DP} = \frac{1261.067}{Re_{eq}^{0.411} Re_{LS}^{0.57}}$;

- for corrugated chevron plates (Tao and Infante-Ferreira 2019):

$$f_{DP} = \left(4.207 - \frac{2.673}{\alpha^{0.46}}\right)\left(\frac{4.200 - 5.41\ Bd^{1.2}}{Re_{eq}^{0.95}}\right)\left(\frac{P_{sat}}{P_{cr}}\right)^{0.3}$$

where Bd is the dimensionless Bond number, which measures the significance of the buoyancy forces with respect to the surface tension (Amalfi 2016):

$$Bd = \frac{(\rho_L - \rho_G) g D_H^2}{\sigma}$$

$$Re_{eq} = \frac{G\left[(1 - x_{av}) + x_{av}\left(\frac{\rho_L}{\rho_G}\right)^{0.5}\right] D_H}{\mu_L}$$

The pressure drop is thus given by:

$$\Delta P_c = 2N f_{DP} \frac{G^2}{\rho_{av}} \frac{L}{D_H} \pm \rho\, g\, L$$

where:

– G is the mass flux;

– L is the flow length;

– D_H is the hydraulic diameter;

– ρ_{av} is the average density, calculated by treating the two-phase flow like a flow of a homogeneous fluid of average vapor quality, x_{av}. ρ_{av} is given by: $\frac{1}{\rho_{av}} = \frac{x_{av}}{\rho_G} + \frac{1-x_{av}}{\rho_L}$.

Appendix 4

Fouling of Heat Exchangers

A4.1. Introduction

In industry, fluids (liquids or gases) that need to be cooled or heated are often not of high purity; in most cases, they are charged with suspended matter or dissolved salts (Garret-Price 1984; Kuppan 2000). To achieve the desired thermal schedule, these fluids need to circulate in heat exchangers. As these fluids flow through the exchangers, suspended matter slowly deposits on the walls of the tubes or plates of the vessels, constituting *solid films*. These films can be rust, scale, silt, coke, etc. They induce an additional resistance to heat transfer. It is then said that the exchanger becomes fouled (gets dirty).

This deposition process is called *fouling*. It starts gradually on the transfer areas (tubes or plates) and the thicknesses of the deposits increase throughout the operation of the exchanger. If no action is taken, these fouling deposits become troublesome both for fluid flow (significant increase in pressure drops) and for heat transfer. This can even block flow within the exchanger (Garret-Price 1984; Somerscales 1988).

A4.2. Nature of fouling

Fouling is generated by chemical deposits, or by sedimentations originating from matter suspended in the fluid entering the exchanger: sludge, sand, dead algae, salt and any chemical that becomes insoluble in the carrier fluid at a given temperature (Garrett-Price 1984; Chenoweth 1988; Kazi 2019).

In addition to chemical deposits, fouling can also occur as a result of biological deposits. These are mainly due to unicellular microorganisms generally present in natural waters (river water, sea water and brackish water). They can also be initiated by multicellular organisms that develop despite preventive chemical treatments having been carried out.

A4.3. How to detect fouling in time

When an exchanger is in operation, any fouling deposits cannot be seen. In fact, the latter occur inside the exchanger, at the walls of the tubes or plates. Therefore, fouling can only be detected from the consequences it induces on the operation of an exchanger (Taborek et al. 1972; Kazi 2019).

Moreover, the deposition process is generally extremely slow. The possible consequences can thus only be "felt" after several months or even several years of operation when, prematurely, the exchangers no longer assure the thermal schedules for which they were designed (Peel 1995; Kuppan 2000).

The consequences of fouling are demonstrated by:

– *A decrease in thermal performance* (exchanged power less than the nominal power). This is easily observed when the exchanger no longer carries out its thermal schedule: the outlet temperatures are no longer at the desired levels;

– *An increase in pressure drops*. The power consumed by the circulation pump of the circuit concerned by fouling becomes greater than the design power;

– *Possible fluid leaks* occurring as a result of corrosion of the walls of the tubes or plates. This corrosion is generated beneath the deposits if the fouling continues over a long period without action[1].

A4.4. Taking fouling into consideration

A4.4.1. *Consideration of fouling from a thermal perspective*

Fouling introduces two new heat transfer resistances (Kern 1959; Marner et al. 1985; Peel 1995): an internal resistance, R_i, which corresponds to the thickness of solid deposited inside the metal wall of the tubes or plates, and an external

1 It should be noted that fouling can be triggered by corrosion. This is due to the fact that a corrosion point presents an imperfect surface state that could promote the attachment of particulates. The fouling then becomes widespread from this initial point.

resistance, R_e, which corresponds to the fouling crust deposited on the outside (see Figure A4.1).

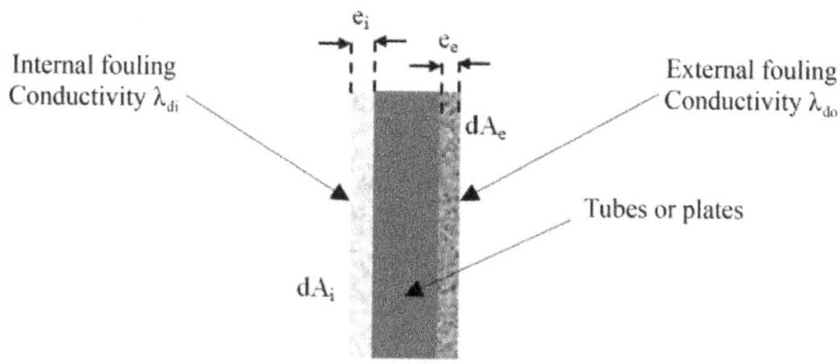

Figure A4.1 *Internal and external fouling. For a color version of this figure, see www.iste.co.uk/benallou/energy6.zip*

Strictly speaking, these thicknesses need to induce conductive thermal resistances, internal (R_i) and external (R_e), of the form $R_i = \frac{e_i}{\lambda_{di} dA_i}$ and $R_e = \frac{e_e}{\lambda_{do}\, dA_e}$ (Peel 1995). In practice, however, a convective description is used, through coefficients h_{di} and h_{de}, called *fouling coefficients*, internal and external respectively. These coefficients make it possible to define *fouling resistances*[2], R_{di} and R_{de} in a convective format: $R_{di} = \frac{1}{h_{di}}$; $R_{de} = \frac{1}{h_{de}}$.

This preference (convective representation rather than conductive) is due to the nature of the data available in the literature on fouling. These data are expressed in terms of *fouling coefficients* h_{di} and h_{de} or *fouling resistances* R_{di} and R_{de} and not in terms of heat conductivities. Orders of magnitude of these coefficients and of the corresponding "resistances" are given, for different fluids, in Table A1.20 (see Appendix 1).

A4.4.2. *Incorporation into calculations*

Being thermal "resistances", the fouling parameters need to be integrated in the calculation of the overall heat transfer coefficient. We will then speak of the *overall coefficient for a fouled (or dirty) heat exchanger*, U_F (see section 5.3.2)[3]:

2 The word "resistance" is put in quotation marks since it is the reciprocal of a heat transfer coefficient rather than a thermal resistance.

3 $\frac{1}{U} = \frac{1}{h_i} + \frac{1}{h_e} + \frac{e}{\lambda_t}$.

$$\frac{1}{U_F} = \frac{1}{U} + R_i + R_e$$

where: U is the overall coefficient for the "clean" heat exchanger (without fouling).

The overall coefficients for fouled heat exchangers are significant in the sizing procedures. As a matter of fact (see Chapter 6), the initial estimation of the exchange area requires an estimation of U_F. Tables A1.21 to A1.23 give the values of U_F to be used in initial calculations of the sizing procedures.

The *heat exchanger cleanliness factor* is often defined by[4]: $D_{CF} = \frac{U_F}{U}$.

The rules of the art recommend a cleanliness factor of around 0.85.

We deduce therefrom that: $1 - D_{CF} = (R_i + R_e)U_F$.

Hence the total fouling resistance: $R_i + R_e = \frac{1-D_{CF}}{U\ D_{CF}}$.

A4.4.3. *Consideration of fouling in terms of cost*

The cost of an exchanger needs to be analyzed considering its two components: the manufacturing cost and the operating/maintenance costs (Garrett-Price et al. 1984).

The manufacturing cost is directly related to the size (transfer area) and the manufacturing complexity, while operation/maintenance includes all of the expenses necessary to run the installation (pumping, utilities, etc.), as well as the maintenance costs (such as cleaning, periodic shutdowns, etc.).

Fouling is incorporated into the exchanger design by allowing a margin at the transfer area; the retained transfer area is greater than that calculated for a "clean" heat exchanger. This increase in surface area allows the exchanger to assure its thermal schedule while it is clean, as well as during the fouling, up to a certain limit. This limit is defined by the fouling resistance values, R_{di} and R_{do}, retained in the design. From an economic point of view, this transfer area margin entails an additional cost, and therefore a greater initial investment for the purchase or creation of the exchanger.

Fouling also implies an *increase in operating costs* resulting from the need for periodic shutdowns in order to conduct cleaning to remove the deposits. In certain

4 $U_F < U \Rightarrow D_{CF} > 1$.

situations where operational constraints require continuous operation of the installation and where shutdown cannot be envisaged without adverse consequences on overall productivity, *a replacement exchanger* is included in the design. This will ensure uninterrupted service, while enabling the necessary periodic cleaning to be performed.

A4.5. How to control fouling

As we have seen, the standard answer to the fouling problem is to oversize exchangers by allowing *a margin on the transfer area* at the design stage. However, this solution is only temporary since it only delays the problem (Knudsen 1984; Chenoweth 1988).

In reality, fouling cannot be avoided when exchangers operate with industrial fluids. It therefore needs to be controlled so as not to lead to a shutdown of the installation. Fouling control strategies should actually be kept in mind starting at the design stage.

A4.5.1. *Actions to be adopted at design stage*

During design, it would be necessary to:

– Opt for exchangers with a large "array"; let us recall that the array type and array pitch are degrees of freedom for the designer (see Chapter 4).

– When the fluids concerned are charged, provide a filter upstream of the exchanger.

– Where applicable, provide coils (springs) inside the tubes, or rubber balls in the shell (Aubin 2015), in the perspective of ensuring continuous cleaning of the exchanger during its operation, while increasing turbulence and thus improving heat transfer (Krueger et al. 2009).

– Envisage chemical treatment of the fluids where necessary.

– Schedule periodic cleaning to remove deposits before this prevents the exchanger from operating.

– In the case of corrosive fouling, provide *cathodic protection* of the exchanger where applicable. Cathodic protection is a technique that reduces corrosion by cathodic polarization. This is done by circulating a low electrical current between an auxiliary anode and the material to be protected, which constitutes the cathode.

A4.5.2. *Actions to be adopted during operation*

The actions to be adopted during exchanger operation would be as follows:

– Implement a pressure drop monitoring system on both the tube side and the shell side in order to be able to detect underperformance in time.

– Perform the scheduled mechanical cleaning of the tubes. In some exchangers, this cleaning can be carried out continuously, without the need to stop operation. This is possible when springs (Aubin 2015) or rubber microspheres are introduced into the tubes (Kuppan 2000);

– Implement the chemical additives schedule, when this has been envisaged (Strauss 1984), during operation by injecting products that act as fouling inhibitors (Kazi 2019), with a view to minimizing fouling (Kuppan 2000; Kazi 2019).

A4.6. Periodic cleaning of heat exchangers

Fouling can be removed by mechanical action or chemical cleaning.

A4.6.1. *Mechanical cleaning*

This type of cleaning requires the exchanger to be shutdown. It is generally performed using brush lances, high-pressure water jet guns, and vapor (for hydrocarbon deposits). The brushes and water guns are generally attached to long rods.

A4.6.2. *Chemical cleaning*

Chemical cleaning makes it possible to access hard-to-reach zones. The cleaning liquid acts by dissolving the foulings through a chemical reaction.

Appendix 5

Nomenclature

A:	Heat transfer area
A_0:	Approximate heat transfer area, used in the sizing algorithms
A_i:	Internal heat transfer area
A_e:	External heat transfer area
A_m:	Mean heat transfer area
a:	Pitch
a_b:	Bypass free area (bypass leakage clearance cross-section area)
A_F:	Free area
a_L:	Longitudinal free area
a_{Min}:	Minimum clearance area at the center of the bundle
A_O:	Free area of a plate-fin heat exchanger
a_{sb}:	Leakage free area between shell and baffle
a_T:	Transverse free area
A_T:	Transversal cross-section area
a_{tb}:	Leakage free area between tubes and the baffle

d_o: Outer diameter of the tubes

d_i: Inner diameter of the tubes

D_E: Equivalent diameter

D_H: Hydraulic diameter of the tubes

D_{Hel}: Helix diameter

D_{lim}: Limit shell diameter to fit all tubes in the shell, as defined in section 2.4.6.5

D_{SB}: Shell-baffle clearance

D_{SB}: Shell-baffle clearance, as defined in section 2.6.4

D_{sh}: Shell diameter

D_{eq}: Shell equivalent diameter

e_c: Spacing between two baffles

e_E: Effective gap. Effective spacing of the tubes, as defined in Table 2.4

e_L: Longitudinal spacing of the tubes

e_T: Transverse spacing of the tubes

F: Array factor (see Tables A2.3 and A2.5)

F_c: Cold fluid flowrate

F_h: Hot fluid flowrate

F_{sh}: Mass flowrate of the fluid circulating in the shell

F_t: Mass flowrate of the fluid circulating in the tubes

f: Friction factor

G: Mass flowrate per unit area

G_L: Longitudinal unit mass flowrate

G_M: Average unit mass flowrate

G_{Max}: Maximum unit mass flowrate

G_{mv}: Average unit flowrate of vapor in exchangers with phase change

G_T: Transverse unit mass flowrate

G_{TM}: Maximum transverse flowrate

h_i: Internal convective heat transfer coefficient (inside tubes)

h_{ideal}: Ideal transfer coefficient on the shell side

h_e: External convective heat transfer coefficient (outside tubes)

h_d: Fouling coefficient

j: Colburn factor

J_C: Configuration factor

J_L: Leakage factor

J_R: Flow regime factor

J_S: Correction factor for nonuniform baffle spacings

L: Length of heat exchanger

L_{TB}: Tube-baffle clearance

L_{TB}: Tube-baffle clearance, as defined in section 2.6.4

N_B: Number of baffles

ND: Nominal diameter

N_{LB}: Number of longitudinal baffles in the shell

N_p: Number of tube-side passes

$N_{p\,Max}$: Maximum value of tube-side passes

NPS: Nominal Pipe Size

N_{sh}:	Number of shell-side passes
N_{ss}:	Number of pairs of sealing strips possibly attached to each baffle
n_t:	Number of tubes per pass
n_c:	Number of baffles
N_t:	Total number of tubes in the shell $N_t = N_p\, n_t$
N_{TB}:	Number of transverse baffles
N_{tcc}:	Effective number of cross-flow layers between baffles
Nu:	Nusselt number
P:	Wetted perimeter through which the heat exchange takes place
Pr:	Prandtl number
Q:	Energy transferred (thermal schedule) in the heat exchanger
Re:	Reynolds number
Re_d:	Reynolds number based on the outer diameter of the tubes, d_o
Re_E:	Reynolds number based on the equivalent diameter
R_{fi}:	Internal fouling thermal resistance
R_{fe}:	External fouling thermal resistance
Re_{avg}:	Average Reynolds number calculated using G_{avg}
Re_{Max}:	Maximum Reynolds number calculated using G_{Max}
R_{sh}:	Shell radius
S:	Free cross-section area perpendicular to the flow
SCH:	Tube schedule
St:	Stanton number

U:	Overall heat transfer coefficient of the heat exchanger
U_F:	Overall heat transfer coefficient for a fouled exchanger
v:	Fluid flow velocity
v_{sh}:	Flow velocity in the shell, far from the baffles
W:	Mass flowrate on the shell side
X:	Thermal efficiency of the heat exchanger, as defined in section 5.7.3
Y:	Correction factor, as defined in section 5.7.3
Z:	Ratio of heat capacities, as defined in section 5.7.3

Greek letters

ΔP_{sh}:	Shell-side pressure drop
ΔP_t:	Tube-side pressure drop
ΔP_t^{Adm}:	Permitted tube-side pressure drop
$\Delta\theta_{eff}$:	Effective temperature difference
$\Delta\theta_{ML}$:	Logarithmic Mean of temperature differences, as defined in section 5.7.3
δ_c:	Shell thickness
δ_F:	Free height above the baffle
δ_t:	Tube thickness
ε:	Cutting ratio of a baffle
λ_t:	Tube wall thermal conductivity
ε:	Cutting ratio of the baffle
ϕ:	Heat flux to be exchanged
θ:	Central angle, as defined in section 2.6.2

θ_c:	Cold fluid temperature
θ_h:	Hot fluid temperature
ρ_{sh}:	Density of the fluid circulating in the shell
ω:	Turbine rotational speed

References

Abadie, T., Houvin, G., Lafitte, S. (2010). Modélisation de l'ébullition nucléée dans les réacteurs à eau pressurisée. Report, EDF, Toulouse.

Afgan, N. and Schlunder, E. (1947). *Heat Exchanger Design and Theory Sourcebook.* McGraw-Hill/Scripta, Washington.

Amalfi, R.L., Vakili-Farahani, F., Thome, J.R. (2016). Flow boiling and frictional pressure gradients in plate heat exchangers. Part 2: Comparison of literature methods to database and new prediction methods. *International Journal of Refrigeration*, 61, 185–203.

Aounallah, Y., Kenning, O.B.R., Whalley, P.B., Hewitt, G.P. (1982). Boiling heat transfer in annular flow. *International Heat Transfer Conference*, 4, 193–199.

Ariu, V. (2014). Heat exchanger analysis for innovative molten salt fast reactor. Master's Thesis, Swiss Federal Institute of Technology, Lausanne.

Aubin, N. (2015). Tube inserts for heat exchangers for energy conservation. In *Proceedings of the International Conference on Heat Exchanger Fooling and Cleaning*, Dublin.

Badger, W.L. (1937). Heat transfer coefficients for condensing dowtherm films. *Industrial and Engineering Chemistry*, 29, 910.

Bandel, J. and Schliinder, E.U. (1974). Frictional pressure drop and convective heat transfer of gas-liquid flow in horizontal tubes. *Proceeding of the 5th Interrnational Conference on Heat Transfer*, 5, 190–194.

Bannon, J.M., Piersall, C.N. Jr., Pucci, P.F. (1965). Heat transfer and flow friction characteristics of perforated nickel plate-fin type heat transfer surfaces. Report, United States Naval Postgraduate School, Monterey.

Barbosa, J.R. Jr., Ribeiro, G.B., de Oliveira, P.A. (2012). A state-of-the-art review of compact vapor compression refrigeration systems and their applications. *Heat Transfer Engineering*, 33, 356–374.

Baviere, R. (2005). Étude de l'hydrodynamique et des transferts de chaleur dans des microcanaux. PhD Thesis, Université Joseph-Fourier, Grenoble.

Bell, K.J. (1960). Exchanger design based on the Delaware research program. *Petroleum and Petrochemical Engineering Journal*, 32, 26–40.

Bell, K.J. (1963). Final report of the co-operative research program on shell and tube heat exchangers. Report, University of Delaware Engineering Experiment Station, Delaware.

Bell, K.J. (1978). Estimate S and T exchanger design fast. *Oil and Gas Journal*, 76, 59–68.

Bell, K.J. (1983). Thermal and hydraulic design of heat exchangers. In *Heat Exchanger Design Handbook*, Schlünder, E.U. (ed.). Hemisphere, Washington.

Bell, K.J. and Mueller, A.C. (1984). Wolverine Tube Heat Transfer Data Book II. Report, Honeywell UOP, Illinois.

Benallou, A. (2018). *Energy Transfers by Conduction*. ISTE Ltd, London, and John Wiley & Sons, New York.

Benallou, A. (2019). *Energy Transfers by Convection*. ISTE Ltd, London, and John Wiley & Sons, New York.

Benallou, A. (2021). Heat exchangers for electronic equipment cooling. In *Heat Exchangers*, Gómez, L.C., Flores, V.M.V., Procopio, M.N. (eds). IntechOpen. doi: 10.5772/intechopen. 100732.

Bergelin, O.P., Brown, G.A., Doberstein, S.C. (1952). Heat transfer. *The American Society of Mechanical Engineers*, 74, 953.

Bhaiya, M. (2009). Solid fluid thermal analysis of a heat exchanger. PhD Thesis, Indian Institute of Technology, Delhi.

Binnie, A.M., Poole, E.G.C. (1937). Theory of the single-pass cross-flow heat interchanger. *Proceedings, Cambridge Philosophical Society*, 33, 403–411.

Bowman, R.A. (1936). Mean temperature difference correction in multipass exchangers. *Industrial and Engineering Chemistry*, 28, 541–544.

Bowman, R.A., Mueller, A.C., Nagle, W.M. (1940). Mean temperature difference in design. *Transactions of the American Society of Mechanical Engineers*, 62(4), 283.

Briggs, D.E. and Young, E.H. (1965). Convection heat transfer and pressure drop of air flowing across triangular pitch banks of finned tubes. *Chemical Engineering Progress Symposium*, 41, 1–10.

Briggs, D.E., Katz, D.L., Young, E.H. (1963). How to design finned-tube heat exchangers. *Chemical Engineering Progress*, 59(11), 49–59.

Brown, W.T. Jr. (1967). A study of flow surface boiling. PhD Thesis, Massachusetts Institute of Technology, Cambridge.

Brownell, L.E. and Young, E.H. (1959). *Process Equipment Design*. John Wiley & Sons, New York.

Burnett, J. (2014). Advances in vapor compression electronics cooling. *Electronics Cooling*, 28–32.

Caccia, M., Tabandeh-Khorshid, M., Itskos, G., Strayer, A.R., Caldwell, A.S., Pidaparti, S., Singnisai, S., Rohskopf, A.D., Schroeder, A.M., Jarrahbashi, D., Kang, T., Sahoo, S., Kadasala, N. et al. (2018). Ceramic-metal composites for heat exchangers in concentrated solar power plants. *Nature*, 562, 406–409.

Cahiers Alfa Laval (1987). Spécial thermique. *Alfa Laval Industries*, 17, 17–18.

Carpenter, E.F. and Colburn, A.P. (1951). The effect of vapor velocity on condensation inside tubes. In *Proceedings of General Discussions on Heat Transfer*, 20–26.

Chen, J.C. (1966). A correlation for boiling heat transfer to saturated fluids in convective flow. *Industrial and Engineering Chemistry Process Design*, 5, 322.

Chen, Y.N. and Weber, M. (1970). Flow induced vibrations in tube bundle heat exchangers with cross and parallel flow. In *Symposium on Flow-Induced Vibration in Heat Exchangers. American Society of Mechanical Engineers*, 57–77.

Chenoweth, J.M. (1988). General design of heat exchangers for fouling conditions. In *Fouling Science and Technology*, Melo, L.F., Bott, T.R., Bernardo, C.A. (eds). Kluwer Publishers, Netherlands.

Chilton, T.H., Drew, T.B., Jebens, R.H. (1944). Heat transfer coefficients in agitated vessels. *Industrial and Engineering Chemistry*, 36, 510.

Chowdhury, K., Linkmeyer, H., Bassiouny, M.K., Martin, H. (1985). Analytical studies on the temperature distribution in spiral plate heat exchangers. *Chemical Engineering Progress*, 19, 183–190.

Colburn, A.P. (1933a). Mean temperature difference and heat transfer coefficient in liquid heat exchangers. *Industrial and Engineering Chemistry*, 25, 873–877.

Colburn, A.P. (1933b). A method of correlating forced convection heat transfer data and a comparison with fluid friction. *Journal of Electronics Cooling and Thermal Control*, 29, 174–210.

Colgan, E.G., Furman, B., Gaynes, M., Graham, W.S., LaBianca, N.C., Magerlein, J.H., Polastre, R.J., Rothwell, M.B., Bezama, R.J., Choudhary, R., Marston, K.C., Toy, H., Wakil, J. et al. (2007). A practical implementation of silicon microchannel coolers for high power chips. *IEEE Transactions on Components and Packaging Technologies*, 30(2), 218–225.

Collier, J.G. (1981). *Convective Boiling and Condensation*. McGraw-Hill, London.

Cook, E.M. (1964). Comparison of equipment for removing heat from process streams. *Chemical Engineering Progress*, 71, 137–142.

Cooper, A. and Usher, J.D. (1983). Plate heat exchangers. In *Heat Exchanger Design Handbook 3*, Bell, K.J., Chisholm, D., Cooper, A. et al. (eds). Hemisphere, Washington.

Debab, A., Chergui, N., Bekrentchir, K., Bertrand, J. (2011). An investigation of heat transfer in a mechanically agitated vessel. *Journal of Applied Fluid Mechanics*, 4(2), 43–50.

Deng T., Zhu, Z., Li, X., Ma, T., Wang, Q. (2020). Experimental study on electrochemical etching for titanium printed circuit heat exchanger channels. *Journal of Materials Processing Technology*, 282.

Dengler, D.E. and Addoms, J.N. (1956). Heat transfer mechanism for vaporization of water in a vertical tube. *Chemical Engineering Progress Symposium*, 52(18), 95–103.

Donohue, D. (1949). Heat transfer and pressure drop in heat exchangers. *Industrial and Engineering Chemistry*, 41, 2499–2511.

Edelstein, S., Perez, A.J., Chen, J.C. (1984). Analytic representation of convective boiling functions. *AIChE. Journal*, 30(5), 840–841.

El Gharbi, N., Kheiri, A., El Ganaoui, M., Blanchard, R. (2015). Numerical optimization of heatexchangers with circular and non-circular shapes. *Case Studies in Thermal Engineering*, 6(C), 194–203.

Fischer, F.K. (1938). Mean temperature difference correction in multipass exchangers. *Industrial and Engineering Chemistry*, 30, 377–383.

Fletcher, P. (1987). Heat transfer coefficients for stirred batch reactor design. *Chemical Engineering Progress*, 33, 33–37.

Fraas, A.P. (1989). *Heat Exchanger Design.* John Wiley & Sons, New York.

Fujii, T. (1991). *Theory of Laminar Film Condensation*. Springer, New York.

Fumimaru, O., Shuji, K., Tokuro, M. (1988). Condensation of binary vapors of immiscible liquids. *International Journal of Heat and Mass Transfer*, 31(2), 245–250.

Gardner, K.A. (1945). Variable heat transfer rate correction in multipass exchangers, shell side film controlling. *Transactions of the ASME*, 67, 31–38.

Garret-Price, B.A., Smith, S.A., Watts, R.L., Knudsen, J.G. (1984). Industrial fouling: Problem characterization, economic assessment, and review of prevention, mitigation, and accommodation techniques. Report, Pacific Northwest Laboratory, Washington.

Gilmour, C.H. (1952). Short cut to heat exchanger design. *Chemical Engineering*, 144–148.

Gilmour, C.H. (1953). Short cut to heat exchanger design. *Chemical Engineering*, 203–229.

Gilmour, C.H. (1954). Short cut to heat exchanger design. *Chemical Engineering*, 190–213.

Gray, D.L., Webb, R.L. (1989). Heat transfer and friction correlations for plate fin and tube heat exchangers having plain fins. *Heat Transfer 1986, Proc. Eighth Int. Heat Transfer Conference*, 6, 2745–2750.

Guerrieri, S.A. and Talty, R.D. (1956). A study of heat transfer to organic liquids in single-tube, natural circulation vertical tube boilers. *Chemical Engineering Progress Symposium*, 52(18), 69–77.

Haaland, S.E. (1983). Simple and explicit formulas for the friction factor in turbulent pipe flow. *Journal of Fluids Engineering*, 89–90.

Hayes N., Jokar, A., Ayub, Z.H. (2011). Study of carbon dioxide condensation in chevron plate exchangers; heat transfer analysis. *International Journal of Heat and Mass Transfer*, 54, 1121–1131.

Hesselgreaves, J.E. (2001). *Compact Heat Exchangers*. Pergamon Press, Oxford.

Horacek, B., Kim, J., Kiger, K. (2004). Spray cooling using multiple nozzles: Visualization and wall heat transfer measurements. *IEEE Transactions on Device and Materials Reliability*, 4(4), 614–625.

Horacek, B., Kiger, K., Kim, J. (2005). Single nozzle spray cooling heat transfer mechanisms. *International Journal of Heat and Mass Transfer*, 48(8), 1425–1438.

Hugo, J.M. (2012). Transferts dans les milieux cellulaires à forte porosité : application à l'optimisation structurale des échangeurs à ailettes. PhD Thesis, Mechanics and Energy Specialty, Université d'Aix-Marseille, Marseille.

Hutchinson, E. (1945). On the measurement of the thermal conductivity of liquids. *Faraday Society*, 41, 87–92.

Incropera, F.P. and DeWitt, D.P. (1966). *Fundamentals of Heat and Mass Transfer*. John Wiley & Sons, New York.

International Technology Roadmap for Semiconductors (2007). International Roadmap Committee. Report, Semiconductor Industry Association, Washington.

Jaw, L. (1964). Temperature relations in shell-and-tube exchangers having one-pass split-flow shells. *The American Society of Mechanical Engineers*, 8, 408–416.

Jens, W.H. and Lottes, P.A. (1951). An analysis of heat transfer, burnout, pressure drop, and density data for high pressure water. Argonne National Laboratory Report ANL-4627, Chicago.

Jenson, S. (1960). Assessment of heat transfer data. *Chemical Engineering Progress Symposium*, 960(56), 195–201.

Jokar, A., Hosni, M.H., Eckels, S.J. (2006). Dimensional analysis on the evaporation and condensation of refrigerant R-134a in minichannel plate heat exchangers. *Applied Thermal Engineering*, 26, 2287–2300.

Kakaç, S. and Liu, H. (2002). *Heat Exchangers Selection Rating and Thermal Design*. CRC Press, New York.

Kannengieser, O. (2009). Étude de l'ébullition sur plaque plane en microgravité, application aux réservoirs des fusées Ariane V. PhD Thesis, Institut Polytechnique de Toulouse, Toulouse.

Kays, W.M. and London, A.L. (1998). *Compact Heat Exchangers*. McGraw-Hill, New York.

Kays, W.M. and Perkins, H.C. (1972). Radiation. In *Handbook of Heat Transfer*. Rohsenow, W.M., Hartnett, J.P. (eds). McGraw-Hill, New York.

Kazi, S.N. (2019). Fouling and fouling mitigation of calcium compounds on heat exchangers by novel colloids and surface modifications. *Reviews in Chemical Engineering*, 36(6). doi: 10.1515/revce-2017-0076.

Kercher, D., Lee, J.B., Brand, O., Allen, M., Glezer, A. (2006). Microjet cooling devices for thermal management of electronics. *IEEE Trans. Comp. Packag. Technol.*, 26, 359–366.

Kern, D.Q. (1950). *Process Heat Transfer*. McGraw-Hill, New York.

Kern, D.Q. and Kraus, A.D. (1972). *Extended Surface Heat Transfer*. McGraw-Hill, New York.

Kern, D.Q. and Seaton, R.E. (1959). A theoretical analysis of thermal surface fouling. *British Chemical Engineering*, 4, 258.

Kim, I.H. (2012). Experimental and numerical investigations of thermal-hydraulic characteristics for the design of a Printed Circuit Heat Exchanger (PCHE) in HTGRs. PhD Thesis, Korea Advanced Institute of Science and Technology.

Kim, E.S., Oh, C.H., Sherman, S. (2008). Simplified optimum sizing and cost analysis for compact heat exchanger in VHTR. *Nuclear Engineering and Design*, 238, 2635–2647.

Kirkbride, C.G. (1933). Heat transfer by condensing vapor on vertical tube. *Trans. AIChE*, 30, 170–186.

Kissel, J.H. (1973). Flow induced vibrations in heat exchangers – A practical look. *Machine Design*, 11, 104–107.

Klimenko, V. (1984). Processes of two-phase heat transfer with cryogenic fluids and development of optimal computational methods. PhD Thesis, Moscow Power Engineering Institute.

Klimenko, V. (1988). A generalized correlation for two-phase forced flow heat transfer. *International Journal of Heat and Mass Transfer*, 31(3), 541–552.

Knudson, J.G. (1984). Fouling of heat exchangers: are we solving the problem? *Chemical Engineering Progress*, 80, 63.

Knudsen, J.G. and Katz, D.L. (1958). *Fluid Dynamics and Heat Transfer*. McGraw-Hill, New York.

Krueger, A. and Pouponnot F. (2009). Heat exchanger performance enhancement through the use of tube inserts in refineries and chemical plants. In *Proceedings of the International Conference on Heat Exchanger Fooling and Cleaning VIII*, Schladming. Springer, Berlin.

Kuppan, T. (2000). *Heat Exchanger Design Handbook*. Marcel Dekker, New York.

Linnhoff, B. (1982). *User Guide on Process Integration for the Efficient Use of Energy*. Institution of Chemical Engineers, London.

Lohrisch, F.W. (1963). What are optimum exchanger conditions?. *Hydrocarbon Process. Petrol. Refiner*, 42(5), 177–180.

Longo, G.A, Righetti, G., Zilio, C. (2015). A new computational procedure for refrigerant condensation inside herringbone-type Brazed Plate Heat Exchangers. *International Journal of Heat and Mass Transfer*, 82, 530–536.

Lord, R.C., Minton, P.E., Slusser, R.P. (1970). *Design of Heat Exchangers*. McGraw-Hill, New York.

Ma, T., Zhang, P., Deng, T., Ke, H., Lin, Y., Wang, Q. (2021). Thermal hydraulic characteristics of printed circuit heat exchangers used for floating natural gas liquefaction. *Renewable and Sustainable Energy Reviews*, 137.

Mana, M. (2016). Optimisation énergétique des échangeurs tubulaires. Master's Thesis, Université de Tlemcen, Chetouane.

Mancin, S., Zilio, C., Righetti, G., Rossetto, L. (2013). Mini vapor cycle system for high density electronic cooling applications. *International Journal of Refrigeration*, 36(4), 1191–1202.

Manglik, R.M. and Bergles, A.E. (1995). Heat transfer and pressure drop correlations for the rectangular offset-strip-fin compact heat exchanger. *Exp. Thermal and Fluid Sci.*, 10, 171–180.

Marcinichen, J.B., Lamaison, N., Thome, J.R. (2016). Electronic micro-evaporator cooling systems and flow control. In *Encyclopedia of Two-Phase Heat Transfer and Flow*, Thome, J.R., Kim, J. (eds). World Scientific Publishing, Singapore.

Marner, W.J. and Suitor, J.W. (1987). Fouling with convective heat transfer. In *Handbook of Single-Phase Convective Heat Transfer*, Kakaç, S., Shah, R.K., Aung, W. (eds). John Wiley & Sons, New York.

Marriott, J. (1979). Where and how to use plate heat exchangers. *Process Heat Exchange, Chemical Engineering Magazine*, 156–162.

McAdams, W.H. (1954). *Heat Transmission*. McGraw-Hill, New York.

McKetta, J.J. (1992). *Heat Transfer Design Methods*. Marcel Dekker Inc, New York.

Meysenc, L. (1998). Étude des micro-échangeurs intégrés pour le refroidissement des semi-conducteurs de puissance. PhD Thesis, Institut National Polytechnique de Grenoble, Grenoble.

Milnes, P.D. (2011). Phase separation in two phase microfluidic heat exchangers. PhD Thesis. Stanford University, Stanford.

Modi, J. (1995). Quick manual design of shell and tube heat exchangers. Master's Thesis, Oklahoma State University, Oklahoma.

Mohr, Y. and Runge, R. (1977). Forced convection boiling of пеоп in horizontal tubes. In *Heat Transfer in Boiling*, Hahne, E., Grigull, U. (eds). Hemisphere, Washington.

Monteiro-Guimarães, L.G., Dos Santos-Guzella, M., Cabezas-Gómez, L., Neves Teixeira, F. (2015). Numerical determination of the LMTD correction factor for shell-and-tube 1-2 heat exchangers. *Applied Mechanics and Materials*, 455–459, 789–790.

Moody, L.F. (1944). Friction factors for pipe flow. *Transactions of the American Society of Mechanical Engineers*, 66, 671.

Moon-Hyun, C. and Kyun-Tae, K. (1990). Assessment of the new and existing correlations for laminar and turbulent film condensations on a vertical surface. *International Communications in Heat and Mass Transfer*, 17(4), 431–444.

Morton, D.S. (1962). Heat exchangers dominate process heat transfer. *Chemical Engineering*, 170–176.

Mueller, A.C. (1972). Heat exchangers. In *Handbook of Heat Transfer*, Kays, W.M., Perkins, H.G. (eds). McGraw-Hill, New York.

Murali, S., Bhaskar, D., Rao, Y. (2008). A simple tube sheet layout program for heat exchangers. *American Journal of Engineering and Applied Sciences*, 1(2), 131–135.

Nagle, W.M. (1933). Mean temperature differences in multipass heat exchangers. *Industrial and Engineering Chemistry*, 25, 604–609.

Nusselt, W. (1911). Der Warmeiibergang im Kreuzstrom. *Zeitschriftdes Vereines Deutscher Ingenieur*, 55, 2021–2024.

Nusselt, W. (1930). Eine neue Formel fur den Warmedurchgang im Kreuzstrom. *Technische Mechanik und Thermodynamik*, 1, 417–422.

Oh, C.H. and Kim, E.S. (2008). Heat exchanger design options and tritium transport study for the VHTR system. Report, Idaho National Laboratory, Idaho.

Ohadi, M., Choo, K., Dessiatoun, S., Cetegen, E. (2013). *Next Generation Microchannel Heat Exchangers*. Springer, Berlin.

Oldshue, J.Y. and Gretton, A.T. (1954). Helical coil heat transfer in mixing vessel. *Chemical Engineering Progress*, 50, 615.

Özisik, M.N. (1985). *Heat Transfer: A Basic Approach*. McGraw-Hill, New York.

Peel, D., Wbeeldon, P., Virdee, G.S. (1995). Estimation and control of fouling in heat exchangers. In *IFAC Algorithms and Architectures for Real-Time Control, Ostend*. Elsevier, Amsterdam.

Perry, R.H. and Chilton, C.H. (1973). *Chemical Engineers' Handbook*. McGraw-Hill, New York.

Phelan, P.E., Chiriac, V.A., Lee, T.-Y.T. (2002). Current and future miniature refrigeration cooling technologies for high power microelectronics. *IEEE Trans. Compon. Packag. Technol.*, 25, 356–365.

Phelan, P.E., Catano, J., Michna, G., Gupta, Y., Tyagi, H., Zhou, R., Wen, J., Prasher, R.S., Jensen, M., Peles, Y. (2010). Energy efficiency of refrigeration systems for high-heat-flux microelectronics. *Journal of Thermal Science and Engineering Applications*, 2(3).

Picón-Núñez, M., Polleya, G.T., Riesco-Ávilab, J.M. (2012). Design space for the sizing and selection of heat exchangers of the compact type. *Chemical Engineering Transactions*, 29, 217–222.

Picón-Núñez, M., Rodríguez-Ibarra, J.E., Alvarado-Briones, B. (2014). Design of heat exchangers in complex arrangements for use in preheat trains. *Chemical Engineering Transactions*, 39, 301–306.

PRIATHERM (2018). Design your cooling system using forced convection in just three easy steps [Online]. Available at: https://priatherm.com/design-your-cooling-system-using-forced-convection-in-just-three-easy-steps/.

Qiu, C., Chen, S., Shen, Y., Gan, Z. (2023). Numerical and experimental study on a tube-in-tube heat exchanger working at liquid-hydrogen temperature with a large temperature span. *International Journal of Heat and Mass Transfer*, 208.

Ravelo-Mendivelso, K.Y., Villate-Fonseca, M.T., Hernández-Vásquez, J.D., Miranda-Samper, O.M., Pacheco-Torres, P.J., Campuzano, M.J. (2023). Thermal and hyrdodynamic performance analysis of a shell and tube heat exchanger using the AHP multicriteria method. *International Journal of Technology*, 14(3), 522–535.

Reid, D.R. and Taborek, J. (1994). Selection criteria for plain and segmented finned tubes for heat recovery systems. *ASME J. Eng. Gas Turbines and Power*, 116, 406.

Rich, D.G. (1973). The effect of fin spacing on the heat transfer and friction performance of multi-row, smooth plate fin-and-tube heat exchangers. *ASHRAE Transactions*, 79(2), 137–145.

Robinson, K.K. and Briggs, D.E. (1966). Pressure drops of air flowing across triangular pitch banks of finned tubes. *Chem. Eng. Prog. Symp. Ser.*, 62, 177–184.

Rohsenow, W.M., Weber, G.H., Ling, A.T. (1956). The effect of vapor velocity on lami-nar and turbulent film condensation. *The American Society of Mechanical Engineers*, 78(8), 1637–1643.

Rohsenow, W.M., Hartnett, J.P., Cho, Y.I. (1998). *Handbook of Heat Transfer*. McGraw-Hill, New York.

Roszak, C. and Véron, M. (1929). *Nouvelles études sur la chaleur*. Dunod, Paris.

Sabharwall, P., Kim, E.S., Siahpush, A., Anderson, N., Glazoff, M., Phoenix, B., Mizia, R., Clark, D., McKellar, M., Patterson, M.W. (2011). Feasibility study of secondary heat exchanger concepts for the advanced high temperature reactor. Report, Idaho National Laboratory, Idaho.

Saeed, M., Berrouk, A.S., Siddiqui, M.S., Awais, A.A. (2020). Effect of printed circuit heat exchanger's different designs on the performance of supercritical carbon dioxide Brayton cycle. *Applied Thermal Engineering*, 179.

Saha, M.N. and Srivastava, B.N. (1935). *Treatise on Heat*. The Indian Press, Kolkata.

Sandifer, J.B. (1992). Guidelines for flow-induced vibration prevention in heat exchangers. *Welding Research Council Bulletin*, 372, 1–27.

Sarmiento, A.P.C., Milanez, F.H., Mantelli, M.B.H. (2021). Theoretical models for compact printed circuit heat exchangers with straight semi-circular channels. *Applied Thermal Engineering*, 184, 115435.

Sathyabhama, A. (2019). Boiling of saturated water on grooved surface. *Thermal Science*, 23(2), 1095–1104.

Saunders, E.A.D. (1988). *Heat Exchangers: Selection. Design and Construction*. Longman Scientific & Technical, Essex.

Schlunder, E.U. (1982). *Heat Exchanger Design Handbook.* Hemisphere, Washington.

Schumann, E.A. Jr. (1939). The calculation of heat transfer in cross-flow exchangers. Report, Heat Transfer Professional Group Annual Meeting, Philadelphia, 4–8.

Shah, R.K. (1984). What's new in heat exchanger design? *Mechanical Engineering*, 50–60. 71.

Shah, R.K. and Focke, W.W. (1988). Plate heat exchangers and their design theory. In *Heat Transfer Equipment Design*, Shah, R.K., Subbarao, D.C., Mashelekar, R.M. (eds). Hemisphere, Washington, 227–255.

Shah, R.K. and Sekulic, D.P. (1998). Heat exchangers. In *Handbook of Heat Transfer*, Rohsenow, W.M., Hartnett, J.P., Cho, Y.I. (eds). McGraw-Hill, New York.

Shah, R.K. and Sekulic, D.P. (2003). *Fundamentals of Heat Exchanger Design*. John Wiley & Sons, New York.

Shah, R.K. and Willmott, A.J. (1981). *Thermal Energy Storage and Regeneration*. Hemisphere McGraw-Hill, Washington.

Shannon, M.A. (2002). Micro- to mesoscopic vapor-compression cycle heat pumps: Progress and challenges. In *Next Generation Thermal Management Materials and Systems*, Dallas.

Shannon, M.A., Phillpott, M.L., Miller, N.R., Bullard, C.W., Beebe, D.J., Jacobi, A.M., Hrnjak, P.S., Saif, T., Aluru, N., Sehitoglu, H., Rockett, A., Economy, J. (1999). Integrated Mesoscopic Cooler Circuits (IMCCS). *Intl. Mechanical Engineering Congress and Exhibition*, Nashville. *Proc. ASME-AES*, 39, 75–82.

Shicheng, J., Yibin, J., Liping, P., Yu, Z., Wenxuan, Y. (2014). Investigation of spray cooling performance on different heat transfer surfaces. *Proceedings of the International Conference on Logistics, Engineering, Management and Computer Science.*

Shon, B.H., Jung, C.W., Kwon, O.J., Choi, C.K., Kang, Y.T. (2018). Characteristics on condensation heat transfer and pressure drop for a low GWP refrigerant in brazed plate heat exchanger. *International Journal of Heat and Mass Transfer*, 122, 1272–1282.

Smith, D.M. (1934). Mean temperature difference in cross flow. *Engineering*, 138, 479–481, 606–607.

Somerscales, E.F.C. (1998). Fouling, in two-phase flow heat exchangers: Thermo hydraulic fundamentals and design. In *Two-Phase Flow Heat Exchangers*, Kakaç, S., Bergles, A.E., Fernandes, E.O. (eds). John Wiley & Sons, New York.

Srinivasan, P.S., Nandapurkar, S.S., Holland, S.S. (1970). Friction factors for coils. *Transactions of the Institution of Chemical Engineers*, 48, 156–161.

Stevens, R.A., Fernandes, J., Woolf, J.R. (1957). Mean temperature difference in one, two, and three pass crossflow heat exchangers. *Transactions of the ASME*, 79, 287–297.

Strauss, S.D. and Puckorious, P.R. (1984). Cooling-water treatment for control of scaling, fouling and corrosion. *Power*, 128.

Su, T., Mon, T., Khin, A., Lin, M., Sandar, M. (2008). Heat exchanger design. *International Journal of Mechanical and Mechatronics Engineering*, 2(10), 1151–1158.

Taborek, J. (1983). *Heat Exchanger Design Handbook*. Hemisphere, Washington.

Taborek, J., Hewitt, G.F., Afgan, N. (1983). *Heat Exchangers: Theory and Practice*. Hemisphere, Washington.

Tang, L., Cui, L., Sundén, B. (2020). Optimization of fin configurations and layouts in a printed circuit heat exchanger for supercritical liquefied natural gas near the pseudo-critical temperature. *Applied Thermal Engineering*, 172, 25.

Tao, X. and Infante-Ferreira, C.A. (2019). Heat transfer and frictional pressure drop during condensation in plate heat exchangers: Assessment of correlations and a new method. *International Journal of Heat and Mass Transfer*, 135, 996–1012.

TEMA (1999). Standards of Tubular Exchangers Manufacturers Association. Tubular Exchanger Manufacturers Association, New York.

Ten Broeck, H. (1938). Multipass exchanger calculations. *Industrial and Engineering Chemistry Research*, 30, 1041–1042.

Thibault, D. (2009). Étude du refroidissement par impact de jets à travers une paroi mince et avec un écoulement cisaillant amont: application aux aubes de turbines. PhD Thesis, École Nationale Supérieure de Mécanique et d'Aérotechique, Poitiers.

Thome, J.R. (2016). *Encyclopedia of Two-Phase Heat Transfer and Flow*. World Scientific Publishing, Singapore.

Tinker, T. (1958). Shell side characteristics of shell and tube heat exchangers. *Trans. ASME*, 1, 36–52.

Tsuzuki, N., Kato, Y., Ishiduka, T. (2007). High performance printed circuit heat exchanger. *Applied Thermal Engineering*, 27(10), 1702–1707.

Tuckerman, D.B. and Pease, R.F.W. (1981). High performance heat sink for very-large-scale integrated circuits. *IEEE Electron Dev. Lett.*, 2(5), 126–129.

Turton, R., Ferguson, D., Levenspiel, O. (1986). Charts for the performance and design of heat exchangers. *Chemical Engineering Journal*, 23(3), 81–88.

Underwood, A.J.V. (1933). Graphical computation of logarithmic-mean temperature difference. *Industrial Chemist*, 9, 167–170.

Underwood, A.J.V. (1934). The calculation of the mean temperature difference in multipass heat exchangers. *Journal Institution of Petroleum Technologists*, 20, 145–158.

Wales, R.E. (1981). Mean temperature difference in heat exchangers. *Chemical Engineering*, 88(4), 77-81.

Walker, G. (1982). *Industrial Heat Exchangers: A Basic Guide*. Hemisphere, Washington.

Walker, G. (1983). *Industrial Heat Exchangers: A Basic Guide*. Hemisphere/McGraw-Hill, Washington.

Walker, W.M. and Reising, G.F.S. (1968). Flow induced vibrations in cross-flow heat exchangers. *Chemical and Process Engineering*, 95–103.

Wang, Z. and Zhao, Z. (1993). Analysis of performance of steam condensation heat transfer and pressure drop in plate condensers. *Heat Transfer Engineering*, 4(14), 32–41.

Wang, C.C., Chi, K.U., Chang, C.J. (2000). Heat transfer and friction characteristics of plain fin-and-tube heat exchangers – Correlation. *International Journal of Heat and Mass Transfer*, 43, 2692–2700.

Wang, W.-Q., Qiu, Y., He, Y.-L., Shi, H.Y. (2019). Experimental study on the heat transfer performance of a molten-salt printed circuit heat exchanger with airfoil fins for concentrating solar power. *International Journal of Heat and Mass Transfer*, 135, 837–846.

Webb, R.L. (1987). Enhancement of single-phase heat transfer. In *Handbook of Single-Phase Heat Transfer*, Kakac, S., Shah, R.K., Aung, W. (eds). John Wiley & Sons, New York.

Webb, R.L. (1994). *Principles of Enhanced Heat Transfer*. John Wiley & Sons, New York.

Webb, R.L. (2005). Next generation devices for electronic cooling with heat rejection to air. *Journal of Heat Transfer*, 127(1), 2–10.

Yoon, S.J., Sabharwall, P., Kim, E.S. (2013). Analytical study on thermal and mechanical design of printed circuit heat exchangers. Report, Idaho National Laboratory/Department of Energy/Office of Nuclear Energy Under DOE Idaho Operations Office Contract DE-AC07-05ID14517, Idaho.

Yoon, S.J., Sabharwall, P., Kim, E.S. (2014). Numerical study on crossflow printed circuit heat exchanger for advanced small modular reactors. *International Journal of Heat and Mass Transfer*, 70, 250–263.

Yovanovich, M.M., Khan, W.A., Culham, J.R. (2005). Convection heat transfer from tube banks in crossflow: Analytical approach. In *43rd Aerospace Sciences Meeting & Exhibit*, Reno, 10–13.

Zhang, H., Che, F., Lin, T., Zhao, W. (2019a). *Modeling, Analysis, Design, and Tests for Electronics Packaging beyond Moore*. Elsevier Science, New York.

Zhang, J., Kærn, M.R., Ommen, T., Elmegaard, B., Fredrik, H. (2019b). Condensation heat transfer and pressure drop characteristics of R134a, R1234ze (E), R245fa and R1233zd (E) in a plate heat exchanger. *International Journal of Heat and Mass Transfer*, 128, 136–149.

Zuber, N., Tribus, M. (1958). Further remarks on the stability of boiling heat transfer. Report, 58-5, U.C.L.A., Los Angeles.

Zukauskas, A. (1972). Heat transfer from tubes in cross-flow. *Adv. Heat Transf.*, 8, 93–160.

Zukauskas, A. (1987). Convective heat transfer in cross flow. In *Handbook of Single Phase Convective Heat Transfer*, Kakaç, S., Shah, R.K., Aung, W. (eds). John Wiley & Sons, New York.

Zukauskas, A. and Ulinskas, R. (1988). *Heat Transfer in Tube Banks in Crossflow*. Hemisphere, Washington.

Index

A, B

C, D

E, F

G, H

L, M, N

O, P, Q

R, S, T

Other titles from

iSTE

in

Energy

2023

ALVAREZ-HERAULT Marie-Cécile, GOUIN Victor, CHARDIN-SEGUI Trinidad, MALOT Alain, COIGNARD Jonathan, RAISON Bertrand, COULET Jérôme
Distribution System Planning: Evolution of Methodologies and Digital Tools for Energy Transition

BEN KHADER BOUZID Monia, CHAMPENOIS Gérard
Electrical and Mechanical Fault Diagnosis in Wind Energy Conversion Systems

JANNOT Yves, MOYNE Christian, DEGIOVANNI Alain
Heat Transfer 1: Conduction and Convection
Heat Transfer 2: Radiation and Exchangers

ROBYNS Benoît, LENGLET Claude, BARRY Hervé, BOZZO-REY Malik
Smart Users for Energy and Societal Transition

2022

ABOUELATTA Mohamed, SHAKER Ahmed, GONTRAND Christian
Smart Power Integration

BOUTAUD Benoit
Energy Autonomy: From the Notion to the Concepts

GONTRAND Christian
Electromagnetism: Links to Special Relativity

2021

ROBYNS Benoît, DAVIGNY Arnaud, FRANÇOIS Bruno, HENNETON Antoine, SPROOTEN Jonathan
Electricity Production from Renewable Energies (2nd edition)

2020

BOISGIBAULT Louis, AL KABBANI Fahad
Energy Transition in Metropolises, Rural Areas and Deserts

SOUALHI Abdenour, RAZIK Hubert
Electrical Systems 1: From Diagnosis to Prognosis
Electrical Systems 2: From Diagnosis to Prognosis

2019

BENALLOU Abdelhanine
Energy Transfers by Convection
(Energy Engineering Set – Volume 3)
Energy Transfers by Radiation
(Energy Engineering Set – Volume 4)
Mass Transfers and Physical Data Estimation
(Energy Engineering Set – Volume 5)

LACHAL Bernard
Energy Transition

2018

BENALLOU Abdelhanine
Energy and Mass Transfers: Balance Sheet Approach and Basic Concepts
(Energy Engineering Set – Volume 1)
Energy Transfers by Conduction
(Energy Engineering Set – Volume 2)

JEMEÏ Samir
Hybridization, Diagnostic and Prognostic of Proton Exchange Membrane Fuel Cells: Durability and Reliability

RUFFINE Livio, BROSETA Daniel, DESMEDT Arnaud
Gas Hydrates 2: Geoscience Issues and Potential Industrial Applications

VIALLET Virginie, FLEUTOT Benoit
Inorganic Massive Batteries
(Energy Storage – Batteries and Supercapacitors Set – Volume 4)

2017

BROSETA Daniel, RUFFINE Livio, DESMEDT Arnaud
Gas Hydrates 1: Fundamentals, Characterization and Modeling

LA SCALA Massimo
From Smart Grids to Smart Cities: New Challenges in Optimizing Energy Grids
(Advanced Smart Grids Set – Volume 2)

MOLINA Géraldine, MUSY Marjorie, LEFRANC Margot
Building Professionals Facing the Energy Efficiency Challenge

SIMON Patrice, BROUSSE Thierry, FAVIER Frédéric
Supercapacitors Based on Carbon or Pseudocapacitive Materials
(Energy Storage – Batteries and Supercapacitors Set – Volume 3)

2016

ALLARD Bruno
Power Systems-on-Chip: Practical Aspects of Design

ANDRÉ Michel, SAMARAS Zissis
Energy and Environment
(Research for Innovative Transports Set - Volume 1)

DUFOUR Anthony
Thermochemical Conversion of Biomass for the Production of Energy and Chemicals

SOUSTELLE Michel
Phase Transformations
(Chemical Thermodynamics Set – Volume 5)
Thermodynamics of Surfaces and Capillary Systems
(Chemical Thermodynamics Set – Volume 7)

2015

MONCONDUIT Laure, CROGUENNEC Laurence, DEDRYVÈRE Rémi
Electrodes for Li-ion Batteries: Materials, Mechanisms and Performance
(Energy Storage – Batteries and Supercapacitors Set – Volume 2)

ROBYNS Benoît, FRANÇOIS Bruno, DELILLE Gauthier,
SAUDEMONT Christophe
Energy Storage in Electric Power Grids

ROSSI Carole
Al-based Energetic Nanomaterials: Design, Manufacturing, Properties and Applications
(Nanotechnologies for Energy Recovery Set – Volume 2)

TARASCON Jean-Marie, SIMON Patrice
Electrochemical Energy Storage
(Energy Storage – Batteries and Supercapacitors Set – Volume 1)

SOUSTELLE Michel
Phases Modeling Tools: Application to Gases
(Chemical Thermodynamics Set – Volume 1)
Modeling of Liquid Phases
(Chemical Thermodynamics Set – Volume 2)
Thermodynamic Modeling of Solid Phases
(Chemical Thermodynamics Set – Volume 3)
Chemical Equilibria
(Chemical Thermodynamics Set – Volume 4)
Ionic and Electrochemical Equilibria
(Chemical Thermodynamics Set – Volume 6)

2014

DE LARMINAT Philippe
Climate Change: Identification and Projections

2012

BECKERS Benoit
Solar Energy at Urban Scale

FOULADGER Javad
Electrothermics

ROBYNS Benoît, DAVIGNY Arnaud, FRANÇOIS Bruno, HENNETON Antoine, SPROOTEN Jonathan
Electricity Production from Renewable Energies

2011

GAO Fei, BLUNIER Benjamin, MIRAOUI Abdellatif
Proton Exchange Membrane Fuel Cell Modeling

MONMASSON Eric
Power Electronic Converters: PWM Strategies and Current Control Techniques

MULTON Bernard
Marine Renewable Energy Handbook

RAZIK Hubert
Handbook of Asynchronous Machine with Variable Speed

2010

BRUNET Yves
Energy Storage

JUFER Marcel
Electric Drives: Design Methodology

2009

SABONNADIÈRE Jean-Claude
Low Emission Power Generation Technologies and Energy Management
Renewable Energy Technologies

2008

MEUNIER Gérard
The Finite Element Method for Electromagnetic Modeling

Printed and bound by CPI Group (UK) Ltd, Croydon, CR0 4YY
16/09/2026
14963075-0006